STINK BUGS

Biorational Control Based on Communication Processes

STINK BUGS
Biorational Control Based on Communication Processes

Editors

Andrej Čokl

Department of Organisms and Ecosystems Research
National Institute of Biology
Ljubljana
Slovenia

Miguel Borges

Laboratório de Semioquímicos
Embrapa Recursos Genéticos e
Biotecnologia Brasilia, DF
Brazil

CRC Press
Taylor & Francis Group
Boca Raton London New York

CRC Press is an imprint of the
Taylor & Francis Group, an **informa** business
A SCIENCE PUBLISHERS BOOK

Cover illustration reproduced by kind courtesy of Cecilia Rodrigues Vieira.

CRC Press
Taylor & Francis Group
6000 Broken Sound Parkway NW, Suite 300
Boca Raton, FL 33487-2742

Printed on acid-free paper
Version Date: 20170223

International Standard Book Number-13: 978-1-4987-3280-2 (Hardback)

Library of Congress Cataloging-in-Publication Data

Names: Čokl, Andrej, editor. | Borges, Miguel (Entomologist), editor.
Title: Stink bugs : biorational control based on communication processes / editors, Andrej Čokl, Department of Entomology, National Institute of Biology, Ljubljana, Slovenia, Miguel Borges, Laboratório de Semioquímicos, Embrapa Recursos Genéticos e Biotecnologia Brasilia, DF, Brazil.
Description: Boca Raton, FL : CRC Press, 2017. | "A science publishers book." | Includes bibliographical references and index.
Identifiers: LCCN 2016058118| ISBN 9781498732802 (hardback : alk. paper) | ISBN 9781498732826 (e-book : alk. paper)
Subjects: LCSH: Stinkbugs. | Stinkbugs--Control. | Animal communication. | Chemical senses. | Chemoreceptors.
Classification: LCC QL523.P5 S75 2017 | DDC 595.7/54--dc23
LC record available at https://lccn.loc.gov/2016058118

Visit the Taylor & Francis Web site at
http://www.taylorandfrancis.com

and the CRC Press Web site at
http://www.crcpress.com

Dedication

We dedicate this book to the next generation of researchers and would like to say thanks to our teachers, friends and colleagues who share our curiosity about the fantastic world of stink bugs.

Preface

Life on our planet depends on insects. Their success depends, among others things, on highly developed communications, which have been described at various levels in most groups. Because the mechanisms used by the estimated millions of species to exchange information are highly diverse and in very different surroundings, knowledge about communication remains, in most cases, focused on a specific insect group or species. Stink bugs are diversified insects, with wonderful morphological and colour variations between species. They can be herbivores or carnivores and specialists or generalists, and they have a relatively long life cycle, representing a group of insects for which a body of broad knowledge on communication during mating behaviour is increasing. Through evolution, stink bugs developed a most effective intra- and interspecific communication that is in tune with numerous host plants and natural enemies. Because of these characteristics, stink bugs easily adapt and survive in various environmental conditions, such as natural and agricultural systems where several species represent economically important pests.

Increased amounts of new information on stink bug communication and their economic impact in the Neotropics and worldwide, together with an increasing value of the concept of biorational control, compelled us to summarize this knowledge in this book. During long-term joint research on stink bug communication and behaviour, performed by highly specialized groups in Brazil, Slovenia and worldwide, we learned that addressing a single modality gives information on mechanisms that underlie specific processes but ignores basic questions about what the species really drives toward in the field to search and select a mate. Wherever possible, we discuss laboratory results and field observations of species in their natural surroundings. Understanding stink bug communication requires important information on species phylogeny, classification and biology, knowledge on host plant-insect relationships and interactions at various trophic levels. Together with world specialists in systematics, ecology, physiology and communication from Brazil, USA and Europe, we present integrative updated knowledge about the stink bug communication process as the scientific basis for target-oriented monitoring and effective biorational control.

The book is written in a manner to be understood by the broader scientific and public community and to be used as a tool for development of various biological control methods. We are aware that no scientific work comes to an end and that any new information is incomplete and premature. Nevertheless, it has value as a trigger for the human curiosity that drives basic science and induces ideas about how to implement them in praxis. This is one of the reasons we are most grateful to our teachers and why we dedicate this book to all that admire life and insects. We do so with special care for our students, present and future, hoping they will get a chance to proceed with further and deeper investigations of various phenomena that underlie mechanisms driving communication processes in stink bugs and other insects.

Contents

List of Contributors

Maria Carolina Blassioli-Moraes
Laboratório de Semioquímicos, Embrapa Recursos Genéticos e Biotecnologia, Avda. W5 Norte (Final) 71070-917, Brasilia, DF, Brazil.
Email: carolina.blassioli@embrapa.br

Miguel Borges
Laboratório de Semioquímicos, Embrapa Recursos Genéticos e Biotecnologia, Avda. W5 Norte (Final) 71070-917, Brasilia, DF, Brazil.
Email: miguel.borges@embrapa.br

Douglas Henrique Bottura Maccagnan
Laboratory of Entomology, Universidade Estadual de Goiás, Campus Iporá, Avda R2, 76200-000, Iporá, GO, Brazil.
Email: douglas.hbm@ueg.br

Stefano Colazza
Dipartimento di Scienze Agrarie e Forestali, Università degli Studi di Palermo, Viale delle Scienze 13, Edificio 5, 90128 Palermo, Italy.
Email: stefano.colazza@unipa.it

Ted E. Cottrell
United States Department of Agriculture, Agricultural Research Service Southeastern, Fruit & Tree Nut Research Laboratory, Dunbar Road, Byron, GA 31008, USA.
Email: ted.cottrell@ars.usda.gov

Andrej Čokl
Department of Organisms and Ecosystems Research, National Institute of Biology, Večna pot 111, SI-1000 Ljubljana, Slovenia.
Email: andrej.cokl@nib.si

Jocelia Grazia
Universidade Federal do Rio Grande do Sul, Departamento de Zoologia, Av. Bento Gonçalves 9500, Prédio 43435, Sala 223, 91501-970 Porto Alegre, RS, Brasil.
Email: jocelia@ufrgs.br

Salvatore Guarino
Dipartimento di Scienze Agrarie e Forestali, Università degli Studi di Palermo, Viale delle Scienze 13, Edificio 5, 90128 Palermo, Italy.
Email: salvatore.guarino@unipa.it

Raul Alberto Laumann
Laboratório de Semioquímicos, Embrapa Recursos Genéticos e Biotecnologia, Avda. W5 Norte (Final) 71070-917, Brasilia, DF, Brazil.
Email: raul.laumann@embrapa.br

Tiago Lucini
Department of Zoology, Federal University of Paraná, PO Box 19020, Curitiba, PR 81531-980, Brazil.
Email: tiago_lucini@hotmail.com

Diego Martins Magalhães
Universidade de Brasília, Instituto de Ciências Biológicas, Departamento de Zoologia, Campus Universitário Darcy Ribeiro, CEP 70910-900 Brasília-DF, Brazil.
Email: magalhaes.dmm@gmail.com

Antônio Ricardo Panizzi
Laboratory of Entomology, Embrapa Wheat, PO Box 3081, Passo Fundo RS, 99001-970, Brazil.
Email: antonio.panizzi@embrapa.br

Ezio Peri
Dipartimento di Scienze Agrarie e Forestali, Università degli Studi di Palermo, Viale delle Scienze 13, Edificio 5, 90128, Palermo, Italy.
Email: ezio.peri@unipa.it

Cristiano F. Schwertner
Universidade Federal de São Paulo, Departamento de Ciências Biológicas, Rua Artur Riedel, 275, Eldorado, 09972270 Diadema, SP, Brasil.
Email: schwertner@unifesp.br

Nataša Stritih
Department of Organisms and Ecosystems Research, National Institute of Biology, Večna pot 111, SI-1000 Ljubljana, Slovenia.
Email: Natasa.Stritih@nib.si

Glynn P. Tillman
United States Department of Agriculture, Agricultural Research Service, Crop Protection & Management Research Laboratory, PO Box 748, Tifton, GA 31793, USA.
Email: glynn.tillman@ars.usda.gov

Meta Virant-Doberlet
Department of Organisms and Ecosystems Research, National Institute of Biology, Večna pot 111, SI-1000 Ljubljana, Slovenia.
Email: meta.virant@nib.si

Alenka Žunič-Kosi
Department of Organisms and Ecosystems Research, National Institute of Biology, Večna pot 111, SI-1000 Ljubljana, Slovenia.
Email: alenka.zunič-kosi@nib.si

Introduction

Miguel Borges[1,*] and *Andrej Čokl*[2]

The idea of writing this book originated from the broad and deep knowledge on stink bug communication, their economically important impact, especially in Neotropics, and in the concept of biorational control, presented and discussed in the book "Biorational Control of Arthropod Pests: Application and Resistance Management" edited by Isaac Ishaaya and A. Rami Horowitz (2009). In this book various possibilities for the manipulation of insect signalling in order to monitor and control insect pest species were described. The aim of the present book is to focus attention specifically on stink bugs of the subfamily Pentatominae in order to get a deeper insight into their group communication processes as the scientific basis for target-oriented monitoring and effective biorational control. The suborder Heteroptera, with about 37,000 described and perhaps another 25,000 not yet described species, represents one of the most abundant and diverse insect groups with incomplete metamorphosis (Panizzi et al. 2000). Pentatomidae, with an estimated 4570 species worldwide (McPherson and McPherson 2000), represent together with Lygaeidae, the third largest family within Heteroptera, being surpassed by Reduviidae and Miridae. Schuh and Slater (1995) divided the family Pentatomidae into the subfamilies; Asopinae, Cyrtocorinae, Discocephalinae, Edessinae, Pentatominae, Phyllocephalinae, Podopinae and Serbaninae. The recent classification of the Neotropical pest species of Pentatomidae was greatly influenced by the work that Rolston and collaborators carried out more than 25 years ago. However, the classification proposed by Rolston was much more utilitarian rather than reflecting the phylogeny of the group (Rider et al. in press), making the recognition of the evolutionary history of the group difficult. The

[1] Semiochemicals Laboratory, Embrapa Genetic Resources and Biotechnology, Avda W5 Norte (Final), 71070-917, Brasilia, DF, Brazil.

[2] Department of Organisms and Ecosystems Research, National Institute of Biology, Večna pot 111, SI-1000 Ljubljana, Slovenia.
Email: Andrej.Cokl@nib.si

* Corresponding author: miguel.borges@embrapa.br

current subfamilial classifications of the Pentatomidae (see Chapter 1) are based on the works of Gapud (1991) and Grazia et al. (2008), which strongly supported the majority of the subfamilies as monophyletic groups. The tribal classifications of the subfamily Pentatominae are derived from Rider (1998–2015), and provide a more realistic context for comparative studies. Concerning the generic level, only minor changes in the classification of the Neotropical species have been recorded. All the New World species of *Acrosternum* (e.g., *A. impicticorne, A. hilare, A. marginatum*) were included in the subgenus *Chinavia* by Rolston (1983). More recently, based on the study of genital characters, Schwertner (2005) and Schwertner and Grazia (2006, 2007) considered *Chinavia* to be a distinct monophyletic group, more related to *Nezara* than to *Acrosternum*, thus bringing support to *Chinavia* as a valid genus. Consequently, all species from the New World formerly included in *Acrosternum* should currently be incorporated in *Chinavia* (e.g., *Chinavia impicticornis, Chinavia hilaris, Chinavia marginata*). It is interesting to note that the chemical analysis of the sex pheromones of the *Chinavia* and *Nezara* species agrees with this study that is, the ratio between the two components (*cis* and *trans*-(Z)-bisosabolene-epoxide) of the blend of the sex-pheromones produced by the *Chinavia* spp. males is closer to that of *Nezara* than to that of *Acrosternum* (see Chapters 1, 5, Moraes et al. 2008).

The most economically important heteropteran species belonging to the Cimicomorpha and Pentatomorpha infraorders (Schaefer and Panizzi 2000) and different aspects of the biology and control of Heteroptera of economic importance are extensively described in the books edited by Carl W. Schaefer and Antonio Ricardo Panizzi (2000) and McPherson and McPherson (2000).

The aim of the present book is to give an up-to date compilation of the information available about communication in Pentatominae and Asopinae (Heteroptera: Pentatomidae) stink bugs and provide new ideas on their biorational control, based on broad knowledge of their mating behaviour and communication. Intra- and interspecific communication runs in stink bugs by chemical, vibrational and visual signals used simultaneously during the mating behaviour and host foraging, which will be presented and deliberated upon in Chapter 4. Their chemical communication is presented in several chapters of this book. Chapter 5 considers the progress made during the last decade in chemical communication, comprising of intra- and interspecific chemical communication by using stink bugs' pheromones and allomones and the use of their semiochemicals as kairomones by natural enemies. Chapter 8 deals with the communication between plants and stink bugs and their natural enemies comprising of chemical communication as well as the landscape (visual) influence of plants on stink bugs' host choice. This latter topic is also discussed in Chapter 2 together with the influence of the nutritional parameters of plants on stink bugs and the importance of the landscape on their foraging behaviour. The seasonal variation and geographical distribution of stink bug populations gives important background information which is necessary to study the different phenomena connected with communication. The application of chemical communication is covered in Chapter 11, where the advances of the pheromone for the monitoring of phytophagous stink bugs are discussed.

Vibrational communication is summarized in Chapters 6, 7 and 11, embracing three different aspects: in Chapter 6 the fundamentals of the substrate-borne vibratory communication in stink bugs are presented, Chapter 7 discusses the effects of biotic and abiotic factors on vibratory communication and Chapter 11 deals with advances obtained by utilising vibratory signals for stink bug monitoring and control. Furthermore, data on species phylogeny, general biology and reproductive behaviour are presented in Chapter 1. The stink bugs feed on the sap-fluid, developing fruits and seeds of the host plants. Only species of the subfamily Asopinae, as an apomorphy within the family are predators. Their chemical communication is discussed in Chapter 9, and their vibrational communication is expounded upon in Chapter 3. The last chapter (Chapter 12) of this book gives an overview and a summary of the main results obtained from the studies with stink bug communications and discusses the application of these results to develop new tools for biorational control.

Widely-used broad-spectrum insecticides have serious consequences on the environment and at long-term, the insecticides have its efficiency decreased on the target species. Arthropods become resistant through insecticide application, and the negative impact on plants and beneficial arthropods causes their secondary outbreaks and repeated population growth. Additionally, users are endangered by the residues in food, and the farmers are endangered due to their handling of insecticides. The use of low target-specific insecticides interferes with other control techniques and in many cases all of the above mentioned side-effects combine and lead to unexpected damage at different levels. Development of alternative control measures became more and more important with the increasing constraints of intensive insecticide use. Along this time Stern and co-authors (1959) proposed the "Integrated Control" concept that provided fundamental basis for the "Integrated Pest Management" (IPM) strategy.

The IPM programs combine tactics, like the use of predators, parasites and competitors, implementation of different cultures that decrease pest population and methods to directly kill pests or change their habitat to make it unsuitable for them. Target-specific chemical control combined with previously mentioned measures can be used after relevant monitoring. The goal of guidelines-established use of pesticides is to impose minimal risk on environment, human health and non-target and beneficial organisms.

Biorational approach provides an opportunity to upgrade the IPM strategies (Ishaaya 2003, Horowitz and Ishaaya 2004, Ishaaya et al. 2005).

The term "biorational" has been used inconsistently in different contexts. The variety of its definitions and descriptions are reviewed and discussed by Horowitz and co-authors (2009). They propose its descriptive use as an adjective characterizing compatibility with living systems within specific contexts. In such a way an insecticide may be biorational in one or not in another system. In this book, we use the term "biorational" following the Horowitz and co-authors' proposal (2009) as an adjective: processes that have little or no consequence for the environment and non-target species when applied in a specific manner or ecological context, with the goal to have lethal or other suppressive or behaviour modifying action on the target organism, augmenting the control system.

The reader will get an excellent overview on chemical and vibrational communication, but will note that visual and contact mechanical communication are scarcely covered in this book, since there is still little information available on this subject as far as stink bugs are concerned. In addition, chemical communication mainly discusses the biological and behavioural aspects, since there are two excellent recent reviews (Weber et al. in press, Millar 2005) covering more chemical aspects, like synthesis and structural elucidation. The update given here is to be the base for further studies in the future, to develop new tools to control and monitor herbivorous pests in order to reach a more biorational agricultural system.

Acknowledgments

We thank Dr. C.F. Schwertner for providing information in the taxonomy section. Also we would like to thank the National Council for Scientific and Technological Development (CNPq), the Brazilian Corporation of Agricultural Research (EMBRAPA), the Research Support Foundation of the Federal District (FAP-DF) and the Slovene Research Agency.

References

Gapud, V.P. 1991. A generic revision of the subfamily Asopinae, with consideration on its phylogenetic position in the family Pentatomidae and superfamily Pentatomoidea (Hemiptera: Heteroptera). Philippines Entomol. 8: 865–961.

Grazia, J., C.F. Schwertner and A. Ferrari. 2006. Description of five new species of Chinavia Orian (Hemiptera, Pentatomidae, Pentatominae) from western and northwestern South America. Denisia 19: 423–434.

Grazia, J., R.T. Schuh and W.C. Wheeler. 2008. Phylogenetic relationships of family groups in Pentatomoidea based on morphology and DNA sequences (Insecta: Heteroptera). Cladistics 24: 932–976.

Grazia, J., A.R. Panizzi, C. Greve, C.F. Schwertner, L.A. Campos, T.A. Garbelotto et al. 2015. Stink bugs (Pentatomidae). pp. 681–756. *In*: Panizzi, A.R. and J. Grazia (eds.). True Bugs (Heteroptera) of the Neotropics. Springer, Dordrecht, GE.

Horowitz, A.R. and I. Ishaaya. 2004. Biorational insecticides mechanisms, selectivity and importance in pest management programs. pp. 1–28. *In*: Horowitz, A.R. and I. Ishaaya (eds.). Insect Pest Management- Field and Protected Crops. Springer, Berlin, Heidelberg, New York.

Horowitz, A.R., P.C. Ellsworth and I. Ishaaya. 2009. Biorational pest control an overview. pp. 1–20. *In*: Ishaaya, I. and A.R. Horowitz (eds.). Biorational Control of Arthropod Pests: Application and Resistance Management. Springer, Dordrecht, Heidelberg, London, New York.

Ishaaya, I. 2003. Biorational insecticides mechanism and application. Arch. Insect Biochem. Physiol. 54: 144.

Ishaaya, I., S. Kontsedalov and A.R. Horowitz. 2005. Biorational insecticides: mechanism and cross-resistance. Arch. Insect Biochem. Physiol. 58: 192–199.

Ishaaya, I. and A.R. Horowitz. 2009. Biorational Control of Arthropod Pests Application and Resistance Management. Issac Ishaaya and A. Rami Horowitz (eds.). Springer, Berlin.

McPherson, J.E. and R.M. McPherson. 2000. Stink Bugs of Economic Importance in America and North of Mexico. CRC Press, Boca Raton, London, New York, Washington D.C.

Millar, J.G. 2005. Pheromones of true bugs. pp. 37–84. *In*: Schulz, S. (ed.). The Chemistry of Pheromones and other Semiochemicals II—Topics in Current Chemistry, Volume 240. Springer-Verlag Berlin, Heidelberg, Germany.

Panizzi, A.R., J.E. McPherson, D.G. James, M. Javahery and R.M. McPherson. 2000. Stink bugs (Pentatomidae). pp. 421–474. *In*: Schaefer, C.W and A.R. Panizzi (eds.). Heteroptera of Economic Importance. CRC Press, Boca Raton, Florida, USA.

Rider, D.A. 1998–2015. Pentatomoidea home page. North Dakota State University, Fargo, ND, USA. Available at: https://www.ndsu.edu/pubweb/~rider/Pentatomoidea/ [Accessed in 13 May 2016].

Rolston, L.H. 1974. Revision of the genus *Euschistus* in Middle America (Hemiptera, Pentatomidae, Pentatomini). Entomologica Americana 48: 1–102.

Rolston, L.H. 1978. A new subgenus of *Euschistus* (Hemiptera: Pentatomidae). Journal of the New York Entomological Society 86: 102–120.

Rolston, L.H. and F.J.D. McDonald. 1979. Keys and diagnoses for the families of western hemisphere Pentatomoidea, subfamilies of Pentatomidae and tribes of Pentatominae (Hemiptera). Journal of the New York Entomological Society 87: 189–207.

Rolston, L.H., F.J.D. McDonald and D.B. Thomas Jr. 1980. A conspecuts of Pentatomini genera of the western hemisphere. Part I (Hemiptera: Pentatomidae). Journal of the New York Entomological Society 88: 120–132.

Rolston, L.H. 1981. Ochlerini, a new tribe in Discocephalinae (Hemiptera: Pentatomidae). J. New York Entomol. Soc. 89: 40–42.

Rolston, L.H. 1982. A revision of Euschistus Dallas subgenus Lycipta Stal (Hemiptera: Pentatomidae). Proceedings Entomological Society of Washington 84: 281–296.

Rolston, L.H. 1983. A revision of the genus *Acrosternum* Fieber, subgenus *Chinavia* Orian, in the western hemisphere (Hemiptera: Pentatomidae). Journal of the New York Entomological Society 91: 97–176.

Rolston, L.H. 1984. Key to the males of the nominate subgenus of *Euschistus* in South America, with descriptions of three new species (Hemiptera: Pentatomidae). Journal of the New York Entomological Society 92: 352–364.

Schaefer, C.W. and A.R. Panizzi. 2000. Economic importance of Heteroptera: A general view. pp. 3–8. *In*: Schaefer, C.W. and A.R. Panizzi (eds.). Heteroptera of Economic Importance. CRC Press, Boca Raton, London, New York, Washington D.C.

Schuh, R.T. and J.A. Slater. 1995. True bugs of the world (Hemiptera: Heteroptera). Classification and natural history. Cornell University Press, Ithaca, New York, U.S.A.

Schwertner, C.F. 2005. Phylogeny and classification of the green stink bugs of *Nezara group* of genera (Hemiptera, Pentatomidae, Pentanominae). PhD Thesis, Universidade Federal do Rio Grande do Sul, RS, Brazil. [in Portuguese, abstract in English] Available at: http://www.lume.ufrgs.br/handle/10183/1/.

Schwertner, C.F. and J. Grazia. 2006. Description of six new species of *Chinavia* (Hemiptera, Pentatomidae, Pentatominae) from South America. Iheringia, Série Zoologia 96: 237–248.
Schwertner, C.F. and J. Grazia. 2007. O gênero *Chinavia* Orian (Hemiptera, Pentatomidae, Pentatominae) no Brasil, com chave pictórica para os adultos. Revista Brasileira de Entomologia 51: 416–435.
Stern, V.M., R.F. Smith, R. Van den Bosch and K.S. Hagen. 1959. The integration of chemical and biological control of the spotted alpha aphid. The integrated control concept. Hilgardia 29: 81–101.
Weber, Donald C., A. Khrimian, M.C. Blassioli-Moraes and J.G. Millar. Semiochemistry of Pentatomoidea. Chapter 15 *In*: McPherson, J. et al. (eds.). Biology of Invasive Stink Bugs and Related Species. CRC Press, in press.

CHAPTER 1

Stink Bug Classification, Phylogeny, Biology and Reproductive Behavior

Jocelia Grazia[1,*] and *Cristiano F. Schwertner*[2]

Introduction

The family Pentatomidae encompasses a monophyletic group of true bugs (Grazia et al. 2008), representing the fourth largest family within Heteroptera and one of the most diverse groups of hemimetabolous insects (Fig. 1.1). The pentatomids are known as stink bugs and green bugs, and contain about 900 genera and more than 4,700 species in the world. Nine subfamilies are recognized (Grazia et al. 2008, Rider 1998–2015), half of them with restricted distribution (Table 1.1).

The size ranges from small to medium, the body from more or less oval to elongated, the antennae are usually 5-segmented (4-segmented, e.g., *Cyrtocoris* White and *Peromatus* Amyot and Serville, or 3-segmented, e.g., *Omyta* Spinola), the scutellum always surpassing the middle of the abdomen, subtriangular to spatulate in shape. Some species may have most of the abdomen covered by the scutellum (e.g., some Asopinae, Pentatominae and Podopinae), a convergence characteristic found in other Pentatomoidea families, most notably in the Scutelleridae. Several species are bright-colored or have conspicuous patterns, and pentatomids are the most common and abundant true bugs that release an unpleasant odor (hence the name stink bugs).

[1] Universidade Federal do Rio Grande do Sul, Departamento de Zoologia, Av. Bento Gonçalves 9500, Prédio 43435, Sala 223, 91501-970 Porto Alegre, RS, Brasil. Email: jocelia@ufrgs.br

[2] Universidade Federal de São Paulo, Departamento de Ecologia e Biologia Evolutiva, Rua Artur Riedel 275, Eldorado, 09972-270 Diadema, SP, Brasil. Email: schwertner@unifesp.br

* Corresponding author

Figure 1.1. Examples of stink bug diversity and behavior. 1.1a, *Grazia tincta*, egg mass (ecloded) and 1st instar nymphs' aggregation. 1.1b, *Chinavia ubica*, sucking on a soybean leaf stem. 1.1c, *Carpocoris pudicus* (courtesy of D. Takiya). 1.1d, *Dinocoris gibbus*, females showing maternal care to eggs and 1st instar nymph.

Table 1.1. Classification of the family Pentatomidae, with the number of genera and species and distribution of each recognized taxon. The classification follows Grazia et al. (2015).

Subfamily	# Gen	# spp	Distribution
Aphylinae	2	3	Australian
Asopinae	66	299	World
Cyrtocorinae	4	11	Neotropical
Discocephalinae	76	315	Neotropical
Edessinae	7	306	Neotropical
Pentatominae	621	3336	World
Phyllocephalinae	45	213	Afrotropical, Australian, Oriental, Palearctic
Podopinae	64	249	World
Stirotarsinae	1	1	Neotropical

Most of the species are phytophagous, feeding on the sap-fluid, developing fruits and seeds of the hosts. Only the species of the subfamily Asopinae are predators, an apomorphy within the family. Several species are related to agroecosystems, and the family includes some of the most important crop pests among true bugs.

Current Classification

Traditionally, seven infraorders are recognized within Heteroptera, each representing a monophyletic lineage (Schuh and Slater 1995, Weirauch and Schuh 2011). The family Pentatomidae is included in the infraorder Pentatomomorpha, which contains more than 16,000 species of true bugs, and together with the infraorder Cimicomorpha comprises a clade also known as Geocorise, including almost ninety percent of the diversity within Heteroptera. Among pentatomomorphans, the superfamily Pentatomoidea represents a monophyletic group, including Pentatomidae and related families, most of them with similar facies and blurred classification history (Rider et al., in press). The monophyly of both, the superfamily Pentatomoidea and the family Pentatomidae has been recovered in every study including molecular data (i.e., Li et al. 2005, Xie et al. 2005, Li et al. 2012, Yuan et al. 2015), morphological data (i.e., Gapud 1991, Henry 1997, Grazia et al. 2008, Yao et al. 2013) or both (i.e., Grazia et al. 2008).

The taxonomic history of the family Pentatomidae had been in a state of flux until recently (Rider 2000, Cassis and Gross 2002), although the current subfamilial classification is now more or less stable (Rider et al., in press). The modern concept of the family was first arranged by Gross (1975b), and further developed by Rolston and McDonald (1979). However, these two works differed widely in the proposed infrafamilial classification and were strongly regional in scope, hampering an effective unifying classification. Gapud's (1991) proposal includes the recognition of nine subfamilies within Pentatomidae, a classification strongly supported by the results of Grazia et al. (2008) and followed here (Table 1.1). Some authors considered a tenth subfamily, Serbaninae, as a part of the Pentatomidae, a result not supported by Grazia et al. (2008). A tabular key to identify each of the subfamilies referenced in this chapter is presented in Table 1.2.

The subfamily **Aphylinae** Bergroth was established as a separated group because *Aphylum syntheticum* Bergroth was considered to be an isolated taxon combining characters of the pentatomoid family-groups Scutellerinae, Graphosomatinae, Plataspinae and Pentatominae (Bergroth 1906, Schouteden 1906). Reuter (1912) raised it to family rank, a classification accepted by Schuh and Slater (1995) and Cassis and Gross (2002), but Rider (2006) and Grazia et al. (2008) returned the taxon to subfamily rank within Pentatomidae. Grazia et al. (2008) choose to treat Aphylinae at subfamily rank in recognition of their many shared similarities with the Pentatomidae *sensu stricto*, as mentioned above.

Asopinae has been recognized as a suprageneric group at least since Amyot and Serville (1843), who named the group Spissirostres. Taxonomy of the subfamily was revised by Thomas (1992, 1994). De Clerq (2000) reviewed the most studied species of asopines (main and secondary) on a world

Table 1.2. Tabular identification key to the subfamilies of the Pentatomidae (characters in **bold** represent diagnostic characters that allow unequivocally the identification of the subfamily).

	Aphylinae	Asopinae	Cyrtocorinae	Discocephalinae	Edessinae	Pentatominae	Phyllocephalinae	Podopinae	Stirotarsinae
Body shape	**Oval**	Obovate	Quadrangular	Obovate	Obovate to elongated	Obovate to elongated	Elongated	Oval to elongated	Obovate
Surface of the body	Punctuated	Punctuated, sometimes scarcely	Strongly punctuated	Variable	Punctuated, sometimes scarcely	Variable	Variable	Punctuated to strongly punctuated	Strongly punctuated
Number of antennal segments	5	5	4	5 or 4	5 or 4	5, 4, or 3	5	5	5
Rostral development	Not Crassate	**Crassate**	Not Crassate	Not Crassate	Not Crassate	Not Crassate	Not Crassate	Not Crassate	Not Crassate
Length of rostrum	Reaching mesosternum	Extending at least to mesocoxae	Extending at least to mesocoxae	Extending at least to mesocoxae	**Attaining mesocoxae**	Variable, but always surpassing procoxae	**Attaining procoxae**	Variable, but always surpassing procoxae	Attaining metacoxae
Location of the labium insertion (ventral view)	At the same line of the eyes	**Above the line of the eyes**	At the same line of the eyes	**At the same or below the line of the eyes**	At the same line of the eyes	At the same line or above the line of the eyes	At the same line of the eyes	At the same line or above the line of the eyes	**Above the line of the eyes**
First and part of second rostral segment inside the bucculae	No	No	No	No	No	No	**Yes**	No	No

Shape of the scutellum	**Ovate, reaching the apex of the abdomen**	Triangular or elongated, but never reaching the apex of the abdomen	Elongated, never reaching the apex of the abdomen	Triangular to elongated, but never reaching the apex of the abdomen	Triangular, never reaching the apex of the abdomen	Triangular to elongated, but never reaching the apex of the abdomen	Triangular, never reaching the apex of the abdomen	**Elongated, never reaching the apex of the abdomen**	Triangular, never reaching the apex of the abdomen
Dorsal thorn-like projection of the scutellum	Absent	Absent	**Present**	Absent	Absent	Absent	Absent	Absent	Absent
Exponium	**Present**	Absent	Absent	Absent	Absent	Absent	Absent	Absent	Absent
Metasternum development	Never developed	Never developed	Never developed	Never developed	**Developed, anterior margin projected over mesosternum**	Elevated or not, never projected over the mesosternum	Never developed	Never developed	Never developed
Number of tarsal segments	3	3	**2**	3	3	3 or 2	3	3	**2**
Longitudinal carina on tarsi	Absent	Absent	Absent	Absent	Absent	Absent	Absent	Absent	**Present**
Excavation of the last tarsal segment in females	Absent	Absent	Absent	**Present**	Absent	Absent	Absent	Absent	Absent
Paired trichobothria	Present or Absent	Present	Present	Present	Present	Present	Present	**Trichobothria single**	Present

basis, and discussed information available for each one. Discussions on phylogenetic relationships were done by Gapud (1991), Hasan and Kitching (1993), Gapon and Konstantinov (2006), Gapon (2010) and Gapud (2015).

The subfamily **Cyrtocorinae** Distant, endemic to the Neotropics was first recognized as a group by Amyot and Serville (1843). It was revised by Packauskas and Schaefer (1998) and raised to family rank, although Gapud (1991) had separated Cyrtocorinae (as a pentatomid subfamily) from the rest of Pentatomidae on the basis of the genitalic characters of both sexes. As for the Aphylinae, Grazia et al. (2008) chose to treat Cyrtocorinae at subfamily rank in recognition of their many shared similarities with the Pentatomidae *sensu stricto*, as mentioned above.

Classification of the **Discocephalinae** is still at an early stage. The group was proposed as a separated family by Fieber (1860) to include *Discocephala* Laporte, *Dryptocephala* Laporte and *Platycarenus* Fieber, currently representatives of the tribe Discocephalini. The tribal arrangement adopted today is the same since Rolston (1981).

The subfamily **Edessinae** was proposed by Amyot and Serville (1843) to be included within the group Brevirostres (together with the Phyllocephalinae), and contained seven genera until recently: *Edessa* F., *Brachystethus* Laporte, *Peromatus* Amyot and Serville, *Olbia* Stål, *Pantochlora* Stål, *Doesburgedessa* Fernandes, and *Paraedessa* Silva and Fernandes (Rolston and McDonald 1979, Barcellos and Grazia 2003, Fernandes 2010, Silva et al. 2013). Other two genera were recently described, based on species formerly included in *Edessa* (Santos et al. 2015, Correia and Fernandes 2016). Rider et al. (in press) recognized other six genera in this subfamily, including *Lopadusa* Stål, *Mediocampus* Thomas, *Neopharnus* Van Duzee, *Praepharnus* Barber and Bruner, *Pharnus* Stål and *Platistocoris* Rider. Revisionary studies are recognizing different groups of species in *Edessa* (e.g., Fernandes and Van Doesburg 2000, Fernandes and Campos 2011, Silva and Fernandes 2012). The heterogeneity of *Edessa* can be confirmed by its confusing taxonomical and nomenclatural history that culminates with Kirkaldy's catalogue (1909) where he listed the subgenera but considered himself unable to include the species in them.

The subfamily **Pentatominae** is a heterogeneous taxon that includes several genera and groups of genera not recognized as belonging to any of the other subfamilies (Grazia et al. 2015, Rider et al., in press), and does not have a definition based on unique characteristics. Therefore, authors differ about the classification of the subfamily (Cassis and Gross 2002). It was first recognized by Stål (1865) to include five generic groups of Amyot and Serville (1843), Halydes, Pentadomides, Pododides, Raphigastrides and Sciocorides. Currently, the subfamily may be divided into more than 40 tribes (Rider 1998–2015, Rider et al., in press), most of them never tested in a phylogenetic context. Schuh and Slater (1995), for example, included eight tribes in Pentatominae, from which only Pentatomini is registered in the Neotropical Region. Rider et al. (in press) considers there to be 42 tribes in Pentatominae.

Phyllocephalinae was recognized as a group of genera by Amyot and Serville (1843). The current tribal classification is based mostly on Linnavuori (1982), Ahmad and Kamaluddin (1988, 1990) and Kamaluddin and Ahmad (1988), and four tribes are recognized (Rider et al., in press).

Podopinae was first recognized by Stål (1876) as a group of genera. Lethierry and Severin (1893) gave the status of subfamily to the group. The current classification was established by Dávidová-Vilimová and Stys (1994), and included five species groups (Rider et al., in press): *Brachycerocoris* group, *Deroploa* group, *Graphosoma* group, *Podops* group and *Tarisa* group.

More recently, Rider (2000) proposed **Stirotarsinae** for the monotypic genus *Stirotarsus* Bergroth based on the unique antennal, rostral and tarsal characters, along with the relative rare ostiolar, tibial and spiracular characters. Bergroth (1911) included *Stirotarsus* among Asopinae based primarily on the dilated structure of the tibiae. Gapud (1991) excluded it from the Asopinae and tentatively placed *Stirotarsus* in the Dinidoridae. Thomas (1992) confirmed the exclusion from Asopinae, but indicated that the determination of its proper placement would require further studies.

Pentatomidae Phylogeny: Where Are We?

The phylogenetic relationships among pentatomid subfamilies and tribes are almost completely ignored. In a series of papers, Leston (e.g., 1953, 1954a, 1954b) was the first to document and discuss the monophyly and phylogenetic relationships within Pentatomidae. McDonald (1966), Gross (1975b) and Linnavuori (1982) expanded these studies and suggested apomorphic characters for the recognition of several monophyletic groups within the family. Gross (1975b) and Linnavuori (1982) also discussed possible phylogenetic relationships among those groups. However, all these studies were limited in scope (e.g., based in regional fauna) and were carried out without any explicit methodology.

The first contribution on pentatomid relationships using phylogenetic methodology was that of Gapud (1991); he analyzed 41 characters in 13 terminal taxa following the prior general schemes that had divided Pentatomoidea into 11 families. Gapud (1991) considered the clade including ((Lestonidae + Plataspidae) (Thyreocoridae + Cydnidae) + Pentatomidae) as the most derived within pentatomoids, the family Pentatomidae supported by six synapomorphies including mostly male and female genitalia.

More recently, in a phylogenetic analysis using both morphology and molecular sequences, Grazia et al. (2008) supported the conclusions of Gapud, recovering Pentatomidae as monophyletic. Morphological characters of the female genitalia (loss of gonapophyses 8 and first rami, gonapophyses 9 reduced and fused to gonocoxites 9, gonangulum absent, and ductus receptaculi dilated and invaginated, forming three distinct walls) and 28S partition gave strong support to these results.

Although well-supported as a natural group, the monophyly and relationships of the subfamilies and tribes within Pentatomidae are mostly unknown. Some of the subfamilies have strong support from morphological characters (e.g., Aphylinae, Asopinae, Cyrtocorinae, Phyllocephalinae), but only Discocephalinae and Edessinae have been explicitly tested in a phylogenetic context (Barcellos and Grazia 2003, Campos and Grazia 2006). Also most of the recognized groups were never included in the studies using molecular data. The morphological analysis in Grazia et al. (2008) recovered the following relationship (Cyrtocorinae (Aphylinae + Pentatomidae *sensu stricto*)).

Although the group were never tested by a phylogenetic analysis, the unique feeding habits and the associated modifications strongly corroborate the monophyly of the subfamily **Asopinae**. Some groups of genera are recognized within the subfamily (e.g., Rider et al., in press), nevertheless the phylogenetic relationships within these taxa are poorly understood.

Campos and Grazia (2006) recovered the monophyly of the **Discocephalinae**, supported by two exclusive synapomorphies, first segment of labium reaching the prosternum, and the dorsal surface of the male proctiger membranous at basal third, and one homoplastic synapomorphy, the metasternum with the median carina. Although a first cladistic approach is available for Ochlerini (Campos and Grazia 2006), hypotheses of the relationships among the genera of Discocephalini are based on the morphological interpretations by previous authors (e.g., Ruckes 1961, Becker 1977, Rolston 1990), but are lacking at all for several genera. Recently Garbelotto (personal communication) in a cladistic analysis of the tribe Discocephalini, based on morphological characters, questioned the monophyly of Discocephalinae and indicated the need for a new arrangement for the tribe Discocephalini. Relationships with other groups of Pentatomidae are still unknown.

The subfamily **Edessinae** as conceived in Barcellos and Grazia (2003) represents a monophyletic group supported by several characteristics, but its composition at supra-specific level is still not clear due to *Edessa* that is hiding the real diversity of the subfamily. The classification proposed by Rider et al. (in press) needs to be tested. The main synapomorphy supporting Edessinae is the well-developed metasternum, elevated and projected anteriorly to the mesosternum (Barcellos and Grazia 2003).

Hasan and Kitching (1993) provided a cladistic analysis of some of the tribes of **Pentatominae,** but did not propose a revised classification for the group. The group as currently conceived is probably polyphyletic, at least based on the evidence available so far (e.g., Gapud 1991, Hasan and Kitiching 1993, Campos and Grazia 2006, Grazia et al. 2008). The relationship among tribes is unknown, and most of the tribes do not have unique diagnostic characteristics to define them, which make the establishment of the exact composition of these groups difficult. Schaefer and Ahmad (1987) and Schwertner and Grazia (2012) proposed phylogenetic hypotheses within the tribes Lestonocorini and Procleticini, respectively. Memon et al. (2011) in a phylogeny of the South Asian halyine established the monophyly of the tribe, although its taxonomy and systematic position need revision (Wall 2004, Barão et al. 2012).

Studies on relationships of the taxa within **Phyllocephalinae** are still lacking, but some relationship with other groups of stink bugs with short rostrum (e.g., Edessinae) was suggested (Gapud 1991, Hasan and Kitching 1993). The monophyly of the group was never tested, although it was strongly corroborated by the unique morphology of the head, rostrum and male genitalia (Linnavuori 1982, Hasan and Kitching 1993, Konstantinov and Gapon 2005).

The monophyly of the **Podopinae** has been questioned by some authors in the past (e.g., Davidová-Vilímová and McPherson 1995), but currently most authors have accepted it as a natural taxon (Rider et al., in press). Relationships within the subfamily recently became available, and indicate the *Podops* group to be a sister group to a clade including all other groups of genera (Rider et al., in press).

Biology and Reproductive Behavior

Stink bugs are typical sucking insects and phytophagy is the ancestral condition in the family (Schuh and Slater 1995, Schaefer 2009) (Fig. 1.1b). They are generally recorded feeding on different parts of the hosts (e.g., leaves, stems, branches and vascular system), but the main feeding sites are the reproductive parts, like the developing fruits and seeds (Schuh and Slater 1995, Panizzi 1997, 2000, Schaefer 2009). Those species of the subfamily Asopinae which are predators of other insects and small arthropods, are usually slow and soft-bodied individuals, e.g., larvae of Lepidoptera, Diptera, and Hymenoptera (Schuh and Slater 1995, De Clercq 2008). For both phytophagous and predatory species, polyphagy seems to be the rule, but some may have a more restricted diet (e.g., oligophagous) or may be limited to specific plant parts (phytophagous) or habitats (predators) (De Clercq 2000, Schaefer 2009, De Clercq 2008). Among the phytophagous stink bugs, some species show a preference for certain plant families (Panizzi 1997, 2000, Yonke 1990) which suggests that some hosts (or part of that host) may be more important to early nymphs developing, while late instars and adults may be less demanding in nutritional terms (or more "generalists").

The life cycle of phytophagous pest species has been studied more extensively (e.g., *Aelia* spp., *Chinavia* spp., *Edessa* spp., *Euchistus* spp., *Nezara* spp.), and the accounts available in the literature about stink bugs' biology are mostly based on them (Panizzi and Slansky 1985, Panizzi 1997, McPherson and McPherson 2000, Panizzi et al. 2000, Schaefer 2009, Panizzi et al. 2012 and references in there). In early spring, overwintered adult females seek out suitable hosts and typically deposit their eggs on wild host plants. Often these overwintering populations are found along field borders, particularly along tree lines near their overwintering sites. As soon as the reproductive peak of the host-plant decreases, the adults disperse looking for better feeding sites. Later-developing cultivated plants become more attractive when the initial wild hosts dry down, and their proximity allows easy access for stink bug colonization in crops. Adults of predatory species emerge and start to feed and reproduce using different prey.

Mating behavior (pre-copula and copula) is recorded in several species and seems to be the rule among pentatomids. Precopulatory events in Pentatomidae may be divided into two components: long and mid-range attraction mediated by pheromones and sound emission, and short-range courtship characterized by coordinated physical interaction (Borges et al. 1987, Blassioli-Moraes et al. 2014, Laumann et al. 2013). The composition of sex pheromones has been identified for several species of economic importance (e.g., Leal et al. 1998, Borges and Aldrich 1994) and males are responsible for the emission of such signals; structural signatures of sex pheromones in pentatomids can be observed exclusively in closely related species (Moraes et al. 2008). Even though pheromones seem to be species-specific in composition and/or ratio of compounds (Aldrich 1991), females may be unable to identify conspecific males.

Calling songs in Pentatomidae are emitted through a contraction and retraction of the abdominal tergites and their contact with plant substrates (Ryan and Walter 1992, Čokl and Virant-Doberlet 2003). The amount of sex pheromones exhaled by males appears to be regulated, at least in part, in response to female calling songs (Miklas et al. 2003).

Short-ranged courtship includes an initial phase in which the males touch the female bodies with their antennae (= antennation). Following that, the males usually use their heads to touch the female abdomen, frequently attempting to lift the female abdomen (Owusu-Manu 1980, Drickamer and McPherson 1992). The females may abandon courtship after this behavior (McLain 1981, Borges et al. 1987). If the female is receptive, she may stay motionless or elevate her abdomen, while the male turns in the opposite direction. Male pygophore is then extruded and rotated 180°; genital coupling is achieved with the individuals in necessarily end-to-end position (Borges et al. 1987, Blassioli-Moraes et al. 2014). The process of mate choice may be influenced by multiple factors, and males and females usually choose mates with the higher body size. Previous copula experience also seems important. All species studied to date show a polygamous mating system with both males and females copulating multiple times.

Nearly all species studied to date show very long copula, ranging from a few hours to six days, and this strategy has been associated with increased reproductive success in males of Pentatomidae (McLain 1991, Wang and Millar 1997, Rodrigues et al. 2008).

Pentatomid eggs are laid in clusters with tight rows, usually on the underside of leaves but different parts of the host plant can also be used (e.g., Fig. 1.1d). Eggs with translucent chorion are more common in Pentatomidae than eggs with pigmented chorion (Fig. 1.1a, d). The number of aero-micropylar processes varies from 10 to 86, but is rarely used for specific recognition, due to its overlapping in congeneric species. The number of eggs per mass is relatively constant within the Pentatomidae species, but may vary between species, between individuals and even between clutches, and is related to the number of ovarioles in females and the oviposition strategy of each species. In most of the species small egg masses are laid, usually with 14 eggs which is the number of ovarioles more frequently found in Pentatomidae (Pendergrast 1957); however, some species may lay egg masses of up to 300 eggs, e.g., *Nezara viridula* (Panizzi et al. 2000).

There are five nymphal instars, and the first instars only need moisture for survival. Esquivel and Medrano (2014) recorded feeding during the first instar. The duration of the development from the egg to adulthood is about 48 to 80 days. Life history traits are strongly influenced by abiotic conditions (temperature and humidity) and food quality, and between one to seven generations can occur each year. Emerging nymphs are gregarious and remain on/near the egg mass, then begin to feed and disperse as they grow. Diapause in the adult stage is common in high latitudes and regions with marked seasonality, and swarming is recorded in some species. Parasitoids of eggs (Hymenoptera: Platygastridae) and adults (Diptera: Tachinidae) are extensively studied in those species which are of economic importance.

One of the striking aspects of the pentatomids' biology is the parental care shown by several species of different subfamilies and tribes, which seems to have evolved independently several times within the Heteroptera. Within Pentatomidae, this behavior is better known in the tribe Discocephalini (Discocephalinae), with females protecting the eggs and young nymphs against parasitoids and predators (e.g., Eberhard 1975, Santos and Albuquerque 2001). Maternal care has been recorded in the Discocephalini genera *Antiteuchus* Dallas, *Dinocoris* Burmeister (Fig. 1.1d), *Eurystethus* Mayr, and *Mecistorhinus* Dallas. Also, at least three other lineages showed maternal care behavior: *Cyrtocoris* spp. (Cyrtocorinae; Brailovsky et al. 1988), an unidentified species of *Chlorocoris* (Pentatominae; Goula

2008) and *Bromocoris souefi* (Pentatominae: Halyini; Monteith 2006). More recently, unique cases of paternal care (when males take care of the young) in Pentatomidae were reported in the Edessinae species *Edessa nigropunctata* and *Lopadusa augur* (Requena et al. 2010).

Trophobiosis between *Eurystethus microlobatus* Ruckes (Discocephalini) and ants was discovered and described in the Brazilian Cerrado recently, a behavior recorded in other true bug families but previously unknown among pentatomids (Guerra et al. 2011).

Species of economic importance

More than one hundred species of phytophagous stink bugs with economic importance are known worldwide, including 4 subfamilies (Table 1.3). Some species are considered to be secondary minor or local pests, while some can damage extensive areas in certain parts of the world. At least one species, *Nezara viridula* (Fig. 1.2), is considered to be the main pest in all continents and may be considered synantropic, an example of the plasticity that stink bugs can reach as pests (e.g., Panizzi et al. 2000). More recently, the invasive species *Bagrada hilaris* (Fig. 1.3) and *Halyomorpha halys* (Fig. 1.4) in the northern hemisphere and *Euschistus heros* (Fig. 1.5) in South America have put applied entomologists on alert (e.g., Reed et al. 2013, Cesari et al. 2015, Smaniotto and Panizzi 2015).

A detailed account of all species is beyond the scope of this chapter. In this section, we will provide an annotated list of references concerning the classification and identification of the major species of economic importance. The tabular key presented in Table 1.2 allows the identification of all subfamilies discussed in this section. A key to the identification of all tribes of Pentatominae is available in Rider et al. (in press).

Subfamily Discocephalinae

Tribe Ochlerini

The two genera of discocephalines that include species of economic importance are part of the tribe Ochlerini, a group that has been studied intensively recently (Campos and Grazia 2006, Garbelotto et al. 2013, Roell and Campos 2015). The tribe is supported as a monophyletic group (Campos and Grazia 2006, Roell and Campos 2015). Garbelotto et al. (2013) included both genera in the *Herrichella* clade, with *Lincus* representing a more basal lineage within the clade. Rolston (1992) provided a key that allows identification of most genera.

The genus *Lincus* is the most speciose genus within the tribe, with 35 spp. described (Rolston 1983b, Grazia et al. 2015). Besides *L. lobuliger* (Fig. 1.6), several species are possible vectors of phytomonas to palms

Table 1.3. Classification of the stink bugs of economic importance in the world, including distribution, main crop and pest status. The classification follows Rider (1998–2015). Biogeographical regions: AU = Australian; AF = Afrotropical; NA = Nearctic; NT = Neotropical; O = Oriental; P = Palearctic; WORLD = distributed in all regions; M = main pest; S = secondary or minor pest.

Subfamily/Tribe	Species	Distribution	Main Crop(s)	Status
DISCOCEPHALINAE				
Ochlerini	*Lincus lobuliger* Breddin	NT	Palms	M
	Macropygium reticulare (Fabricius)	NT	Palms	S
EDESSINAE	*Edessa meditabunda* (Fabricius)	NT	Alfalfa, Pea, Soybean, Potato, Tomato	M
	Edessa rufomarginata (De Geer)	NT	Potato, Tobacco, Tomato	S
PENTATOMINAE				
Aelini	*Aelia acuminata* (Linnaeus)	O, P	Grains	S
	Aelia americana Dallas	NA	Grains	S
	Aelia furcula Fieber	P	Grains	M
	Aelia germari Küster	P	Grains	M
	Aelia melanota Fieber	P	Grains	M
	Aelia rostrata Boheman	P	Grains	M
	Aelia virgata (Herrich-Schaeffer)	P	Grains	S
Agonoscelini	*Agonoscelis rutila* (Fabricius)	AU	Soybean, Horehound	S
Antestiini	*Antestiopsis clymeneis* (Kirkaldy)	AF	Coffee	S
	Antestiopsis crypta Greathead	AF	Coffee	S
	Antestiopsis falsa Schouteden	AF	Coffee	S
	Antestiopsis facetoides Greathead	AF	Coffee	S
	Antestiopsis intricata (Ghesq. & Carayon)	AF	Coffee	S
	Antestiopsis thunbergii (Gmelin) [*A. orbitalis* (Westwood) syn. jr.]	AF	Coffee	M

Table 1.3 contd. ...

...Table 1.3 contd.

Subfamily/Tribe	Species	Distribution	Main Crop(s)	Status
	Plautia affinis (Dallas)	P	Mulberry, Grape, Vegetable crops	S
	Plautia stali Scott	P	Vegetable crops	M
Axiagastini	*Axiagastus campbelli* Distant	AU	Coconut, Palms	S
Bathycoelini	*Bathycoelia natalicola* Distant	O	Litchi, Macadamia	M
	Bathycoelia talassina (Herrich-Schaffer)	O	Cocoa	M
Cappaeini	*Halyomorpha halys* (Stål)	NA, O, P	Apple, Corn, Cotton, Grape, Maple, Ornamental Cherry, Peach, Pear, Soybean, Tomato	M
	Halyomorpha marmorea Fabricius	O	Areca nut	S
	Halyomorpha mista Uhler	P	Vegetable crops	S
Carpocorini	*Carpocoris fuscispinus* (Boheman)	P	Grains (wheat, canola)	S
	Dichelops furcatus (Fabricius)	NT	Corn, Oat, Soybean, Wheat	S
	Dichelops melacanthus (Dallas)	NT	Corn, Soybean, Wheat	S
	Dolycoris baccarum (Linnaeus)	P	Grains (wheat, sunflower)	M
	Dolycoris penicillatus Horváth	P	Grains (wheat), Pears, Tomatoes	S
	Euschisthus conspersus Uhler	NA	Apple, Apricot, Cotton, Pear, Tomato	S
	Euschisthus heros (Fabricius)	NT	Cotton, Soybean	M
	Euschistus picticornis (Stål)	NT	Soybean	S
	Euschisthus servus (Say)	NA	Apple, Alfalfa, Corn, Cotton, Peach, Pear, Pecan, Sorghum, Soybean	M

Table 1.3 contd. ...

...Table 1.3 contd.

Subfamily/Tribe	Species	Distribution	Main Crop(s)	Status
	Euschisthus tristigmus (Say)	NA	Apple, Pear, Peach, Pecan, Soybean	S
	Euschisthus variolarius (Palisot de Beauvois)	NA	Corn, Peach, Pear, Soybean	S
	Mormidea maculata Dallas	NT	Rice	S
	Mormidea notulifera Stål	NT	Rice	M
	Mormidea v-luteum (Lichtenstein)	NT	Rice	M
	Oebalus insularis Stål	NT	Rice	S
	Oebalus poecilus (Dallas)	NT	Barley, Oat, Rice, Wheat	M
	Oebalus pugnax (Fabricius)	NA	Rice, Sorghum, Wheat	M
	Oebalus ypsilongriseus (De Geer)	NT	Rice	M
	Tibraca limbativentris (De Geer)	NT	Rice, Soybean, Tomato, Wheat	M
Eysarcorini	*Aspavia armigera* (Fabricius)	AF	Cowpea, Soybean, Rice	S
Halyini	*Apodiphus amygdale* (Germar)	PA	Almond, Apple, Pistachio nuts	M
Myrocheini	*Dictyotus coenosus* (Westwood)	AU	Vegetable crops	S
Nezarini	*Chinavia acuta* (Dallas)	AF	Soybean	S
	Chinavia armigera (Stål)	NT	Soybean	S
	Chinavia hilaris (Say)	NA	Corn, Cotton, Pecan, Soybean, Tobacco, Tomato	M
	Chinavia impicticornis (Stål)	NT	Bean, Cowpea, Soybean	S
	Chinavia marginata (Palisot de Beauvois)	NT	Bean, Soybean	S
	Chinavia nigridorsata Breddin [*C. bella* syn. jr.]	NT	Bean, Pea, Soybean	S
	Chinavia ubica (Rolston)	NT	Soybean	S
	Chlorochroa ligata (Say)	NA	Sorghum, Cotton, Pecan	S

Table 1.3 contd. ...

...Table 1.3 contd.

Subfamily/Tribe	Species	Distribution	Main Crop(s)	Status
	Chlorochroa uhleri Stål	NA	Alfalfa, Cotton, Tomato, Wheat	
	Nezara antennata Scott	P	Soybean	S
	Nezara viridula (Linnaeus)	WORLD	Corn, Cotton, Cowpea, Lima bean, Macadamia, Pecan, Rice, Sorghum, Soybean, Tomato, Wheat	M
	Palomena angulosa Motschulsky	P	Alfalfa, Bean, Potato	M
	Palomena prasina (Linnaeus)	P	Hazelnut	M
Pentatomini	*Arvelius albopunctatus* (De Geer)	NT, NA	Tomatoes	S
Piezodorini	*Piezodorus guildinii* (Westwood)	NT	Alfalfa, Bean, Pear, Soybean	M
	Piezodorus hybneri (Gmelin)	AF, O, P	Alfalfa, Bean, Pear, Soybean	M
Rhynchocorini	*Biprorulus bibax Breddin*	AU	Citrus	M
	Cuspicona simplex Walker	AU	Potatoes, Tomatoes	S
Strachiini	*Bagrada hilaris* (Burmeister)	AF, O, NA	Vegetable crops (Crucifers)	M
	Eurydema oleacea (Linnaeus)	P	Vegetable crops (Crucifers)	
	Eurydema ornata (Linnaeus)	P	Vegetable crops (Crucifers)	
	Eurydema pulchra (Westwood)	O, P	Vegetable crops (Crucifers)	M
	Eurydema rugosa Motschulsky	P	Vegetable crops (Crucifers)	M
	Murgantia histrionica (Hahn)	NA, NT	Vegetable crops (Crucifers)	S
	Stenozygum coloratum (Klug)	AF, PA	Caper plants	
Unplaced	*Thyanta custator* (Fabricius)	NA	Alfalfa, Beans, Corn, Sorghum, Soybean, Wheat	S

Table 1.3 contd. ...

...*Table 1.3 contd.*

Subfamily/Tribe	Species	Distribution	Main Crop(s)	Status
	Thyanta pallidovirens Stål	NA	Soybean, Tomatoes, Nuts, Vegetable crops	S
	Thyanta perditor (Fabricius)	NT	Soybean, Sorghum, Rice	S
PODOPINAE	*Scotinophora coarctata* (Fabricius)	O, P	Rice	M

(Maciel et al. 2015). A partial key to identify the species of *Lincus* is found in Rolston (1983b), and should be complemented with the more recent species described (see Maciel et al. 2015 for references). The genus *Macropygium* is monotypic, including only *M. reticulare* (Rolston 1990, Grazia et al. 2015) (Fig. 1.7), although the morphological variability found among populations suggests that new species should be recognized.

Subfamily Edessinae

Edessa spp.

The genus *Edessa* is the most speciose within Edessinae, and its monophyly was supported by Barcellos and Grazia (2003). More recently some groups of species formerly included in *Edessa* have been recognized as distinct genera (Santos et al. 2015, Correia and Fernandes 2016). Several species groups have been reviewed (e.g., Fernandes and Van Doesburg 2000, Silva et al. 2006, Silva and Fernandes 2012).

Identification of *E. meditabunda* (Fig. 1.8) can be difficult, since the main characteristic used in its recognition (the dull brow hemelytra) is also found among other unrelated species of *Edessa*. There is no recent key published that allows the identification of this species among other edessines, and the only reliable character to recognize the species is the male genitalia. The group *rufomarginata* was revised recently (Silva et al. 2004, 2006), and an identification key to the species in Portuguese is available.

Subfamily Pentatominae

Aelia spp.

The genus *Aelia* Fabricius (Fig. 1.9) includes 16 species, the majority of which are distributed in the Palearctic region, with one species, *Aelia americana* Dallas, occurring in the USA (Froeschner 1988, Derjanschi and Péricart 2005, Rider 2006). The Palearctic species are a part of a complex of pests of

Figure 1.2–1.17. Facies dorsal of selected stink bugs of economic importance. 1.2, *Nezara viridula*. 1.3, *Bagrada hilaris*. 1.4, *Halyomorpha halys*. 1.5, *Euschistus heros*. 1.6, *Lincus lobuliger*. 1.7, *Macropigyum reticulare*. 1.8, *Edessa meditabunda*. 1.9, *Aelia acuminata*. 1.10, *Antestiopsis thunbergi*. 1.11, *Plautia affinis*. 1.12, *Chinavia hilaris*. 1.13, *Chinavia acuta*. 1.14, *Piezodorus hybneri*. 1.15, *Biprorulus bibax*. 1.16, *Eurydema ornatum*. 1.17, *Scotinophora coarctacta*.

wheat and barley also known as sunn pests (Javahery et al. 2000, Panizzi et al. 2000).

The tribe Aelini also includes two other genera, the monotypic genus *Aeliopsis* Bergevin and the Holarctic genus *Neotiglossa* Kirkby (13 spp.). However the inclusion of the first has been recently questioned, and the monophyly and relationship of the tribe need to be addressed (Rider et al., in press). Keys to the genera of Aeliini and to the most of the Palearctic species are found in Derjanschi and Péricart (2005).

Tribe Antestiini

The tribe Antestiini Atkinson includes 17 genera and more than 170 species from the Old World (Rider et al., in press). The taxonomy of the whole group is still poorly established, and identification at the species level is difficult. The tribe probably does not represent a monophyletic group.

A key to the Afrotropical genera of Antestiini can be found in Linnavuori (1982). Keys to the Afrotropical species of *Antestia*, *Antestiopsis* (Fig. 1.10) and *Plautia* (Fig. 1.11) are found in Greathead (1966), Linnavuori (1973), and Linnavuori (1982). The Palearctic species of *Plautia* were revised by Liu and Zheng (1994) and the Oriental species by Ahmad and Rana (1996).

Halyomorpha spp.

The genus *Halyomorpha* Mayr is currently part of the tribe Cappaeini Atkinson, and includes more than 35 species, distributed originally in the Afrotropical, Oriental and Palearctic regions (Linnavuori 1982, Rider 2006). The species *Halyomorpha halys* (Fig. 1.4) is native to Asia and has recently spread across most countries of the Northern Hemisphere, both in Europe and North America (Hoebeke and Carter 2003, Lee et al. 2013).

The taxonomy of *Halyomorpha* has been dealt with only on a regional basis, and the identification of the species in its native distribution can be difficult. Species recorded in the Afrotropical region were reviewed in Linnavuori (1982), while species of the Oriental region were reviewed in Ahmad and Zaidi (1989). A key to separate *H. halys* from similar species in Central Europe was presented by Wyniger and Kment (2010).

Tribe Carpocorini

The tribe Carpocorini is the most speciose tribe within Pentatomidae, with almost 130 genera and more than 500 species worldwide (Rider et al., in press). The species of economic importance are included in 12 genera (Table 1.3), and can be separated into two broad groups of genera, the *Carpocoris* group (including *Carpocoris* Kolenati and *Dolycoris* Mulsant and

Rey) and the *Euschistus* group (the remaining genera). These two groups also include several other genera of Carpocorini and potentially represent separated lineages within the tribe (Rider et al., in press). The genera included in the *Carpocoris* group are distributed in the Nearctic, Oriental and Palearctic regions (Froeschner 1988, Rider 2006), while the genera of the *Euschistus* group are restricted to the New World (Rolston and McDonald 1984, Grazia et al. 2015).

The identification of *Carpocoris* and *Dolycoris* were issued by different authors, usually on a regional basis (e.g., Tamanini 1959a,b, Wyniger and Kment 2010). Ribes and Pagola-Carte (2009) presented an updated key to the Euromediterranean species of *Carpocoris*, while Tamanini (1959a) and Hasan and Afzal (1990) presented keys for the identification of the species of *Dolycoris*. The genera included in the *Euschistus* group can be identified using the works of Rolston and McDonald (1984) and Grazia et al. 2015. Identification at the species level can be done following Rolston (1974, 1978, 1984), Grazia (1978), and Fernandes and Grazia (1998).

The Nezara group

The *Nezara* group was first recognized by Linnavuori (1982) to include species of *Aethemenes* Stål, *Acrosternum* Fieber, *Chinavia* Orian (Fig. 1.12 and 1.13) and *Nezara* Amyot and Serville (Fig. 1.2) from the Afrotropical region. More recently, Schwertner (2005) supported the monophyly of the group, including *Neoacrosternum* Day, *Porphyroptera* and *Pseudoacrosternum* Day (= *Aesula* Stål) as well. Besides the Afrotropical distribution, endemic species are also found in the Nearctic, Neotropical, Oriental, and part of the Palearctic (China and Japan) regions. The species *Nezara viridula* (Linnaeus) has worldwide distribution.

All the genera of the *Nezara* group are currently placed in the tribe Nezarini, which includes at least 26 genera and more than 250 species of green stink bugs that remain green after death (Rider et al., in press). The monophyly of the tribe needs to be tested.

The validity of the genus *Chinavia* has only recently been fully appreciated. Orian (1965) proposed the genus *Chinavia* to include nine species from the Afrotropical region all removed from *Acrosternum*. The author also mentioned that probably all Western Hemisphere species of *Acrosternum* should be re-examined in the light of his propostion, once the species of the latter genus appeared to be confined to the Palearctic-Mediterranean region. Linnavuori (1972) did not agree with Orian and synonymized both genera, but Roche (1977) considered *Chinavia* to be a valid genus. Without reference to this latter work, Rolston (1983a) highlighted that the diversity of *Acrosternum* was confused to include all species in a single taxon, and considered *Chinavia* to be a subgenus, including all Western Hemisphere species of *Acrosternum* in it. More recently, Schwertner

(2005) and Schwertner and Grazia (2006, 2007), based on the study of all species included in both *Acrosternum* and *Chinavia*, considered the latter a distinct monophyletic group, more related to *Nezara* than to *Acrosternum*, thus bringing support to *Chinavia* as a valid genus. Currently, *Chinavia* includes species distributed in the Afrotropical, Nearctic and Neotropical regions. The genus *Acrosternum* is mostly confined to the Palearctic region, with only two species reaching the Afrotropical region and one species in the Oriental region.

A key to all the species of *Nezara* is available in Ferrari et al. (2010). Keys for identification of *Chinavia* species can be found in Linnavuori (1972, 1982)—Afrotropical and Rolston (1983a)—Western Hemisphere, with additions of new species by Rider and Rolston (1986), Rider (1987), Eger (1988), Frey-da-Silva and Grazia (2001), Schwertner and Grazia (2006) and Grazia et al. (2006). Schwertner and Grazia (2007) presented a pictorial key to the species of *Chinavia* recorded in Brazil.

Piezodorus spp. (Fig. 1.14)

The genus has worldwide distribution, with 12 species recorded in all biogeographical regions (Linnavuori 1982, Froeschner 1988, Cassis and Gross 2002, Rider 2006, Grazia et al. 2015, Salini and Viraktamath 2015). Monophyly of the *Piezodorus* is well-supported (Ahmad 1995), but the evolution of its species is still poorly studied. The tribe Piezodorini also includes the genera *Anaximenes* Stål, *Chaubattiana* Distant and *Pausias* Jakovlev, but monophyly of the group was never tested (Rider et al., in press).

The identification of the important economic species is usually straightforward, the characters used to separate those species are mostly from general morphology. A key to all the species of *Piezodorus* is presented by Ahmad (1995).

Biprorulus bibax (Fig. 1.15)

This monotypic genus is recorded in Eastern Australia and is currently included in the tribe Rhynchocorini, a well-supported group of 18 genera and more than 100 species distributed mainly in the Australian and Oriental regions (Rider et al., in press); three species of the nominal genus *Rhynchocoris* are also found in Southern China (Rider 2006). Gross (1975a, 1976) reviewed part of the tribe, but did not include the genus *Biprorulus* in his keys. There is no recent paper dealing with the taxonomy and identification of *B. bibax*, although the typical morphology of this species is easily recognized among rhynchocorines of economic importance (Fig. 1.15). A web-key to the identification of all Australian genera of Pentatomidae is available at http://keys.australianmuseum.net.au/stink_intro.htm.

Tribe Strachiini

This group includes 20 genera and 142 species distributed in all biogeographical regions (Froeschner 1988, Cassis and Gross 2002, Rider 2006, Grazia et al. 2015, Salini and Viraktamath 2015), but most speciose in the Oriental and Palearctic regions (Rider et al., in press). Two genera contain the major species of economic importance (Table 1.3): *Eurydema* Laporte (Fig. 1.16), with 33 species distributed in the Oriental and Palearctic region; and the genus *Bagrada* Stål with more than 15 species, distributed originally in the Old World (Afrotropical, Oriental and Palearctic regions). The species *Bagrada hilaris* (Fig. 1.3) was recently reported as an invasive pest in the US (Palumbo and Natwick 2010, Reed et al. 2013) and has caused severe damage to Cole crops in the Western States of the country. Panizzi et al. (2000) referenced this species as a secondary pest in India (under the name *Bagrada cruciferarum* Kirkaldy), but it is also known to cause damage in other countries (Reed et al. 2013).

Keys to identify *Bagrada* and *Eurydema* among other Strachiini of different regions are available in Linnavuori (1982)—Afrotropical, Derjanschi and Péricart (2005)—Palearctic and Salini and Viraktamath 2015)—Oriental. The species of economic importance can be identified using the keys in Derjanschi and Péricart (2005); identification of the species in the Afrotropical, Oriental and Nearctic are straightforward, as only *Bagrada hilaris* (in all three regions) and *Eurydema pulchra* (Oriental region) are recorded.

Subfamily Podopinae

Scotinophara spp.

The genus is the most speciose within the subfamily, with up to 40 spp. described, and is included in the *Podops* group (equivalent to the Tribe Podopini in several references, e.g., Rider 2006) (Rider et al., in press). The species are distributed in the Old World (Afrotropical, Australian, Oriental and Palearctic genera), with *S. coarctacta* (Fig. 1.17) being the most widespread species (Rider 2006).

The *Podops* group represents a morphologically homogeneous group (Rider et al., in press), and especially species of *Podops* Laporte and *Scotinophara* can be confused in the regions where the species are sympatric. A key to identify the genera in the Palearctic region is provided in Péricart (2010), and is also useful for other regions. Species of *Scotinophara* are reviewed in Barrion et al. (2007)—Phillipines and Péricart (2010)— Euromediterranean.

Acknowledgments

We acknowledge the National Research Council (CNPq) for the fellowship to JG (Proc. 302494/2010-3 and 305009/2015-0) and Fundação de Amparo a Pesquisa do Estado de São Paulo (FAPESP) for the funding to CFS (Proc. n. 2014/00729-3). We thank Bruno C. Genevcius for helping with insights and additions to the text on 'Biology and Reproductive Behavior'.

References

Ahmad, I. and S. Kamaluddin. 1988. A new tribe and a new species of the subfamily Phyllocephalinae (Hemiptera: Pentatomidae) from the Indo-Pakistan subcontinent. Orient. Insects 22: 241–258.

Ahmad, I. and R.H. Zaidi. 1989. A revision of the genus *Halyomorpha* Mayr (Hemiptera: Pentatomidae: Pentatominae: Carpocorini) from Indo-Pakistan subcontinent with description of a new species from Potohar region of Pakistan and their cladistic analysis. Proc. Pak. Cong. Zool. 9: 237–153.

Ahmad, I. and S. Kamaluddin. 1990. A new tribe for Phyllocephaline genera *Gellia* Stål and *Tetroda* Amyot and Serville (Hemiptera: Pentatomidae) and their revision. Annotat. Zool. Bot. 195: 1–20.

Ahmad, I. 1995. A review of pentatomine legume bug genus *Piezodorus* Fieber (Hemiptera: Pentatomidae: Pentatominae) with its cladistic analysis. Proc. Pak. Cong. Zool. 15: 329–358.

Ahmad, I. and N.A. Rana. 1996. A review of antestiine genus *Plautia* Stål (Hemiptera: Pentatomidae: Pentatominae) from Indo-Pakistan subcontinent and their cladistic relationships. Pak. J. Entomol. Soc. Karachi. 11: 45–57.

Aldrich, J.R. 1991. Pheromones of good and bad bugs. Entomol. Soc. Qld. News Bull. 19(2): 19–27.

Amyot, C.J.B. and A. Serville. 1843. Histoire naturelle des insectes. Hémiptères. Librairie Encyclopedique de Roret, Paris, France.

Barão, K.R., A. Ferrari and J. Grazia. 2012. Phylogeny of the South Asian Halyini? Comments on Memon et al. (2011): towards a better practice in pentatomidae phylogenetic analysis. Ann. Entomol. Soc. Am. 105(6): 751–752.

Barcellos, A. and J. Grazia. 2003. Cladistics analysis and biogeography of *Brachystethus* Laporte (Heteroptera, Pentatomidae, Edessinae). Zootaxa 256: 1–14.

Barrion, A.T., R.C. Joshi, A.L.A. Barrion-Dupo and L.S. Sebastian. 2007. Systematics of the Philippine rice black bug, *Scotinophara* Stål (Hemiptera: Pentatomidae). pp. 3–179. *In*: Joshi, R.C., A.T. Barrion and L.S. Sebastian (eds.). Rice Black Bugs. Taxonomy, Ecology, and Management of Invasive Species. Philippine Rice Research Institute, Science City of Muñoz, Philippines.

Becker, M. 1977. A review of the genus *Colpocarena* Stål (Heteroptera, Pentatomidae, Discocephalinae). Rev. Bras. Biol. 37(2): 367–373.

Bergroth, E. 1906. Aphylinae und Hyocephalinae, zwei neue Hemipteren-Subfamilien. Zool. Anz. 29: 644–649.

Bergroth, E. 1911. Zur Kenntnis der neotropischen Arminen (Hem. Het.). Wien. Entomol. Ztg. 30(6-7): 117–130.

Blassioli-Moraes, M.C., D.M. Magalhaes, A. Čokl, R.A. Laumann, J.P. Da Silva, C.C. Silva and M. Borges. 2014. Vibrational communication and mating behaviour of *Dichelops melacanthus* (Hemiptera: Pentatomidae) recorded from loudspeaker membranes and plants. Physiol. Entomol. 39(1): 1–11.

Borges, M., P.C. Jepson and P.E. Howse. 1987. Long-range mate location and close-range courtship behaviour of the green stink bug, *Nezara viridula* and its mediation by sex pheromones. Entomol. Exp. Appl. 44(3): 205–212.
Borges, M. and J.R. Aldrich. 1994. Attractant pheromone for nearctic stink bug, *Euschistus obscurus* (Heteroptera, Pentatomidae)—insight into a neotropical relative. J. Chem. Ecol. 20(5): 1095–1102.
Brailovsky, H., L. Cervantes and C. Mayorga. 1988. Hemiptera-Heteroptera de Mexico XL: La familia Cyrtocoridae Distant en la Estacion de Biologia Tropical "Los Tuxtlas" (Pentatomoidea). An. Inst. Biol. Univ. Nac. Autón. Méx. (Zool.) 58: 537–560.
Campos, L.A. and J. Grazia. 2006. Análise cladística e biogeografia de Ochlerini (Hemiptera, Pentatomidae, Discocephalinae). Iheringia, Sér. Zool. 96: 147–163.
Cassis, G. and G. Gross. 2002. Hemiptera: Heteroptera (Pentatomomorpha). pp. xiv+737. *In*: Houston, W. and A. Wells (eds.). Zoological Catalogue of Australia. Csiro Publishing, Melbourne, Australia.
Cesari, M., L. Maistrello, F. Ganzerli, P. Dioli, L. Rebecchi and R. Guidetti. 2015. A pest alien invasion in progress: potential pathways of origin of the brown marmorated stink bug *Halyomorpha halys* populations in Italy. J. Pest Sci. 88: 1–7.
Čokl, A. and M. Virant-Doberlet. 2003. Communication with substrate-borne signals in small plant-dwelling insects. Ann. Rev. Entomol. 48(1): 29–50.
Correia, A.O. and J.A.M. Fernandes. 2016. *Grammedessa*, a new genus of Edessinae (Hemiptera: Heteroptera: Pentatomidae). Zootaxa 4107: 541–565.
Davidová-Vilímová, J. and P. Štys. 1994. Diversity and variation of trichobothrial patterns in adult Podopinae (Heteroptera: Pentatomidae). Acta Universit. Carolinae Biol. 37(1-2) [1993]: 33–72.
Davidová-Vilímová, J. and J.E. McPherson. 1995. History of the higher classification of the subfamily Podopinae (Heteroptera: Pentatomidae), a historical review. Acta Universit. Carolinae Biol. 38: 99–124.
De Clercq, P. 2000. Predaceous stink bugs (Pentatomidae: Asopinae). pp. 737–789. *In*: Schaefer, C.W. and A.R. Panizzi. (eds.). Heteroptera of Economic Importance. CRC Press, Boca Raton, FL, USA.
De Clercq, P. 2008. Stink bugs, predatory (Hemiptera: Pentatomidae, Asopinae). pp. 3042–3045. *In*: Capinera, J.L. (ed.). Encyclopedia of Insects, Vol. 3. Kluwer Academic Publishers, Dordrecht, SH, Netherlands.
Derjanschi, V.V. and J. Péricart. 2005. Hémiptères Pentatomoidea euro-méditerranéens 1. Généralités. Systématique: Prémiere Partie. Faune Fr. 90: 1–494.
Drickamer, L.C. and J.E. McPherson. 1992. Comparative aspects of mating behavior patterns in six species of stink bugs (Heteroptera: Pentatomidae). Great Lakes Entomol. 25(4): 287–295.
Eberhard, W.G. 1975. The ecology and behavior of a subsocial pentatomid bug and two scelionid wasps: strategy and counterstrategy in a host and its parasites. Sm. C. Zool. 205: 1–39.
Eger Jr., J.E. 1988. A new species of *Acrosternum* Fieber, subgenus *Chinavia* Orian, from ecuador (Heteroptera: Pentatomidae: Pentatomini). Fla. Entomol. 71: 120–124.
Esquivel, J.F. and E.G. Medrano. 2014. Ingestion of a marked bacterial pathogen of cotton conclusively demonstrates feeding by first instar southern green stink bug (Hemiptera: Pentatomidae). Environ. Entomol. 43: 110–115.
Fernandes, J.A.M. and J. Grazia. 1998. Revision of the genus *Tibraca* Stål (Heteroptera, Pentatomidae, Pentatominae). Ver. Bras. Zool. 15: 1049–1060.
Fernandes, J.A.M. and P.H. van Doesburg. 2000. The *E. cervus*-group of *Edessa* Fabricius, 1903 (Heteroptera: Pentatomidae: Edessinae). Zool. Med. Leiden 74: 151–165.
Fernandes, J.A.M. 2010. A new genus and species of Edessinae from Amazon Region (Hemiptera: Heteroptera: Pentatomidae). Zootaxa 2662: 53–65.
Fernandes, J.A.M. and L.D. Campos. 2011. A new group of species of *Edessa* Fabricius, 1803 (Hemiptera: Heteroptera: Pentatomidae). Zootaxa 3019: 63–68.

Ferrari, A., C.F. Schwertner and J. Grazia. 2010. Review, cladistics analysis and biogeography of *Nezara* Amyot and Serville (Hemiptera: Pentatomidae). Zootaxa 2424: 1–41.
Fieber, F.X. 1860. Die europäischen Hemiptera. Halbflügler (Rhynchota Heteroptera). Nach der analytischen Methode bearbeitet. Gerold, Wien, Austria.
Frey-da-Silva, A. and J. Grazia. 2001. Novas espécies de *Acrosternum* subgênero *Chinavia* (Heteroptera, Pentatomidae, Pentatomini). Iheringia, Sér. Zool. 90: 107–126.
Froeschner, R.C. 1988. Family Pentatomidae Leach, 1815. The stink bugs. pp. 544–607. *In*: Henry, T.J. and R.C. Froeschner (eds.). Catalog of the Heteroptera, or True Bugs, of Canada and the Continental United States. E.J. Brill, Leiden, New York.
Gapon, D.A. and F.V. Konstantinov. 2006. On the structure of the aedeagus in shield bugs (Heteroptera, Pentatomidae): III. Subfamily Asopinae. Entomol. Rev. 86: 806–819.
Gapon, D.A. 2010. *Conquistator*, a new genus for *Podisus mucronatus* Uhler, 1897 (Heteroptera: Pentatomidae: Asopinae) with a re-description of type species. Zoosyst. Rossica 18(2): 264–270.
Gapud, V.P. 1991. A generic revision of the subfamily Asopinae, with consideration on its phylogenetic position in the family Pentatomidae and superfamily Pentatomoidea (Hemiptera-Heteroptera). Philippines Entomol. 8: 865–961.
Gapud, V.P. 2015. The Philippine genus *Stilbotes* Stål and a new tribe of Asopinae (Hemiptera: Pentatomidae). Asia Life Sci. 24(2): 493–498.
Garbelotto, T. de A., L.A. Campos and J. Grazia. 2013. Cladistics and revision of *Alitocoris* with considerations on the phylogeny of the *Herrichella* clade (Hemiptera, Pentatomidae, Discocephalinae, Ochlerini). Zool. J. Linn. Soc. 168: 452–472.
Grazia, J. 1978. Revisão do gênero *Dichelops* Spinola, 1837 (Heteroptera, Pentatomidae, Pentatomini). Iheringia, Sér. Zool. 53: 3–119.
Grazia, J., C.F. Schwertner and A. Ferrari. 2006. Description of five new species of Chinavia Orian (Hemiptera, Pentatomidae, Pentatominae) from western and northwestern South America. Denisia 19: 423–434.
Grazia, J., R.T. Schuh and W.C. Wheeler. 2008. Phylogenetic relationships of family groups in Pentatomoidea based on morphology and DNA sequences (Insecta: Heteroptera). Cladistics 24: 932–976.
Grazia, J., A.R. Panizzi, C. Greve, C.F. Schwertner, L.A. Campos, T.A. Garbelotto et al. 2015. Stink bugs (Pentatomidae). pp. 681–756. *In*: Panizzi, A.R. and J. Grazia (eds.). True Bugs (Heteroptera) of the Neotropics. Springer, Dordrecht, GE.
Greathead, D.J. 1966. A taxonomic study of the species of *Antestiopsis* (Hemiptera: Pentatomidae) associated with *Coffea arabica* in Africa. Bull. Entomol. Res. 56: 515–554.
Gross, G.F. 1975a. A revision of the Pentatomidae (Hemiptera-Heteroptera) of the *Rhynchocoris* group from Australia and adjacent areas. Rec. S. Aust. Mus. 17: 51–167.
Gross, G.F. 1975b. Handbook of the Flora and Fauna of South Australia. Plant-feeding and other Bugs (Hemiptera) of South Australia. Heteroptera—Part 1. Handbooks Committee, South Australian Government, Adelaide, Australia.
Gross, G.F. 1976. Handbook of the Flora and Fauna of South Australia. Plant-feeding and other Bugs (Hemiptera) of South Australia. Heteroptera—Part 2. Handbooks Committee, South Australian Government, Adelaide, Australia.
Guerra, T.J., F. Camarota, F.S. Castro, C.F. Schwertner and J. Grazia. 2011. Trophobiosis between ants and *Eurystethus microlobatus* Ruckes 1966 (Hemiptera: Heteroptera: Pentatomidae) a cryptic, gregarious and subsocial stinkbug. J. Nat. Hist. 45(17-18): 1101–1117.
Hasan, S.A. and M. Afzal. 1990. Studies on the genus *Dolycoris* Mulsant et Rey (Pentatomidae: Carpocorini) with description of four new species from Pakistan. Biologia (Lahore) 36: 57–61.
Hasan, S.A. and I.J. Kitching. 1993. A cladistic analysis of the tribes of the Pentatomidae (Heteroptera). Jap. J. Entomol. 61: 651–669.
Henry, T.J. 1997. Phylogenetic analysis of family groups within the infraorder Pentatomomorpha (Hemiptera: Heteroptera), with emphasis on the Lygaeoidea. Ann. Entomol. Soc. Am. 90(3): 275–301.

Hoebeke, E.R. and M.E. Carter. 2003. *Halyomorpha halys* (Stål) (Heteroptera: Pentatomidae): a polyphagous plant pest from Asia newly detected in North America. Proc. Entomol. Soc. Wash. 105: 225–237.

Javahery, M., C.W. Schaefer and J.D. Lattin. 2000. Shield bugs (Scutelleridae). pp. 475–503. *In*: Schaefer, C.W. and A.R. Panizzi (eds.). Heteroptera of Economic Importance. CRC Press, Boca Raton, London, New York, Washington, D.C.

Kamaluddin, S. and I. Ahmad. 1988. A revision of the tribe Phyllocephalini (Hemiptera: Pentatomidae: Phyllocephalinae) from Indo-Pakistan subcontinent with description of five new species. Orient. Insects 22: 185–240.

Kirkaldy, G.W. 1909. Catalogue of the Hemiptera (Heteroptera) Vol. I: Cimicidae. Felix L. Dames, Berlin.

Konstantinov, F.V. and D.A. Gapon. 2005. On the structure of the aedeagus in shield bugs (Heteroptera, Pentatomidae): 1. Subfamilies Discocephalinae and Phyllocephalinae. Entomol. Rev. 8: 221–235.

Laumann, R.A., A. Kavčič, M.C. Moraes, M. Borges and A. Čokl. 2013. Reproductive behaviour and vibratory communication of the neotropical predatory stink bug *Podisus nigrispinus*. Physiol. Entomol. 38: 71–80.

Leal, W.S., S. Kuwahara, X. Shi, H. Higuchi, C.E.B. Marino, M. Ono et al. 1998. Male-released sex pheromone of the stink bug *Piezodorus hybneri*. J. Chem. Ecol. 24(11): 1817–1829.

Lee, D.H., B.D. Short, S.V. Joseph, J.C. Bergh and T.C. Leskey. 2013. Review of the biology, ecology, and management of *Halyomorpha halys* (Hemiptera: Pentatomidae) in China, Japan, and the Republic of Korea. Environ. Entomol. 42: 627–641.

Leston, D. 1953. On the wing-venation, male genitalia and spermatheca of *Podops inuncta* (F.), with a note on the diagnosis of the subfamily Podopinae Dallas (Hem., Pentatomidae). J. Soc. British Entomol. 4: 129–135.

Leston, D. 1954a. Notes on the ethiopian pentatomoidea (Hem.) XII. On some specimens from southern Rhodesia, with an investigation of certain features in the morphology of *Afrius figuratus* (Germar) and remarks upon the genitalia in Amyotinae. Occ. Pap. Natl. Mus. S. Rhodesia 19: 678–686.

Leston, D. 1954b. Wing venation and male genitalia of *Tessaratoma* Berthold with remarks on Tessaratominae Stål (Hemiptera: Pentatomidae). Proc. R. Entomol. Soc. London (A) 29: 9–16.

Lethierry, L. and G. Severin. 1893. Catalogue général des Hémiptères. Bruxelles, Pentatomidae, 1: x + 286 pp.

Li, H.-M., R.-Q. Deng, J.-W. Wang, Z.-Y. Chen, F.-L. Jia and X.-Z. Wang. 2005. A preliminary phylogeny of the Pentatomomorpha (Hemiptera: Heteroptera) based on nuclear 18S rDNA and mitochondrial DNA sequences. Mol. Phylogenet. Evol. 37: 313–326.

Li, M., Y. Tian, Y. Zhao and W.-J. Bu. 2012. Higher level phylogeny and the first divergence time estimation of Heteroptera (Insecta: Hemiptera) based on multiple genes. PloS One 7: 1–17.

Linnavuori, R.E. 1972. Studies on African Pentatomoidea. Arq. Mus. Bocage (2) 3 (15): 395–434.

Linnavuori, R.E. 1973. Studies on African Heteroptera. Arq. Mus. Bocage (2) 4 (2): 26–69.

Linnavuori, R.E. 1982. Pentatomidae and Acanthosomatidae (Heteroptera) of Nigeria and the Ivory Coast, with remarks on species of the adjacent countries in West and Central Africa. Acta Zool. Fennica 163: 1–176.

Liu, Q.-A. and L.-Y. Zheng. 1994. On the Chinese species of *Plautia* Stål (Hemiptera: Pentatomidae). Entomotaxonomia 16: 235–248.

Maciel, A.S., T.D.A. Garbelotto, I.C. Winter, T. Roell and L.A. Campos. 2015. Description of the males of *Lincus singularis* and *Lincus incisus* (Hemiptera: Pentatomidae: Discocephalinae). Zoologia 32: 157–161.

McDonald, F.J.D. 1966. The genitalia of North American Pentatomoidea (Hemiptera: Heteroptera). Quaest. Entomol. 2: 7–150.

McLain, D.K. 1981. Female choice and the adaptive significance of prolonged copulation in *Nezara viridula* (Hemiptera: Pentatomidae). Psyche 87(3-4): 325–336.

McLain, D.K. 1991. Heritability of size: a positive correlate of multiple fitness components in the southern green stinkbug (Hemiptera: Pentatomidae). Ann. Entomol. Soc. Am. 84(2): 174–178.

McPherson, J.E. and R.M. McPherson. 2000. Stink Bugs of Economic Importance in America North of Mexico. CRC Press, Boca Raton, London, New York, Washington, D.C.

Memon, N., F. Gilbert and I. Ahmad. 2011. Phylogeny of the south Asian halyine stink bugs (Hemiptera: Pentatomidae: Halyini) based on morphological characters. Ann. Entomol. Soc. Am. 104: 1149–1169.

Miklas, N., T. Lasnier and M. Renou. 2003. Male bugs modulate pheromone emission in response to vibratory signals from conspecifics. J. Chem. Ecol. 29: 561–574.

Moraes, M.C.B., M. Pareja, R.A. Laumann and M. Borges. 2008. The chemical volatiles (semiochemicals) produced by neotropical stink bugs (Hemiptera: Pentatomidae). Neotrop. Entomol. 37: 489–505.

Orian, A.J. 1965. A new genus of Pentatomidae from Africa, Madagascar and Mauritius (Hemiptera). Proc. R. Entomol. Soc. London. Series B 34: 25–29.

Owusu-Manu, E. 1980. Observations on mating and egg-laying behaviour of *Bathycoelia thalassina* (Herrich-Schaeffer) (Hemiptera: Pentatomidae). J. Nat. Hist. 14: 463–467.

Packauskas, R.J. and C.W. Schaefer. 1998. Revision of the Cyrtocoridae (Hemiptera: Pentatomoidea). Ann. Entomol. Soc. Am. 91: 363–386.

Palumbo, J.C. and E.T. Natwick. 2010. The bagrada bug (Hemiptera: Pentatomidae): A new invasive pest of cole crops in Arizona and California. Plant Health Progress. http://www.plantmanagementnetwork.org/sub/php/brief/2010/bagrada/. [Accessed 13 May 2016].

Panizzi, A.R. and F. Slansky Jr. 1985. Review of phytophagous pentatomids (Hemiptera: Pentatomidae) associated with soybean in the Americas. Fla. Entomol. 68: 184–214.

Panizzi, A.R. 1997. Wild hosts of pentatomids: Ecological significance and role in their pest status on crops. Annu. Rev. Entomol. 42: 99–122.

Panizzi, A.R. 2000. Suboptimal nutrition and feeding behavior of hemipterans on less preferred plant food sources. An. Soc. Entomol. Brasil 29: 1–12.

Panizzi, A.R., J.E. McPherson, D.G. James, M. Javahery and R.M. McPherson. 2000. Stink bugs (Pentatomidae). pp. 421–474. *In*: Schaefer, C.W. and A.R. Panizzi (eds.). Heteroptera of Economic Importance. CRC Press, Boca Raton, FL, USA.

Panizzi, A.P., A.F. Bueno and da F.A.C. da Silva. 2012. Insetos que atacam vagens e grãos. pp. 335–420. *In*: Hoffmann-Campo, C.B., B.S. Corrêa-Ferreira and F. Moscardi (eds.). Soja, Manejo Integrado de Insetos e outros Artrópodes-Pragas. Embrapa, Brasília, DF, Brasil.

Pendergrast, J.G. 1957. Studies on the reproductive organs of the Heteroptera with a consideration of their bearing on classification. T. Roy. Ent. Soc. London 109: 1–63.

Péricart, J. 2010. Hémiptères Pentatomoidea Euro-Méditerranéens. Volume 3: Podopinae Et Asopinae. Faune Fr. 93: 1–290.

Reed, D.A., J.C. Palumbo, T.M. Perring and C. May. 2013. *Bagrada hilaris* (Hemiptera: Pentatomidae), an invasive stink bug attacking cole crops in the southwestern United States. J. Integr. Pest Manag. 4: C1–C7.

Requena, G.S., T.M. Nazareth, C.F. Schwertner and G. Machado. 2010. First cases of exclusive paternal care in stink bugs (Hemiptera: Pentatomidae). Zoologia 27: 1018–1021.

Reuter, O.M. 1912. Bemerkungen über mein neues Heteropterensystem. Öfvers. F. Vetensk-Soc. Förhandl. 54(A)(6): 1–62.

Ribes, J. and S. Pagola-Carte. 2009. Clave de identificación de las especies euromediterráneas de Carpocoris (Hemiptera: Heteroptera: Pentatomidae). Heteropterus Rev. Entomol. 9: 45–48.

Rider, D.A. 1987. A new species of *Acrosternum* Fieber, subgenus *Chinavia* Orian, from Cuba (Hemiptera: Pentatomidae). J. N. Y. Entomol. Soc. 95: 298–301.

Rider, D.A. 1998–2015. Pentatomoidea home page. North Dakota State University, Fargo, ND, USA. Available at: https://www.ndsu.edu/pubweb/~rider/Pentatomoidea/[Accessed in 13 May 2016].

Rider, D.A. 2000. Stirotarsinae, new subfamily for *Stirotarsus abnormis* Bergroth (Heteroptera: Pentatomidae). Ann. Entomol. Soc. Am. 93: 802–806.

Rider, D.A. 2006. Family pentatomidae. pp. 233–402. *In*: Aukema, B. and C. Rieger (eds.). Catalogue of the Heteroptera of the Palaearctic Region. Vol. 5. Entomological Society, Amsterdam, The Netherlands.
Rider, D.A., C.F. Schwertner, J. Vilímová, D. Rédei, P. Kment and D.B. Thomas. 2017. Higher systematics of pentatomoidea. *In*: McPherson, J. (ed.). Biology of Invasive Stink Bugs and Related Species. CRC Press. (in press).
Rider, D.A. and L.H. Rolston. 1986. Three new species of *Acrosternum* Fieber, subgenus *Chinavia* Orian, from Mexico (Hemiptera: Pentatomidae). J. N. Y. Entomol. Soc. 94: 416–423.
Roche, P.J.L. 1977. Pentatomidae of the granitic islands of Seychelles (Heteroptera). Rev. Zool. Africaine 91: 558–572.
Rodrigues, A.R.S., J.E. Serrno, V.W. Teixeira, J.B. Torres and A.A. Teixeira. 2008. Spermatogenesis, changes in reproductive structures, and time constraint associated with insemination in *Podisus nigrispinus*. J. Insect Physiol. 54: 1543–1551.
Roell, T. and L.A. Campos. 2015. *Candeocoris bistillatus*, new genus and new species of Ochlerini from Ecuador (Hemiptera: Heteroptera: Pentatomidae). Zootaxa 4018: 573–583.
Rolston, L.H. 1974. Revision of the genus *Euschistus* in middle America (Hemiptera, Pentatomidae, Pentatomini). Entomol. Amer. 48: 1–102.
Rolston, L.H. 1978. A revision of the genus *Mormidea* (Hemiptera: Pentatomidae). J. N. Y. Entomol. Soc. 86: 161–219.
Rolston, L.H. 1981. Ochlerini, a new tribe in Discocephalinae (Hemiptera: Pentatomidae). J. N. Y. Entomol. Soc. 89: 40–42.
Rolston, L.H. 1983a. A revision of the genus *Acrosternum* Fieber, subgenus *Chinavia* Orian, in the western hemisphere (Hemiptera: Pentatomidae). J. N. Y. Entomol. Soc. 91: 97–176.
Rolston, L.H. 1983b. A revision of the genus *Lincus* Stål (Hemiptera: Pentatomidae: Discocephalinae: Ochlerini). J. N. Y. Entomol. Soc. 91: 1–47.
Rolston, L.H. 1984. Key to the males of the nominate subgenus of *Euschistus* in South America, with descriptions of three new species (Hemiptera: Pentatomidae). J. N. Y. Entomol. Soc. 92: 352–364.
Rolston, L.H., and F.J.D. McDonald. 1979. Keys and diagnoses for the families of Western Hemisphere Pentatomoidea, subfamilies of Pentatomidae and tribes of Pentatominae (Hemiptera). J. N. Y. Entomol. Soc. 87: 189–207.
Rolston, L.H. and F.J.D. McDonald. 1984. A conspectus of Pentatomini of the western hemisphere. Part 3 (Hemiptera: Pentatomidae). J. N. Y. Entomol. Soc. 92: 69–86.
Rolston, L.H. 1990. Key and diagnoses for the genera of 'broadheaded' discocephalines (Hemiptera: Pentatomidae). J. N. Y. Entomol. Soc. 98: 14–31.
Rolston, L.H. 1992. Key and diagnoses for the genera of Ochlerini (Hemiptera: Pentatomidae: Discocephalinae). J. N. Y. Entomol. Soc. 100: 1–41.
Ruckes, H. 1961. Notes on the *Mecistorhinus-Antiteuchus* generic complex of discocephaline pentatomids (Heteroptera, Pentatomidae). J. N. Y. Entomol. Soc. 69: 147–156.
Ryan, M.A. and G.H. Walter. 1992. Sound communication in *Nezara viridula* (L.) (Heteroptera: Pentatomidae): further evidence that signal transmission is substrate-borne. Experentia 48: 1112–1115.
Salini, S. and C.A. Viraktamath. 2015. Genera of Pentatomidae (Hemiptera: Pentatomoidea) from south India–an illustrated key to genera and checklist of species. Zootaxa 3924: 1–76.
Santos, A.V. and G.S. Albuquerque. 2001. Eficiência do cuidado maternal de *Antiteuchus sepulcralis* (Fabricius) (Hemiptera: Pentatomidae) contra inimigos naturais do estágio de ovo. Neotrop. Entomol. 30: 641–646.
Santos, B.T., V.J. Silva and J.A.M. Fernandes. 2015. Revision of *Ascra* with proposition of the *bifida* species group and description of two new species (Hemiptera: Pentatomidae: Edessinae). Zootaxa 4034: 445–470.
Schaefer, C.W. and I. Ahmad. 1987. A cladistic analysis of the genera of the Lestonocorini (Hemiptera: Pentatomidae: Pentatominae). Proceedings of the Entomological Society of Washington 89: 444–447.

Schaefer, C.W. 2009. *Prosorrhyncha* (Heteroptera and Coleorrhyncha). pp. 839–855. *In*: Resh, V.H. and R.T. Cardé (eds.). Encyclopedia of Insects, 2nd ed. Vol. 5. Academic, Amsterdam, The Netherlands.

Schouteden, H. 1906. Heteroptera. Fam. Pentatomidae. Subfam. Aphylinae. Genera Insectorum, fasc. 47, 4 pp., 1 pl.

Schuh, R.T. and J.A. Slater. 1995. True Bugs of the World (Hemiptera: Heteroptera): Classification and Natural History. Cornell University Press, NY, USA.

Schwertner, C.F. 2005. Phylogeny and classification of the green stink bugs of *Nezara group* of genera (Hemiptera, Pentatomidae, Pentanominae). Ph.D. Thesis, Universidade Federal do Rio Grande do Sul, RS, Brazil. [in Portuguese, abstract in English] Available at: http://www.lume.ufrgs.br/handle/10183/1/.

Schwertner, C.F. and J. Grazia. 2006. Description of six new species of *Chinavia* (Hemiptera, Pentatomidae, Pentatominae) from South America. Iheringia, Sér. Zool. 96: 237–248.

Schwertner, C.F. and J. Grazia. 2007. O gênero *Chinavia* Orian (Hemiptera, Pentatomidae, Pentatominae) no Brasil, com chave pictórica para os adultos. Rev. Bras. Entomol. 51: 416–435.

Schwertner, C.F. and J. Grazia. 2012. Review of the neotropical genus *Aleixus* Mcdonald (Hemiptera: Heteroptera: Pentatomidae: Procleticini), with description of a new species and cladistic analysis of the Tribe Procleticini. Entomol. Amer. 118: 252–262.

Silva, E.J.E., J.A.M. Fernandes and J. Grazia. 2004. Morphological variants in *Edessa rufomarginata* and revalidation of *E. albomarginata* and *E. marginalis* (Heteroptera, Pentatomidae, Edessinae). Iheringia. Sér. Zool. 94: 261–268.

Silva, E.J.E., J.A.M. Fernandes and J. Grazia. 2006. Characterization of the group *Edessa rufomarginata* and description of seven new species (Heteroptera, Pentatomidae, Edessinae). Iheringia. Sér. Zool. 96: 345–362.

Silva, V.J. da and J.A.M. Fernandes. 2012. A new species group in *Edessa* Fabricius, 1803 (Heteroptera: Pentatomidae: Edessinae). Zootaxa 3313: 12–22.

Silva, V.J. da, D.M. Nunes and J.A.M. Fernandes. 2013. *Paraedessa*, a new genus of Edessinae (Hemiptera: Heteroptera: Pentatomidae). Zootaxa 3716: 395–416.

Smaniotto, L.F. and A.R. Panizzi. 2015. Interactions of selected species of stink bugs (Hemiptera: Heteroptera: Pentatomidae) from leguminous crops with plants in the Neotropics. Fla. Entomol. 98: 7–17.

Stål, C. 1865. Hemiptera Africana. Vol. 1. Norstedtiana, Stockholm.

Stål, C. 1876. Enumeratio Hemipterorum. Bidrag till en Förteckning öfver alla hittills kända Hemiptera, Jemte Systematiska Meddelanden. K. Svensk. Vetensk-Akad. Handl. 14: 1–162.

Tamanini, L. 1959a. Caratteri generici di *Dolycoris* Mulsant et Rey e *Eudolycoris* nov. gen. con tavola dicotomica delle entita della sottoregione mediterranea. Mem. Soc. Entomol. Ital. 38: 73–83.

Tamanini, L. 1959b. I *Carpocoris* della regione Paleartica. Tabella per la determinazione delle entita e loro distribuzione. Mem. Soc. Entomol. Ital. 38: 120–142.

Thomas, D.B. 1992. Taxonomic synopsis of the asopinae Pentatomidae (Heteroptera) of the western hemisphere. The Thomas Say Foundation, ESA, Monographs 16: 1–156.

Thomas, D.B. 1994. Taxonomic synopsis of the old World asopine genera (Pentatomidae: Heteroptera). Insecta Mundi 8: 145–212.

Wall, M.A. 2004. Phylogenetic relationships among Halyini (Pentatomidae: Pentatominae) genera based on morphology, with emphasis on the taxonomy and morphology of the *Solomonius*-group. Ph.D. thesis, University of Connecticut, CT.

Wang, Q. and J.G. Millar. 1997. Reproductive behavior of *Thyanta pallidovirens* (Heteroptera: Pentatomidae). Ann. Entomol. Soc. Am. 90: 380–388.

Weirauch, C. and R.T. Schuh. 2011. Systematics and evolution of Heteroptera: 25 years of progress. Ann. Rev. Entomol. 56: 487–510.

Wyniger, D. and P. Kment. 2010. Key for the separation of *Halyomorpha halys* (Stål) from similar-appearing pentatomids (Insecta: Heteroptera: Pentatomidae) occurring in Central Europe, with new Swiss records. Mitt. Sch. Entomol. Ges. 83: 261–270.

Xie, Q., W.-J. Bu and L.-Y. Zheng. 2005. The Bayesian phylogenetic analysis of the 18S rRNA sequences from the main lineages of Trichophora (Insecta: Heteroptera: Pentatomomorpha). Mol. Phylogenet. Evol. 34: 448–451.
Yao, Y., W. Cai, D.A. Rider and D. Ren. 2013. Primipentatomidae fam. nov. (Hemiptera: Heteroptera: Pentatomomorpha), an extinct insect family from the Cretaceous of north-eastern China. J. Syst. Palaeontol. 11: 63–82.
Yonke, T.R. 1990. Order hemiptera. pp. 22–65. *In*: Stehr, F.W. (ed.). Immature Insects, Vol. 2. Kendall/Hunt, Dubuque, Iowa, USA.
Yuan, M.-l., Q.-l. Zhang, Z.-L. Guo, J. Wang and Y.-Y. Shen. 2015. Comparative mitogenomic analysis of the superfamily Pentatomoidea (Insecta: Hemiptera: Heteroptera) and phylogenetic implications. BioMed. Central Genom. 16: 1–15.

CHAPTER 2

Host Plant-Stink Bug (Pentatomidae) Relationships

Antônio R. Panizzi[1,*] and *Tiago Lucini*[2]

Introduction

The family Pentatomidae comprises more than 4,700 species spread around the world, divided into seven subfamilies: Asopinae, Cyrtocorinae, Discocephalinae, Edessinae, Pentatominae, Phyllocephalinae, and Stirotarsinae. Pentatominae represents the most abundant and diverse group within Pentatomidae, with > 2,800 species (Grazia et al. 2015). The subfamily Pentatominae consists entirely of plant feeders; in general, all are polyphagous and feed on both cultivated and non-cultivated plants. The subfamilies Edessinae and Pentatominae include phytophagous pentatomids of economic importance, i.e., insect pests; the first is represented mostly by the neotropical genus *Edessa*, and the second, Pentatominae, contains the majority of species that are pests of cultivated crops belonging to different genera, such as, *Aelia*, *Arvelius*, *Biprorulus*, *Chinavia* (*Acrosternum*), *Dichelops*, *Dolycoris*, *Euschistus*, *Mormidea*, *Nezara*, *Oebalus*, *Palomena*, *Piezodorus*, *Plautia*, and *Tibraca* (Schuh and Slater 1995, Panizzi et al. 2000).

The feeding process of phytophagous stink bugs is very interesting and sophisticated, because it involves piercing-sucking stylets (mouthparts modified from mandibles and maxillae) that are inserted into the plant tissue. For this reason, their feeding behaviors on specific plant tissues are difficult to watch and to interpret correctly as compared to phytophagous

[1] Laboratory of Entomology, Embrapa Wheat, PO Box 3081, Passo Fundo RS, 99001-970, Brazil.
[2] Department of Zoology, Federal University of Paraná, PO Box 19020, Curitiba, PR 81531-980, Brazil.
Email: tiago_lucini@hotmail.com
* Corresponding author: antonio.panizzi@embrapa.br

defoliators such as caterpillars, whose feeding process is easier to directly observe and interpret.

The technique known as EPG (Electropenetrography or Electrical Penetration Graph) (McLean and Kinsey 1964, Tjallingii 1978, Backus and Bennett 2009) reveals the many activities related to the feeding process of any piercing-sucking insect, like stink bugs, otherwise invisible inside plant tissue. Recent EPG studies of stink bug feeding recorded two pentatomids on soybean plants, the brown-winged stink bug, *Edessa meditabunda* (F.) (Lucini and Panizzi 2016) and the red-banded stink bug, *Piezodorus guildinii* (Westwood) (Lucini et al. 2016).

Stink bugs may feed on different parts of the host plants; however, the preference is to feed on fruits and on immature seeds (e.g., Panizzi et al. 2000, McPherson and McPherson 2000, Olson et al. 2011). Moreover, in the absence of a suitable and preferred host plant, stink bugs may explore alternate host plants, such as the less-preferred plants that are available in that particular space and time (Panizzi 2000, Karban and Agrawal 2002, Després et al. 2007); this behavior has been observed and reported for several species, such as the cosmopolitan southern green stink bug, *Nezara viridula* (L.), a seed feeder that was observed to feed on the leaf veins of a less-preferred plant, castor bean, *Ricinus communis* (Panizzi 2000).

Although the majority of stink bug species in the subfamily Pentatominae are thought to be polyphagous, some of these generalist feeders may prefer to explore plants in some particular taxa (preference for one or few kinds of taxa) (Table 2.1). For example, *E. meditabunda* has a preference to feed on plants of Solanaceae (solanaceous plants) and Fabaceae (legumes) (Silva et al. 1968, Rizzo 1971, Lopes et al. 1974); *N. viridula* prefers Brassicaceae (brassicaceous plants) and legumes (Todd and Herzog 1980, Jackai et al. 1990); *Bagrada hilaris* (Burmeister) shows a preference for brassicaceous plants (Reed et al. 2013, Huang et al. 2014), and *P. guildinii* for legumes (Panizzi and Slansky 1985). On the other hand, some pentatomids prefer Poaceae (graminaceous plants), as has been observed for species in the genera *Aelia*, *Oebalus*, and *Mormidea* (see references in Panizzi et al. 2000). Therefore, ostensibly polyphagous heteropterans can at times be oligophagous or even monophagous, depending on the availability of host plants and the time for which they have been exposed to particular plant taxa (local populations with specific feeding habits) (Fox and Morrow 1981); this behavior demonstrating the plasticity of the feeding adaptations of stink bugs illustrates how complex and sophisticated is the biology of phytophagous pentatomids.

The localization of suitable plants is a challenge that any phytophagous insect faces, particularly during adulthood. The search for and selection of host plants by an insect requires not only the choice of one or more species of a plant at random, but also searching for particular plants that may be suitable for feeding, and that ensure the development and survival of

Table 2.1. Major plant taxa explored by phytophagous pentatomids belonging to different subfamilies and genera demonstrating that, in general, although stink bugs are polyphagous, there is a preference for certain taxa (sources: Panizzi et al. 2000, McPherson and McPherson 2000, Smaniotto and Panizzi 2015, Grazia et al. 2015).

Stink bug subfamily	Stink bug genera	Plant family	Common names (plants or group of plants)
Pentatominae	*Aelia*	Poaceae	Wheat, barley, wild grasses
	Antestiopsis	Rubiaceae	Coffee
	Arvelius	Solanaceae	Tomato, potato, wild solanaceous
	Axiagastus	Arecaceae	Coconut
	Bagrada	Brassicaceae	Cabbage, other brassicaceous
	Bathycoelia	Malvaceae	Cocoa
	Biprorulus	Rutaceae	Citrus (orange, lemon, mandarin)
	Carpocoris	Poaceae	Wheat, wild grasses
	Chinavia (*Acrosternum*)	Fabaceae	Legumes (soybean, wild beans)
	Dichelops	Fabaceae Poaceae	Legumes (soybean, wild beans) Corn, wheat
	Dolycoris	Poaceae	Cereals
	Eurydema	Brassicaceae	Cabbage, other brassicaceous
	Euschistus	Fabaceae	Legumes (soybean, wild beans)
	Mormidea	Poaceae	Rice, wild grasses
	Murgantia	Brassicaceae	Cabbage, collard, mustard, turnip, peppergrass, radish, broccoli, cauliflower, kale
	Nezara	Fabaceae	Legumes
		Brassicaceae	Cabbage, collard, mustard, turnip, peppergrass, radish, broccoli, cauliflower, kale
	Oebalus	Poaceae	Rice, wild grasses
	Palomena	Fabaceae Solanaceae	Alfalfa, beans Potato
	Piezodorus	Fabaceae	Legumes (soybean, alfalfa, wild legumes)
	Plautia	Rosaceae	Fruit trees
	Runibia	Solanaceae	Ornamental trees (*Brunfelsia* spp.)
	Scotinophora	Poaceae	Rice

Table 2.1 contd. ...

...*Table 2.1 contd.*

Stink bug subfamily	Stink bug genera	Plant family	Common names (plants or group of plants)
	Thyanta	Poaceae	Wheat, sorghum
		Asteraceae	*Bidens* spp.
		Fabaceae	Legumes
	Tibraca	Poaceae	Rice, wild grasses
	Vitellus	Rutaceae	Citrus
Cyrtocorinae	*Cyrtocoris*	Fabaceae	Soybean, wild legume trees
Discocephalinae	*Lincus*	Arecaceae	Coconut, palm trees
Edessinae	*Edessa*	Solanaceae	Potato, tomato
		Fabaceae	Soybean

its progeny. Several factors are assessed by phytophagous insects while searching for suitable and preferred host plants in order to explore them as a food source; behavioral responses indicate whether they accept or reject a particular host (Bernays and Chapman 1994). Plant features that are important in this process include physical and/or chemical traits (e.g., chemical volatiles, surface waxes, nutrients, secondary compounds), and presence in time and space of plant(s) with suitable nutritional qualities (Strong et al. 1984, Bernays and Chapman 1994, Schoonhoven et al. 2005).

The ability of a phytophagous insect to locate a host plant depends on a series of factors; for instance, its dispersal capability allows it to move into the vicinity of its host plant over great distances (Bernays and Chapman 1994), then the chemical and physical features of the plant (Chew and Renwich 1995) may act as an attractant or a repellent. For example, volatile chemicals produced and released by green plants, commonly known as "green leaf volatiles" may or may not be attractive to the insect. In addition, the substance may negatively affect the performance of one or more species of an insect, but not others (Bernays and Chapman 1994).

In this chapter, we will discuss the modes of feeding of stink bugs on host plants, the role of host plants on stink bug nymphal and adult biology, how the switch in food type from nymph to adult may affect adult biology, the role of less-preferred host plants to sustain populations, and the importance of considering all types of hosts and associated plants to unveil the bioecology of pentatomids in nature.

Feeding Strategies/Tactics and Specific Sites for Food/Water Ingestion

The modes of feeding by phytophagous heteropterans have been presented and discussed by several authors (e.g., Miles 1969, 1972, Miles and Taylor

1994, Hori 2000). The last author described in detail the four ways in which the heteropterans feed: (1) salivary (or stylet) sheath feeding, which includes the formation of a sheath of gelling saliva to anchor/support/lubricate the stylets and is used by most pentatomorphans; (2) lacerate-and-flush feeding, in which the bugs vigorously move the stylets back and forth mechanically destroying the cells for subsequent ingestion; (3) macerate-and-flush feeding, in which the bugs move stylets only slightly but inject saliva (especially pectinase) that macerates (dissolves) the cell contents, which are subsequently ingested with the uptake of secreted saliva; and (4) osmotic pump feeding, in which the bugs secrete and introduce salivary sucrase that increases the osmotic concentration of fluids, especially phloem sap, causing the unloading of nutrients (sugars and amino acids) from the phloem so that these nutrients can be subsequently ingested.

More recently, Backus et al. (2005) proposed a revision of hemipteran feeding strategies, to consolidate the many modes observed in the order into three main strategies: (1) salivary sheath feeding, (2) cell rupture feeding and (3) osmotic pump feeding. The first two strategies include multiple sub-strategies or tactics. Tactics within salivary sheath feeding include various methods of stylet penetration such as intercellular versus intracellular and mandibular-stylets-ahead versus maxillary-stylets-ahead (Backus 1988, Backus et al. 2005). Heteropteran stylet penetration is intracellular and mandibular-stylets-ahead (Backus 1988). Within the cell rupture feeding strategy, there are five tactics identified so far, two of which (macerate-and-flush and lacerate-and-flush) occur in heteropterans, as stated above. Three other tactics are performed by typhlocybine leafhoppers, an auchenorrhynchan/cicadellid subfamily that uniquely performs cell rupture feeding rather than salivary sheath feeding (Backus et al. 2005).

Heteropteran feeding strategies/tactics have been described in the literature in relation to their resulting damage: local lesion at the stylet insertion point; local lesion with the development of secondary symptoms; tissue malformation; and symptoms indicating translocation of some kind of noxious agent, assumed to be salivary. These initial symptoms often lead to the development of further symptoms over time, described as: local wilting and tissue necrosis; abscission of fruits; morphological deformation of fruit/seed; altered vegetative growth, and tissue malformation (see detailed discussion in Hori 2000, and references therein).

Interestingly, pentatomids may use one or more of these strategies/tactics of feeding on the same individual plant. For example, the red-banded stink bug, *P. guildinii*, feeding on the soybean stem, leaf and pod-wall uses the salivary sheath feeding strategy—in this case ingesting from xylem vessels, very probably for hydration. In contrast, when feeding on the soybean seed in the pod (seed locus—seed endosperm) the same insect can use the cell rupturing strategy to obtain nutrients (Lucini et al. 2016);

however, whether the macerate-and-flush or the lacerate-and flush tactic is used is presently unknown. This shift in feeding strategy is not known to occur in any other group in Hemiptera. Clearly, it allows the exploration of different feeding sites on plant organs, probably resulting in improved nutritional balance during the feeding process. Seed endosperm is known to be a nutritionally complete "package" while the xylem vessels contain mostly water.

Plant tissue histology, in which the stylets or salivary sheath introduced into the specific feeding sites on the plant are carefully excised during (or immediately after) feeding, allows identification of exactly which tissue the bugs are exploring in order to take up nutrients and/or water. Recently published articles with two different species of pentatomids demonstrated the feeding sites at the cellular level on different plant organs of a soybean plant (Lucini and Panizzi 2016, Lucini et al. 2016). For instance, the brown-winged stink bug, *E. meditabunda*, feeds on different veinous tissues of the soybean plant (both xylem and phloem, although it could not be determined which of the four phloem cell types [phloem fiber, phloem parenchyma, companion cells or sieve elements] were probed) (Fig. 2.1).

This species of stink bug, different from the majority of species in different genera, does not restrict itself to feeding on reproductive structures (fruits and/or seeds), and actually prefers to feed on stems using a very thin salivary sheath (Silva et al. 2012). These authors demonstrated that when stem-feeding, *E. meditabunda* produces saliva with almost no amylase activity, as compared to the seed-feeding Neotropical brown stink bug, *Euschistus heros* (F.), which does secrete amylase.

The red-banded stink bug, *P. guildinii* almost always shows a salivary sheath surrounding the stylets when feeding on soybean stems (Fig. 2.2 A,

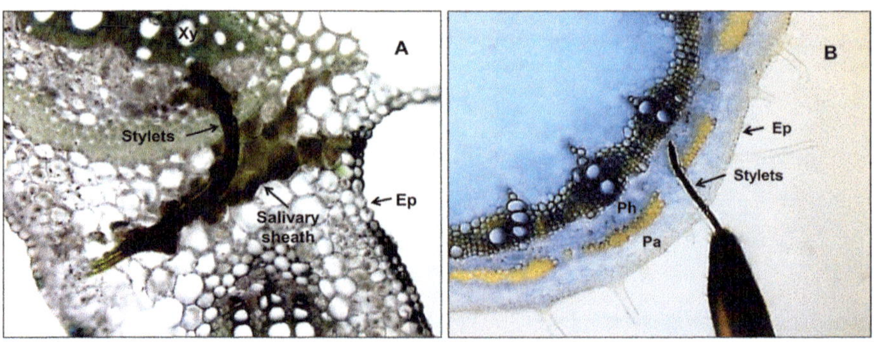

Figure 2.1. Cross-section of the soybean plants containing the stylets of the brown-winged stink bug, *Edessa meditabunda* (F.) excised during feeding on the xylem (A) and on the phloem (B) of the soybean stem, with the gelling salivary sheath secreted during the feeding process Pa: parenchyma, Ep: epidermis, Xy: xylem, Ph: phloem (photos: Tiago Lucini).

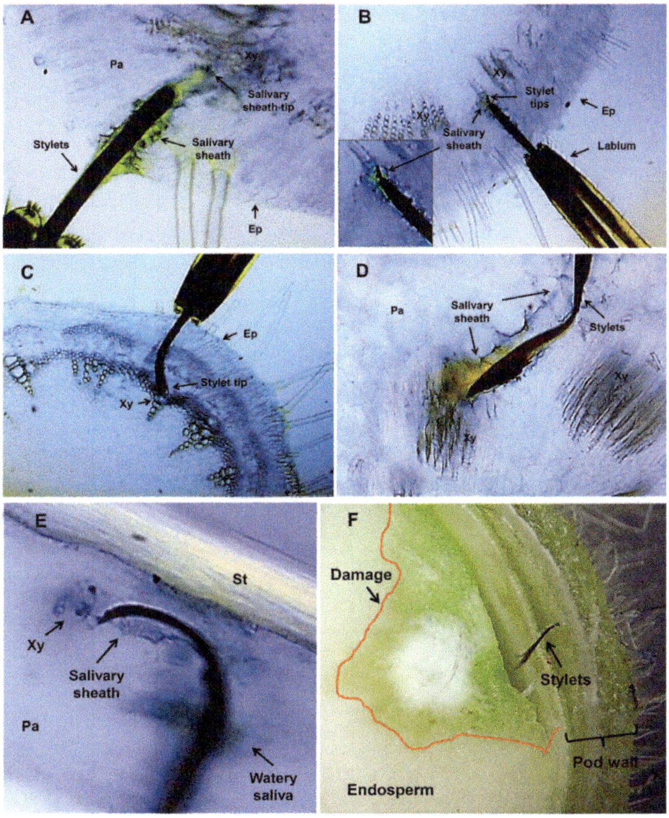

Figure 2.2. Cross-sections of a soybean plant containing stylets and salivary sheaths of *Piezodorus guildinii* on the soybean stem and pod. (A) Salivary sheath ending in parenchyma tissue of the soybean stem (10x). (B) Stylet and salivary sheath tips in parenchyma tissue of the soybean stem, near the likely ingestion cell type (xylem) (5x). (C, D) Stylet tips and/or salivary sheath ending in xylem vessels (longitudinal cut in D) recorded on the soybean stem (5x and 10x, respectively). (E) Stylet tips in xylem vessels recorded on the soybean pod-wall (10x). Note the small salivary sheath near the xylem vessel. The proximal portion of the stylet bundle is surrounded by a fuzzy deposit of saliva that may be watery saliva or out-of-focus sheath saliva. (F) Cross-section of the soybean pod showing damage near the insertion point of the stylets of the stink bug. An orange line was drawn around the damaged area. Ep: stem/pod epidermis; Pa: parenchyma; St: Sclerenchyma; Xy: xylem (from Lucini et al. 2016, with permission of the Entomological Society of America).

B, D, not in Fig. 2.2C). Feeding on the xylem of the stem and on the xylem of the pod-wall is illustrated on Fig. 2.2C and E, respectively. The damaged area in the seed endosperm appears in Fig. 2.2F. These cross-sections containing the stylets reveal not only the specific feeding sites, but also allow us to speculate about the needs of bugs for both nutrients and water, and their resulting nutritional balance.

Host Plants and Associated Plants

In many cases, authors do not make a clear distinction between the host plants and associated plants, and this has been discussed in the literature (e.g., Strong et al. 1984, Bernays and Chapman 1994, Schoonhoven et al. 2005). Often, a plant species on which a particular species of stink bug is collected, feeding or not, is regarded as a host plant; however, this is not always the case. A host plant is defined as a plant that the bug feeds on and that allows reproduction, i.e., the complete development of nymphs that will yield adults able to continue breeding to produce future generations. On the other hand, associated plants are those which the bugs may feed on, use as shelter, lay eggs on, but the nymphs are unable to develop and complete their life cycle; in addition, any resulting adults do not breed.

Among the host plants, there are those which are called preferred host plants, which present high nutritional qualities. When insects feed on them, it is expected that the next generation will be able to reach its maximum reproductive potential, thus maximum fitness, i.e., the maximum reproductive contribution of each individual to the next generation. In those cases, nymphal development and adult reproductive capacity will be the greatest (Panizzi 2007). Accordingly, the preferred host plant is where the pentatomids breed, develop, and reproduce without (or with only minor) restriction. In contrast, on associated plants on which the bug may (out of necessity) feed exclusively, it is unable to reproduce and complete its development (Panizzi 2000).

As an example of the above principles, the southern green stink bug, *N. viridula*, has been recorded on 70 different plant species in the Neotropics, of which 29 species are reported as host plants; the remaining 41 species are considered to be associated plants (Fig. 2.3) (Smaniotto and Panizzi 2015). *N. viridula* does not breed on graminaceous plants; for instance, on wheat the bug may occasionally feed on seed heads during late winter/early spring, but has not been reported to lay eggs on this plant. Immatures of *N. viridula* fed wheat seed heads and mature seeds in laboratory conditions did not complete development (Panizzi 2007).

Other examples include the species *P. guildinii* and *E. meditabunda*, which have been reported on 49 and 40 different plant species in the Neotropics, respectively, with about 50% of these plants being considered as host plants. On the other hand, for other species of pentatomids such as *E. heros*, *Dichelops furcatus* (F.), *Dichelops melacanthus* (Dallas), and *Thyanta perditor* (F.), the majority of the plants (over 70%) have been reported as associated plants (Smaniotto and Panizzi 2015) (Fig. 2.3).

Neotropical green-belly stink bugs, *D. melacanthus* and *D. furcatus*, feed on several different species of cultivated plants soon after they emerge as seedlings because of their habit of overwintering under crop residues on the ground. In addition, these bugs also exploit weed plants. For example,

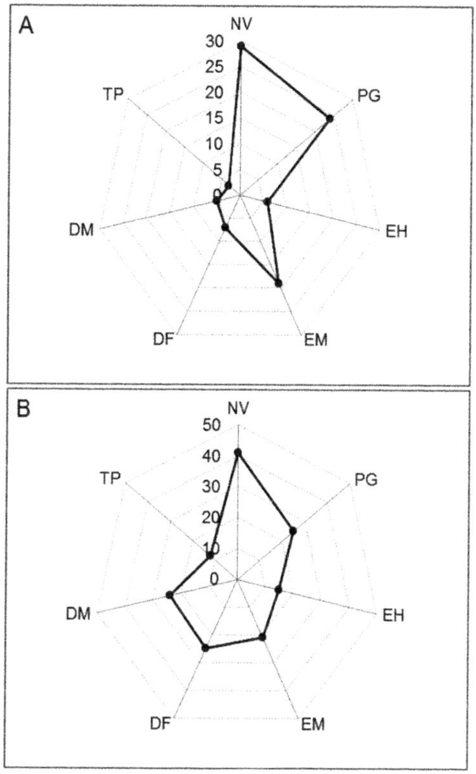

Figure 2.3. Records of host plants and associated plants for different species of Neotropical stink bugs. The dark line represents, (A) Number of plants considered as host plants, i.e., plants on which the stink bugs feed, develop and reproduce, for each species of stink bug; (B) Number of plants considered to be associated plants, i.e., plants on which the stink bugs feed, but do not reproduce, for each species of stink bug. NV = *Nezara viridula*; PG = *Piezodorus guildinii*; EH = *Euschistus heros*; EM = *Edessa meditabunda*; DF = *Dichelops furcatus*; DM = *Dichelops melacanthus*; and TP = *Thyanta perditor*. Figure modified from Smaniotto and Panizzi (2015).

D. melacanthus occurs on soybean in the form of both nymphs and adults, while mostly adults are observed on wheat and on corn during summer and autumn (Fig. 2.4).

On the non-cultivated plants, hairy indigo, *Indigofera hirsuta*, signal grass, *Bachiaria decumbens*, and tropical spider wort, *Commelina benghalensis*, nymphs and adults are observed year-round; on crotalaria (*Crotalaria pallida*) only adults are recovered (Fig. 2.5).

These data illustrate the strong variability in the degree to which nymphs and adults explore cultivated and non-cultivated plants as a source of nutrients/water or for shelter, and how much research effort is needed to completely characterize for a particular stink bug species the role of all associated plants.

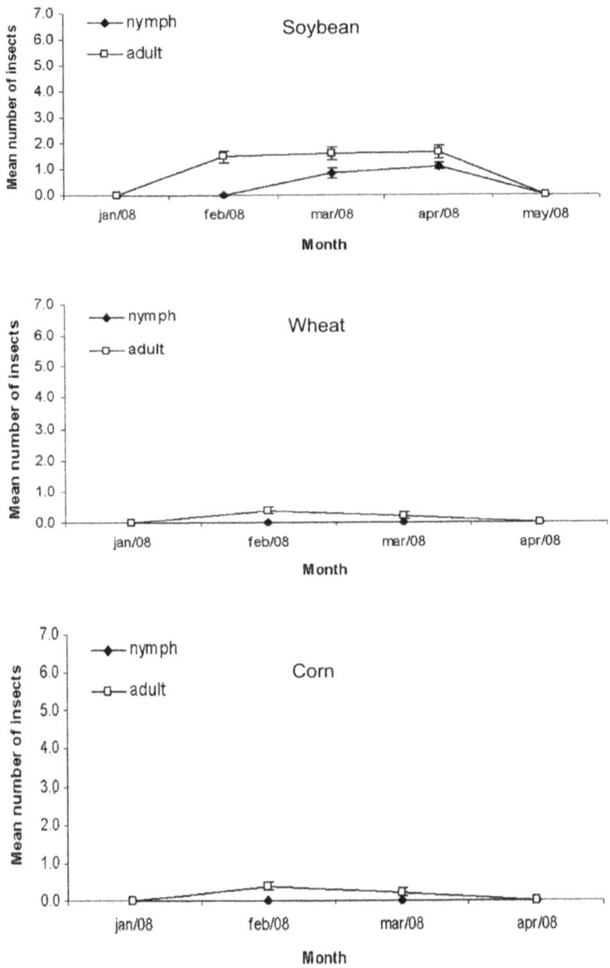

Figure 2.4. Occurrence of *Dichelops melacanthus* on cultivated plants in Londrina, PR, Brazil (from Silva et al. 2013, with permission).

Nymphal Biology on Plants

As eggs hatch, newly ecloded nymphs usually remain aggregated on the top of eggs or egg masses. Heteropteran eggs may be laid in masses or singly, and these egg-laying patterns have been discussed in the literature regarding which type yields the most successful results for nymphal survivorship (e.g., Panizzi 2004). Stink bugs lay egg masses of different sizes and shapes (hexagonal, in double parallel long lines, short 3–4 lines, etc.) and usually nymphs remain on the top of egg shells, or after some time position themselves around the eggs. Laboratory observations of the first-

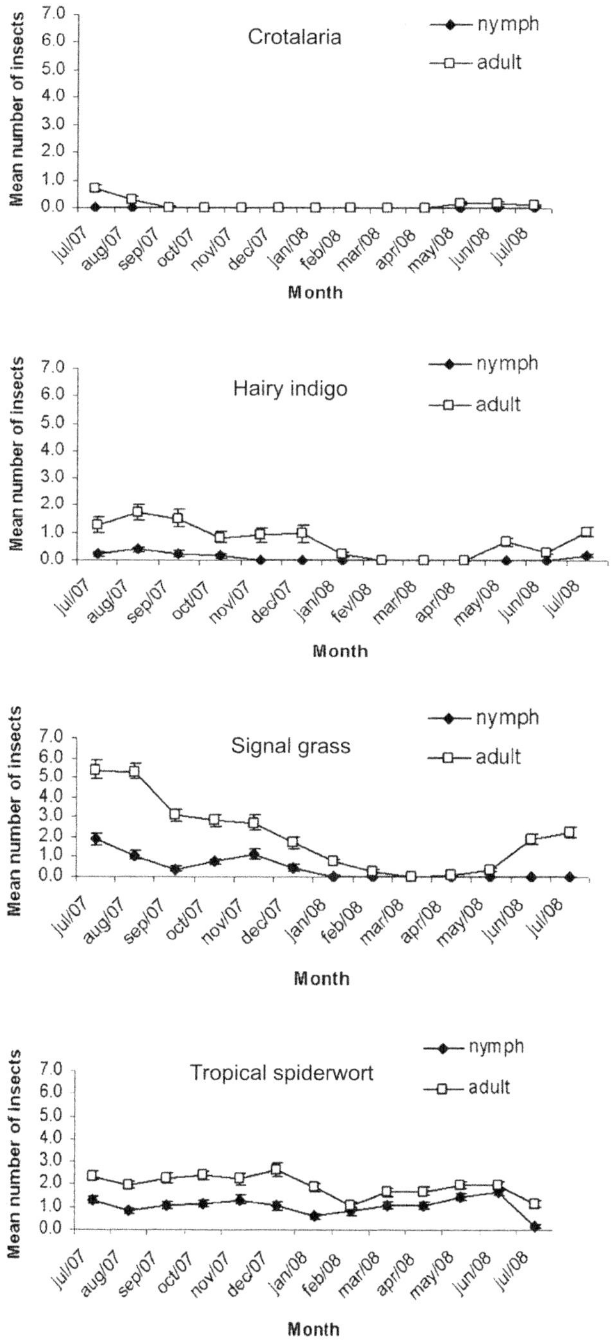

Figure 2.5. Occurrence of *Dichelops melacanthus* on non-cultivated plants in Londrina, PR, Brazil (from Silva et al. 2013, with permission).

instar nymphs of the brown-winged stink bug, *E. meditabunda*, indicated that, immediately after emergence, the nymphs stayed on the top of each chorion for about 3 min; they visited ca. 8 chorions or eggs before moving off the shells or eggs and positioning themselves adjacent to the eggs, which took about 25 min. All nymphs relocated around the egg mass, facing it. This behavior of moving off the chorions or eggs and positioning adjacent to them is speculated to be related to possible symbiont acquisition and/or defense (Calizotti and Panizzi 2014).

Up until recently, the first instars of pentatomids were believed not to feed. However, laboratory studies conducted with the southern green stink bug, *N. viridula*, demonstrated that they ingested a marked bacterial pathogen of cotton from the interior of green bean pods internally inoculated with bacteria (Esquivel and Medrano 2014), thus they are capable of feeding (in the broad sense). However, while the first instars can ingest fluid, it is not known whether such ingestion is obligate or facultative. Previous research has shown that the first instars of many species of pentatomids are able to pass onto the second instar in the absence of food. Thus, they may be ingesting not food but symbionts and other microorganisms present on the egg mass surface and from inside the egg shells during the time they spend on the top of egg shells and before molting to second instar. Therefore, the majority of the studies dealing with nymphal biology on different food sources start from the second instar when food is clearly ingested by the nymphs (see references in Panizzi 1997).

Nymphs of pentatomids feeding on cultivated and on non-cultivated (wild) plants show a great variation in the time of development and survival rate. In general, their performance is better on cultivated than on non-cultivated plants; however, this is not always true. For example, the highly polyphagous *N. viridula* feeding on soybean (pods) shows a second instar-to-adult mean developmental time of 25.4 days (range of 22.3 to 27.8 days), while on non-cultivated plants the mean time is 33.2 days (range of 22.0 to 64.0 days) (Table 2.2); this delay of ca. 8 days on non-cultivated plants reveals the greater unsuitability of these plants as compared to the cultivated ones, in this case, soybean. This may have variable impact on nymphal bioecology, including implications for their synchronized phenology with host plants and greater exposure time to natural enemies, among others. In a similar way, mean nymph mortality (averaged among multiple bugs on individual soybean pods as well as reproductive plants) is 24.8% (range of 2.0 to 60.0%), while on non-cultivated plants the mean mortality is 61.2% (range of 0.0–100.0%) (Table 2.2). Again, on non-cultivated plants *N. viridula* nymphs show a much greater mortality than on their preferred food, soybean.

For a less cosmopolitan and less polyphagous species, such as the red-banded stink bug, *P. guildinii*, nymphs feeding on several wild legumes such as *Sesbania aculeata*, *Indigofera endecaphylla*, and *Indigofera truxillensis*, exhibit

Table 2.2. Developmental time and mortality of nymphs of the southern green stink bug, *Nezara viridula*, feeding on different plants (non-cultivated and cultivated) (adapted from Panizzi 2007, references therein).

Plant species[a]	Days[b]	Mortality (%)
Non-cultivated		
Brassica kaber	26.1–27.5	25.0
C. fasciculata	–	100.0
C. occidentalis	26.7	0.0
Crotalaria lanceolata	27.2–33.9	85.0
C. spectabilis	37.3	26.0
Croton glandulosus	43.5	80.0
Datura stramonium plants[c]	38.7	59.5
Desmodium canum	–	100.0
Desmodium tortuosum	22.0–23.5	65.0
Ebelmoschus esculentus	33.5	10.0
Glycine wightii	25.0–27.5	93.3
Indigofera hirsuta	–	100.0
Leonurus sibiricus	30.4–31.9	25.0
Lepidium virginicum	–	100.0
Ligustrum lucidum	26.9–30.1	38.7
Macroptilium lathyroides plants	33.5	61.7
Melilotus indica plants[c]	47.6	63.7
Physalis virginiana plants[c]	47.5	65.0
Prunus serotine	29.3	78.0
Raphanus raphanistrum	35.4–39.3	56.2
R. raphanistrum	27.5	2.0
Rapistrum rugosum plants[c]	44.1	65.2
Ricinus communis	42.3–42.6	60.2
R. communis plants[c]	50.2	86.5
Sesbania aculeata	–	100.0
S. emerus	20.3–20.8	10.0
S. vesicaria seeds (immature)	20.5–22.2	40.0
S. vesicaria pods	–	100.0
Trifolium repens plants[c]	64.0	98.4
Mean	**33.2**	**61.2**

Table 2.2 contd. ...

...*Table 2.2 contd.*

Plant species[a]	Days[b]	Mortality (%)
Cultivated		
Glycine max	26.2–26.3	60.0
G. max	25.2–27.8	28.9
G. max	25.9–26.0	15.0
G. max	23.0	2.0
G. max	22.9–23.2	22.5
G. max	22.3–23.4	20.0
G. max plants[c]	32.8	25.5
Mean	**25.4**	**24.8**

[a]Unless otherwise indicated, the food offered were fruits. [b]From second stadium to adult. [c]Fruiting plants.

lower total mortality (< 30%) than those feeding on soybean (ca. 70%–range of 57.7 to 88.0%) (references in Panizzi 1997). One should consider that soybean is an exotic plant in the Neotropics, where *P. guildinii* is native, and this may account for its low nymphal performance on such introduced crop, despite its status as a major pest.

These examples and others from the literature demonstrate that, for pentatomid nymphs, food (plant) exploitation is a great challenge. Barriers that are not only physical (e.g., hard tissues, thickness of pod-walls and air space inside pods protecting the seeds, pilosity of different kinds) but also chemical (lack of right compounds—proteins, lipids, amino acids, vitamins, minerals, etc. plus presence of allelochemicals that might be toxic) must be surpassed. In addition, food location might be an extra burden, considering a nymph's limited dispersal ability and the lack of control of food suitability they face in the early stages of development. Host plant choice is performed by their mother, not the nymphs, in locating a site for egg-laying.

Adult Biology on Plants

As nymphs complete their development and adults are produced, they immediately start looking for suitable host plants on which to feed and reproduce. For adults, two variables most commonly used to measure food suitability in laboratory studies are fecundity and longevity. Southern green stink bug *N. viridula* shows almost two times greater fecundity on cultivated plants as compared to non-cultivated plants; in a similar way, adult mean longevity is also greater on cultivated plants as compared to non-cultivated plants, although the difference is not as pronounced for fecundity (Table 2.3).

Table 2.3. Fecundity and longevity of the southern green stink bug, *Nezara viridula*, feeding on different plants (non-cultivated and cultivated) (adapted from Panizzi 2007, references therein).

Plant species[a]	Eggs/female	Longevity (days)
Non-cultivated		
Acanthospermum hispidum plants[b]	0.0	6.9
Brassica kaber	107.4	36.9
Crotalaria lanceolata	29.0	33.7
Datura stramonium plants[b]	30.8	-
Desmodium tortuosum	61.0	35.6
Leonurus sibiricus	91.7	58.8
**Ligustrum lucidum*	256.6	35.8
L. lucidum[c]	261.0	-
Macroptilium lathyroides plants[b]	0.0	-
Physalis virginiana plants[b]	56.8	-
Raphanus raphanistrum	68.8	42.2
Rapistrum rugosum plants[b]	94.3	-
Ricinus communis	0.0	20.5
R. communis plants[b]	95.0	-
Sesbania emerus	273.9	-
Trifolium repens plants[b]	0.0	-
Mean	**89.1**	**33.8**
Cultivated		
Glycine max	67.7	37.9
G. max	99.3	45.1
G. max	110.0	38.0
G. max	139.7	35.0
G. max	203.7	56.0
G. max	116.8	-
G. max plants[b]	124.8	-
Sesamum indicum	297.9	62.0
Mean	**145.0**	**45.7**

[a]Unless otherwise indicated, the food offered were fruits.
[b]Fruiting plants.
[c]Panizzi and Mourão (1999).
*The fecundity value obtained by Coombs (2004) (632 eggs/female) on *Ligustrum lucidum* was not considered because this value is so high that it will biase the mean value presented for the non-cultivated plants (see text for details).

Despite the general greater suitability of cultivated plants as compared to non-cultivated ones (at least, for pest species) the opposite may occur. For example, mean fecundity (eggs/female) for *N. viridula* on privet, *Ligustrum lucidum*, and on the wild legume *Sesbania emerus* is 256.6–261.0, and 273.9 eggs, respectively (Panizzi and Slansky 1991, Panizzi et al. 1996, Panizzi and Mourão 1999). Research work carried out in Australia confirmed the very high fecundity of *N. viridula* on privet (632.0 eggs/female) (Coombs 2004), although the later value is more than two times the values mentioned in previous studies. Because of this very high (outlier) value, we decided not to consider it in the calculation of *N. viridula* mean fecundity on non-cultivated plants presented in Table 2.3. A similarly high fecundity (630.9 eggs/female) was observed for *N. viridula* fed on privet berries, however, soybean seeds and raw peanuts were added as complementary food (Panizzi et al. 1996). In this last case, perhaps a synergistic food effect may have occurred, and the high fecundity obtained cannot be regarded as an effect of privet berries alone.

For the Neotropical red-banded stink bug, *P. guildinii*, the fecundity observed on soybean—overall mean of data presented in Table 6.1 of Panizzi (2007)—is 52.5 eggs/female. More recently, Zerbino et al. (2016) reported the fecundity of *P. guildinii* to be 134.4 and 148.3 eggs/female feeding on soybean and on alfalfa, respectively. These values, although higher than those mentioned in Table 6.1 of Panizzi (2007), are yet smaller as compared to those obtained for certain wild legumes, such as 196.7 on *Indigofera suffruticosa*, 204.8 on *I. hirsuta*, 205.1 on *Sesbania aculeata*, 305.5 on *I. endecaphylla*, and 507.7 eggs/female on *I. truxillensis*. This last value is extremely high when compared to the overall mean of data on non-cultivated plants in Table 6.1 (Panizzi 2007) of 205.0 eggs/female.

Preferred and suitable foods may not be available altogether, or may become scarce, for adults during certain periods of time; in that case, the stink bugs need to find one or more ways to survive. For example, the Neotropical brown stink bug, *E. heros*, and the southern green stink bug *N. viridula*, feed temporarily on stems of the less-preferred plant star bristle, *Acanthospermum hispidum*. This is an atypical behavior, because these bugs are seed feeders; the first species feeds but does not reproduce on the plant, whereas the second species has its longevity drastically reduced (Panizzi and Rossi 1991). It seems that the chemical compounds produced by this plant have a deterrent effect on *N. viridula*, suggesting a toxic effect, which does not seem to occur for *E. heros*. Certain chemical compounds produced by plants can have a deterrent effect, i.e., inhibit feeding and/or oviposition, while in case of other species, these same plant chemicals do not significantly affect their performance (Bernays and Chapman 1994). In addition, under unfavorable environmental conditions (mainly temperature and photoperiod) adults may enter into oligopause/quiescence, as observed for *E. heros* (Panizzi and Niva 1994, Panizzi and Vivan 1997); during this time, the bug does not feed, surviving only on lipids previously accumulated (Panizzi and Hirose 1995).

Impact of Nymph-to-Adult Food-Switch on Stink Bug Biology

Upon reaching the adult stage, phytophagous hemipterans, in general, disperse to feed and reproduce on plant species entirely different from those on which they have developed as nymphs. An exception occurs when they feed on suitable plants of crops cultivated in large areas with plants at different phenological stages of development, allowing different generations of bugs to develop before the plants mature and are harvested. Disregarding this last case, offspring will generally feed and develop on food sources different from those of adults. Therefore, the effect of this food-switch from nymph to adult undoubtedly is a very important component of heteropteran biology; despite that, we suspect that the importance of this switch in food in regulating bug performance has been, in general, underestimated. For the specific case of pentatomids, this was presented and discussed in the 1980's and '90's, in particular for the polyphagous southern green stink bug, *N. viridula* (Kester and Smith 1984, Panizzi et al. 1989, Panizzi and Slansky 1991, Panizzi and Saraiva 1993, Velasco and Walter 1992, 1993).

The performance of adult *N. viridula* was significantly affected when nymphs were fed on two poor food sources, soybean mature seed and pod of *Crotalaria lanceolata*. On these two foods, the longevity of adult females was reduced significantly; in addition, on the second plant, adult females needed a longer time (49 days) to produce only one egg mass and none of the eggs hatched, whereas on the other food sources, the pre-oviposition period ranged from 22 to 24 days. When the nymphs were fed with green bean pod (a good food source) and then the adults were switched to other food sources such as *C. lanceolata*, the longevity of both females and males was increased significantly; similar results were observed when adults were fed soybean and peanut mature seed, as well as beggar weed, *Desmodium tortuosum*, pod (Panizzi and Slansky 1991).

Moreover, the fecundity rates of adult *N. viridula* increased considerably when the nymphs were reared on green bean pod and then switched to a poor food source as adults. These results suggest that, when nymphs are reared on an adequate food source, the deleterious effects of a poor quality diet on adult biology may, at least, be alleviated. On the other hand, poor quality food during nymphal development can directly (negatively) affect adult performance, even when adults are fed good quality food sources (Panizzi and Slansky 1991). Kester and Smith (1984) reported that adults of *N. viridula* that originated from fifth instar nymphs that were fed peanut seeds (poor quality food for nymphs) exhibited greatly reduced fecundity rates.

The nymph-to-adult food-switch syndrome, therefore, can have a positive or negative effect, or even, not have any effect at all on the biology of pentatomids depending on the quality of the food involved in these changes in food sources. These effects also may vary for different species according to their capacity to compensate for unsuitable food sources, which may include both physiological and behavioral adaptations.

The Role of Less-Preferred Plants

Preferred host plants, especially when associated with favorable environmental conditions of temperature and humidity, allow pentatomids to develop and reach their maximum reproductive potential. Although in general polyphagous, these bugs are often faced with the absence or low availability of their preferred host plants. In this case, the bugs must be able to exploit other plants to continue their feeding and breeding (although they may not breed on some of these plants). These plants, known as less-preferred plants, provide some nutrients and also water (Panizzi 2000, Panizzi and Silva 2012).

During the spring and summer seasons, stink bugs find their preferred host plants, usually cultivated crops. However, they spend only part of their lifetime feeding and breeding on these plants; the rest of the time they feed on other types of plants, often less-preferred, or find overwintering sites in where they remain in partial hibernation, not feeding, such as observed for the Neotropical brown stink bug, *E. heros* (Panizzi and Vivan 1997).

The less-preferred plants may be represented by non-cultivated (wild plants, many of them weeds) and also by cultivated plants. Examples of wild, less-preferred plants for Neotropical pentatomids of economic importance are included in Table 2.4. The number of plant species in different plant families on which stink bugs are collected, and on which they are not recorded to reproduce, is certainly impressive. On these plants, stink bugs feed to some degree; they may also use them for shelter, and may even occasionally oviposit on them, however, the nymphs do not complete their development and the resulting adults, in general, do not oviposit on them.

As an example, *E. heros* may feed on the euphorb *Euphorbia heterophylla* (Meneguim et al. 1989, Pinto and Panizzi 1994) and on star bristle, *A. hispidum* (Panizzi and Rossi 1991); other pentatomids such as *N. viridula* and *P. guildinii* feed on the weed *Bidens pilosa* (Lopes et al. 1974, Ferreira and Panizzi 1982); all these plants are wild plants associated with the soybean crop. The green belly stink bug, *D. melacanthus*, also feeds on weeds of the genera *Bidens*, *Sida*, and *Brachiaria*, and also on stems of young plants of corn (Panizzi 2000, Chocorosqui and Panizzi 2004, Manfredi-Coimbra et al. 2005) and wheat (Panizzi 2000, Manfredi-Coimbra et al. 2005), when soybean plants at the reproductive stage with pods are not available.

Table 2.4. Less-preferred feeding plants (associated plants) exploited by some species of stink bug pests in the Neotropics (adapted from Smaniotto and Panizzi 2015).

Stink bug species	Plant family	Plant species
Nezara viridula	Amaranthaceae	*Hebanthe eriantha*
	Anacardiaceae	*Schinus molle*
	Apiaceae	*Foeniculum vulgare*
	Asteraceae	*Acanthospermum hispidum*
		Bidens pilosa
		Cynara cardunculus
		Eupatorium spp.
		Helianthus annuus
		Lactuca sativa
	Bignoniaceae	*Adenocalymma comosum*
		Pyrostegia venusta
	Caricaceae	*Carica papaya*
	Cucurbitaceae	*Cucurbita* sp.
		Momordica charantia
		Sechium edule
	Euphorbiaceae	*Ricinus communis*
	Lauraceae	*Nectandra* sp.
	Smilacaceae	*Smilax brasiliensis*
	Malvaceae	*Abelmoschus esculentus*
		Gossypium hirsutum
		Malva sp.
		Sida sp.
	Meliaceae	*Cedrela fissilis*
	Piperaceae	*Piper* sp.
	Poaceae	*Avena sativa*
		Coix lacryma-jobi
		Oryza sativa
		Sorghum bicolor
		Triticum aestivum
		Zea mays
	Rosaceae	*Prunus persica*
	Rutaceae	*Citrus sinensis*

Table 2.4 contd. ...

...*Table 2.4 contd.*

Stink bug species	Plant family	Plant species
		Citrus sp.
	Solanaceae	*Datura* sp.
		Solanum lycopersicum
		Nicotiana tabacum
		Solanum incarceratum
		Solanum melongena
		Solanum nigrum
		Solanum sisymbrifolium
		Solanum tuberosum
Piezodorus guildinii	Amaranthaceae	*Hebanthe eriantha*
	Apiaceae	*Foeniculum vulgare*
	Asteraceae	*Bidens pilosa*
		Helianthus annuus
	Bignoniaceae	*Adenocalymma comosum*
		Pyrostegia venusta
	Brassicaceae	*Brassica napus*
		Raphanus sativus
	Cactaceae	*Peireskia aculeata*
	Cucurbitaceae	*Sechium edule*
	Euphorbiaceae	*Ricinus communis*
	Lauraceae	*Nectandra* sp.
	Linaceae	*Linum usitatissimum*
	Malvaceae	*Gossypium hirsutum*
	Myrtaceae	*Eugenia uniflora*
		Myrciaria tenella
	Nyctaginaceae	*Bougainvillea glabra*
	Phytolaccaceae	*Phytolacca dioica*
	Rosaceae	*Fragaria ananassa*
	Rubiaceae	*Coffea arabica*
	Sapindaceae	*Serjania fuscifolia*
	Violaceae	*Anchietea salutaris*

Table 2.4 contd. ...

...*Table 2.4 contd.*

Stink bug species	Plant family	Plant species
		Hybanthus atropurpureus
Euschistus heros	Amaranthaceae	*Amaranthus retroflexus*
	Asteraceae	*Acanthospermum hispidum*
		Helianthus annuus
	Brassicaceae	*Brassica napus*
		Brassica oleraceae
	Lauraceae	*Nectandra* sp.
	Malpighiaceae	*Malpighia glabra*
	Malvaceae	*Gossypium hirsutum*
	Ranunculaceae	*Clematis dioica*
	Salicaceae	*Casearia sylvestris*
	Solanaceae	*Nicotiana tabacum*
		Solanum mauritianum
		Solanum megalochiton
		Vassobia breviflora
Edessa meditabunda	Asteraceae	*Dahlia* sp.
		Helianthus annuus
	Cariaceae	*Carica papaya*
	Commelinaceae	*Tradescantia virginiana*
	Cucurbitaceae	*Cucumis melo*
		Cucurbita sp.
		Momordica charantia
		Sechium edule
	Euphorbiaceae	*Manihot esculenta*
	Malvaceae	*Gossypium hirsutum*
		Sida sp.
	Poaceae	*Lolium multiflorum*
		Lolium perene
		Oryza sativa
		Zea mays
	Rutaceae	*Citrus* sp.
	Solanaceae	*Capsicum* sp.

Table 2.4 contd. ...

...*Table 2.4 contd.*

Stink bug species	Plant family	Plant species
		Datura sp.
		Nicotiana tabacum
	Vitaceae	*Vitis vinifera*
Dichelops furcatus	Asteraceae	*Bidens pilosa*
		Conyza bonariensis
		Gochnatia polymorpha
		Helianthus annuus
	Cucurbitaceae	*Citrullus lanatus*
	Fabaceae	*Lotus corniculatus*
		Lupinus albus
		Macroptilium atropurpureum
		Phaseolus vulgaris
		Pisum sativum
		Rhynchosia corylifolia
		Vicia spp.
	Malvaceae	*Gossypium hirsutum*
	Linaceae	*Linum usitatissimum*
	Melastomataceae	*Miconea cinerascens*
	Myrtaceae	*Eugenia uniflora*
		Myrciaria tenella
	Oleaceae	*Olea europaea*
	Poaceae	*Lolium multiflorum*
		Avena sativa
	Rosaceae	*Fragaria x ananassa*
		Prunus myrtifolia
	Scrophulariaceae	*Buddleja thyrsoides*
	Solanaceae	*Nicotiana tabacum*
		Solanum tuberosum
Dichelops melacanthus	Amaranthaceae	*Amaranthus viridis*
		Gomphrena globosa
	Asteraceae	*Bidens pilosa*
		Emilia sonchifolia

Table 2.4 contd. ...

...*Table 2.4 contd.*

Stink bug species	Plant family	Plant species
		Tridax procumbens
	Convolvulaceae	*Ipomoea indica*
	Fabaceae	*Crotalaria pallida*
	Lamiaceae	*Leonotis nepetifolia*
		Leonurus sibiricus
		Stachys arvensis
	Malvaceae	*Malvastrum coromandelianum*
		Sida rhombifolia
	Poaceae	*Avena strigosa*
		Brachiaria decumbens
		Brachiaria plantaginea
		Cenchrus echinatus
		Chloris gayana
		Eleusine indica
		Panicum maximum
		Triticum secale
		Zea mays
	Rubiaceae	*Spermacoce alata*
		Richardia brasiliensis
	Solanaceae	*Solanum americanum*
Thyanta perditor	Asteraceae	*Baccharis trimera*
		Helianthus annuus
	Brassicaceae	*Brassica napus*
		Nasturtium officinale
	Fabaceae	*Crotalaria juncea*
	Linaceae	*Linum usitatissimum*
	Oleaceae	*Ligustrum lucidum*
	Pedaliaceae	*Sesamum indicum*
	Poaceae	*Hordeum vulgare*
		Oryza sativa
		Sorghum vulgare
	Solanaceae	*Solanum paniculatum*

Usually these less-preferred plants are available and located near cultivated fields, not far from where the preferred hosts are grown. In tropical and subtropical areas, the bugs are continuously active; when crop plants mature they move to the less-preferred plants to feed a little and even lay eggs. Eventually, the bugs can complete another generation before the environmental conditions become unfavorable (Panizzi 2007).

True bugs, in general, prefer to feed on fruits and immature seeds (Schuh and Slater 1995), because they are rich in essential nutrients and easy to reach. However, under low availability of these suitable and preferred structures, they feed on other plant organs usually not used as food sources (such as stems and leaves of less-preferred plants), even though their nutritional quality can be low (Panizzi 2000, 2007, Panizzi and Silva 2012). In this case, they change their feeding habits, passing from piercing-and-sucking on typical reproductive structures to feeding on vegetative tissues. Some species, however, prefer to feed on vegetative parts, as has been observed for *Tibraca limbativentris* Stål that feeds on the stems of rice plants (Rizzo 1976), and *E. meditabunda* that feeds on the stems of soybean (Galileo and Heinrichs 1979, Silva et al. 2012).

Changes in the feeding habits of stink bugs may be determined by some other factors such as: less-preferred plants are in vegetative stages of development or they carry seeds and/or fruits not normally used as food by the bugs, or they may produce seeds that might provide some nutrients but may be inaccessible for some reason, such as legumes that have thick pod-walls or pods with air space between the pod-wall and the seeds inside (Panizzi 2007).

Concluding Remarks

This brief discussion, somewhat restricted to the species of bugs from the Neotropics, illustrates how important it is to consider the many implications of the relationships of true bugs, in this case the relationship of stink bugs (pentatomids) with their plants—host plants, as presented here allowing continuous reproduction, or associated plants, those that do not allow reproduction and are used only momentarily. This analysis, we hope, will serve to highlight the many implications of this intricate relationship between stink bugs and plants.

Although the relationships between plants and stink bugs occur all over the different zoogeographical zones, perhaps in no other area of the world is this relationship as intense as it is in the Neotropics. The relatively recent introduction of multiple cropping systems in the Neotropics in the last 10–15 years has caused a boom in the production of diverse commodities, such as those related to grain production (soybean, corn, rice, wheat, and others) and fiber production (cotton). In addition, the widespread adoption by growers of no-tillage cultivation increases the amount of crop residues on the ground, which serve both as food source (fallen seeds) and shelter (debris). Moreover, this system increases the population of weed plants that may serve as "hosts" for stink bugs. In this scenario, there is intense stink bug activity year round, since in the so-called "winter" (dry season near the Equator with relatively high temperature) crops are grown under irrigation and stink bugs are able to colonize them. Such a system creates favorable conditions of food and shelter, not only for the known pest species, but also for the non-pest species of stink bugs that are being intercepted more often now. Species previously considered to be rare or difficult to collect are now being recorded on cultivated plants as well as on associated weeds, probably moving on from their native non-cultivated plants that have been replaced by the plants of the cropping systems. These associations may give rise to new pest species in the coming years.

In conclusion, stink bugs (pentatomids) and their relationships with plants form an interesting and challenging biological system. As we analyze and increase our understanding of this system, we certainly will be able to improve the integrated pest management (IPM) programs for these pest species, in the hope that we can more readily predict outbreaks of potential pest species. For more basic science, further studies will explain what makes a particular species more likely to become associated with a particular plant.

Acknowledgments

We thank Elaine A. Backus (USDA Agricultural Research Service, Parlier, California, USA) for critically reading the manuscript and for improving its readability. We also thank the editors for the invitation to contribute this chapter. Our published work referred to here has been supported by several grants over the years to ARP and by scholarships to several of his students and postdocs by the Brazilian Council of Scientific and Technology Development (CNPq) and Ministry of Education (CAPES). Our published data has been generated with the support of the Embrapa Soybean and Embrapa Wheat research units, Ministry of Agriculture, Livestock and Food Supply (MAPA).

References

Backus, E.A. 1988. Sensory systems and behaviours which mediate hemipteran plant-feeding: a taxonomic overview. J. Insect Physiol. 34: 151–165.

Backus, E.A., M.S. Serrano and C.M. Ranger. 2005. Mechanisms of hopperburn: An overview of insect taxonomy, behavior and physiology. Annu. Rev. Entomol. 50: 125–151.

Backus, E.A. and W.H. Bennett. 2009. The AC-DC correlation monitor: new EPG design with flexible input resistors to detect both R and emf components for any piercing-sucking hemipteran. J. Insect Physiol. 55: 869–884.

Bernays, E.A. and R.F. Chapman. 1994. Host-plant Selection by Phytophagous Insects. Chapman & Hall, New York.

Calizotti, G.S. and A.R. Panizzi. 2014. Behavior of first instar nymphs of *Edessa meditabunda* (F.) (Hemiptera: Pentatomidae) on the egg mass. Fla. Entomol. 97: 277–280.

Chew, F.S. and J.A.A. Renwick. 1995. Host plant choice in *Pieris* butterflies. pp. 214–238. *In*: Carde, R.T. and W.J. Bell (eds.). Chemical Ecology of Insects. Chapman & Hall, New York.

Chocorosqui, V.R. and A.R. Panizzi. 2004. Impact of cultivation systems on *Dichelops melacanthus* (Dallas) (Heteroptera: Pentatomidae) population and damage and its chemical control on wheat. Neotrop. Entomol. 33: 487–492.

Coombs, M. 2004. Broadleaf, *Ligustrum lucidum* Aiton (Oleaceae), a late-season host for *Nezara viridula* (L.), *Plautia affinis* Dallas and *Glaucias amyoti* (Dallas) (Hemiptera: Pentatomidae) in northern New South Wales, Australia. Austr. J. Entomol. 43: 335–339.

Després, L., J.P. David and C. Gallet. 2007. The evolutionary ecology of insect resistance to plant chemicals. Trends Ecol. Evol. 22: 298–307.

Esquivel, J.F. and E.G. Medrano. 2014. Ingestion of a marked bacterial pathogen of cotton conclusively demonstrates feeding by first instar southern green stink bug (Hemiptera: Pentatomidae). Environ. Entomol. 43: 110–115.

Ferreira, B.S.C. and A.R. Panizzi. 1982. Percevejos pragas da soja no Norte do Paraná: abundância em relação à fenologia da planta e hospedeiros intermediários. An. Semin. Nac. Pesq. Soja, Londrina 1: 140–151.

Fox, L.R. and P.A. Morrow. 1981. Specialization: species property of local phenomenon? Science 211: 887–893.

Galileo, M.H.M. and E.A. Heinrichs. 1979. Danos causados à soja em diferentes níveis e épocas de infestação durante o crescimento. Pesq. Agropec. Brasil 14: 279–272.

Grazia, J., A.R. Panizzi, C. Greve, F. Schwertner, L.A. Campos, T.A. Garbelotto et al. 2015. Stink bugs (Pentatomidae). pp. 681–756. *In*: Panizzi, A.R. and J. Grazia (eds.). True Bugs (Heteroptera) of the Neotropics. Springer, Dordrecht.

Hori, K. 2000. Possible causes of disease symptoms resulting from the feeding of phytophagous Heteroptera. pp. 11–35. *In*: Schaefer, C.W. and A.R. Panizzi (eds.). Heteroptera of Economic Importance. CRC Press, Boca Raton.

Huang, T., D.A. Reed, T.M. Perring and J.C. Palumbo. 2014. Host selection behavior of *Bagrada hilaris* (Hemiptera: Pentatomidae) on commercial cruciferous host plants. Crop Protec. 59: 7–13.

Jackai, L.E.M., A.R. Panizzi, G.G. Kundu and K.P. Srivastava. 1990. Insect pests of soybean in the tropics. pp. 91–156. *In*: Singh, S.R. (ed.). Insect Pests of Tropical Food Legumes. Whiley, Chichester.

Karban, R. and A.A. Agrawal. 2002. Herbivore offense. Annu. Rev. Ecol. Evol. Syst. 33: 641–664.

Kester, K.M. and C.M. Smith. 1984. Effects of diet on growth, fecundity and duration of tethered flight of *Nezara viridula*. Entomol. Exp. Appl. 35: 75–81.

Lopes, O.J., D. Link and I.V. Basso. 1974. Pentatomídeos de Santa Maria–Lista preliminar de plantas hospedeiras. Rev. Centr. Ciênc. Rur. 4: 317–322.

Lucini, T. and A.R. Panizzi. 2016. Waveform characterization of the soybean stem feeder *Edessa meditabunda* (F.) (Hemiptera: Heteroptera: Pentatomidae): overcoming the challenge of wiring pentatomids for EPG. Entomol. Exp. Appl. 158: 118–132.

Lucini, T., A.R. Panizzi and E.A. Backus. 2016. Characterization of an EPG waveform library for red banded stink bug, *Piezodorus guildinii* (Hemiptera: Pentatomidae), on soybean plants. Ann. Entomol. Soc. Am. 109: 198–210.

McLean, D.L. and M.G. Kinsey. 1964. A technique for electronically recording aphid feeding and salivation. Nature 202: 1358–1359.

McPherson, J.E. and R.M. McPherson. 2000. Stink Bugs of Economic Importance in America North of Mexico. CRC Press, Boca Raton.

Manfredi-Coimbra, S., J.J. Silva, V.R. Chocorosqui and A.R. Panizzi. 2005. Damage of the green belly stink bug *Dichelops melacanthus* (Dallas) (Heteroptera: Pentatomidae) to wheat. Cienc. Rural 35: 1243–1247.

Meneguim, A.M., M.C. Rossini and A.R. Panizzi. 1989. Desempenho de ninfas e adultos de *Euschistus heros* (F.) (Hemiptera: Pentatomidae) em frutos verdes de amendoim-bravo *Euphorbia heterophylla* (Euphorbiaceae) e em sementes e vagens de soja. Res, Congr, Bras, Entomol, 12, Vol. 1, pp. 43.

Miles, P.W. 1969. Interaction of plant phenols and salivar phenolases in the relationship between plants and Hemiptera. Entomol. Exp. Appl. 12: 736–744.

Miles, P.W. 1972. The saliva of Hemiptera. Adv. Insect Physiol. 9: 183–255.

Miles, P.W. and G.S. Taylor. 1994. "Osmotic pump" feeding by coreids. Entomol. Exp. Appl. 73: 166–173.

Olson, D.M., J.R. Ruberson, A.R. Zeilinger and D.A. Andow. 2011. Colonization preference of *Euschistus servus* and *Nezara viridula* in transgenic cotton varieties, peanut, and soybean. Entomol. Exp. Appl. 139: 161–169.

Panizzi, A.R. 1997. Wild hosts of pentatomids: Ecological significance and role in their pest status on crops. Annu. Rev. Entomol. 42: 99–122.

Panizzi, A.R. 2000. Suboptimal nutrition and feeding behavior of hemipterans on less preferred plant food sources. An. Soc. Entomol. Brasil 29: 1–12.

Panizzi, A.R. 2004. Adaptive advantages for egg and nymph survivorship by egg deposition in masses or singly in seed-sucking Heteroptera. pp. 60–73. *In*: Gujar, G.T. (ed.). Contemporary Trends in Insect Science. Campus Book International, New Delhi.

Panizzi, A.R. 2007. Nutritional ecology of plant feeding arthropods and IPM. pp. 170–222. *In*: Kogan, M. and P. Jepson (eds.). Perspectives in Ecological Theory and Integrated Pest Management. Cambridge University Press.

Panizzi, A.R. and F.Jr. Slansky. 1985. Review of phytophagous pentatomids (Hemiptera: Pentatomidae) associated with soybean in the Americas. Fla. Entomol. 68: 184–215.

Panizzi, A.R., A.M. Meneguim and M.C. Rossini. 1989. Impacto da troca de alimento da fase ninfal para a fase adulta e do estresse nutricional na fase adulta na biologia de *Nezara viridula* (Hemiptera: Pentatomidae). Pesq. Agropec. Bras. 24: 945–954.

Panizzi, A.R. and C.E. Rossi. 1991. The role of *Acanthospermum hispidum* in the phenology of *Euschistus heros* and of *Nezara viridula*. Entomol. Exp. Appl. 59: 67–74.

Panizzi, A.R. and F.Jr. Slansky. 1991. Suitability of selected legumes and the effect of nymphal and adult nutrition in the southern green stink bug (Hemiptera: Heteroptera: Pentatomidae). J. Econ. Entomol. 84: 103–113.

Panizzi, A.R. and S.I. Saraiva. 1993. Performance of nymphal and adult southern green stink bug on an overwintering host and impact of nymph to adult food-switch. Entomol. Exp. Appl. 68: 109–115.

Panizzi, A.R. and C.C. Niva. 1994. Overwintering strategy of the brown stink bug in northern Paraná. Pesq. Agropec. Bras. 29: 509–511.

Panizzi, A.R. and E. Hirose. 1995. Seasonal body weight, lipid content, and impact of starvation and water stress on adult survivorship and longevity of *Nezara viridula* and *Euschistus heros*. Entomol. Exp. Appl. 76: 247–253.

Panizzi, A.R., L.M. Vivan, B.S. Corrêa-Ferreira and L.A. Foerster. 1996. Performance of southern green stink bug (Heteroptera: Pentatomidae) nymphs and adults on a novel food plant (Japanese privet) and other hosts. Ann. Entomol. Soc. Am. 89: 822–827.

Panizzi, A.R. and L.M. Vivan. 1997. Seasonal abundance of the neotropical brown stink bug, *Euschistus heros* in overwintering sites and the breaking of dormancy. Entomol. Exp. Appl. 82: 213–217.

Panizzi, A.R. and A.P. Mourão. 1999. Mating, ovipositional rhythm and fecundity of *Nezara viridula* (L.) (Heteroptera: Pentatomidae) fed on privet, *Ligustrum lucidum* Thunb., and on soybean, *Glycine max* (L.) Merrill fruits. An. Soc. Entomol. Brasil 28: 35–40.

Panizzi, A.R., J.E. McPherson, D.G. James, M. Javahery and R.M. McPherson. 2000. Stink bugs (Pentatomidae). pp. 421–474. *In*: Schaefer, C.W. and A.R. Panizzi (eds.). Heteroptera of Economic Importance. CRC Press, Boca Raton.

Panizzi, A.R. and F.A.C. Silva. 2012. Insect bioecology and nutrition for integrated pest management (IPM). pp. 687–704. *In*: Panizzi, A.R. and J.R.P. Parra (eds.). Insect Bioecology and Nutrition for Integrated Pest Management. CRC Press, New York.

Pinto, S.B. and A.R. Panizzi. 1994. Performance of nymphal and adult *Euschistus heros* (F.) on milkweed and on soybean and effect of food switch on adult survivorship, reproduction and weight gain. An. Soc. Entomol. Brasil 23: 549–555.

Reed, D.A., J.C. Palumbo, T.M. Perring and C. May. 2013. *Bagrada hilaris* (Hemiptera: Pentatomidae), an invasive stink bug attacking cole crops in the southwestern United States. J. Int. Pest Manage. 4: C1–C7.

Rizzo, H.F.E. 1971. Aspectos morfológicos y biológicos de *Edessa meditabunda* (F.) (Hemiptera, Pentatomidae). Rev. Per. Entomol. 14: 272–281.

Rizzo, H.F.E. 1976. Hemípteros de Interés Agrícola. Hemisferio Sur, Buenos Aires.

Schoonhoven, L.M., J.J.A. Van Loon and M. Dicke. 2005. Insect-plant Biology. Oxford University Press, Oxford.

Schuh, R.T. and J.A. Slater. 1995. True Bugs of the World (Hemiptera: Heteroptera). Classification and Natural History. Cornell University Press, Ithaca.

Silva, A.G.D.A., C.R. Gonçalves, D.M. Galvão, A.J.L. Gonçalves, J. Gomes, M.N. Silva et al. 1968. Quarto catálogo dos insetos que vivem nas plantas do Brasil–seus parasitas e predadores. Min Agric, Rio de Janeiro, Parte II, Vol. 1, 622 p.

Silva, F.A.C., J.J. Silva, R.A. Depieri and A.R. Panizzi. 2012. Feeding activity, salivary amylase activity and superficial damage to soybean seed by adult *Edessa meditabunda* (F.) and *Euschistus heros* (F.) (Hemiptera: Pentatomidae). Neotrop. Entomol. 41: 386–390.

Silva, J.J., M.U. Ventura, F.A.C. Silva and A.R. Panizzi. 2013. Population dynamics of *Dichelops melacanthus* (Dallas) (Heteroptera: Pentatomidae) on host plants. Neotrop. Entomol. 42: 141–145.

Smaniotto, L.F. and A.R. Panizzi. 2015. Interactions of selected species of stink bugs (Hemiptera: Heteroptera: Pentatomidae) from leguminous crops with plants in the neotropics. Fla. Entomol. 98: 7–17.

Strong, D.R., J.H. Lawton and S.R. Southwood. 1984. Insects on Plants. Harvard University Press, Cambridge.

Tjallingii, W.F. 1978. Electronic recording of penetration behaviour by aphids. Entomol. Exp. Appl. 24: 721–730.

Todd, J.W. and D.C. Herzog. 1980. Sampling phytophagous Pentatomidae on soybean. pp. 438–478. *In*: Kogan, M. and D.C. Herzog (eds.). Sampling Methods in Soybean Entomology. Springer-Verlag, New York.

Velasco, L.R.I. and G.H. Walter. 1992. Availability of different host plant species and changing abundance of the polyphagous bug *Nezara viridula* (Hemiptera: Pentatomidae). Environ. Entomol. 21: 751–759.

Velasco, L.R.I. and G.H. Walter. 1993. Potential of host-switching in *Nezara viridula* (Hemiptera: Pentatomidae) to enhance survival and reproduction. Environ. Entomol. 22: 326–333.

Zerbino, M.S., N.A. Altier and A.R. Panizzi. 2016. Performance of nymph and adult of *Piezodorus guildinii* (Westwood) (Hemiptera: Pentatomidae) feeding on cultivated legumes. Neotrop. Entomol. 45: 114–122.

CHAPTER 3

Predatory Stink Bugs (Asopinae) and the Role of Substrate-borne Vibrational Signals in Intra- and Interspecific Interactions

*Alenka Žunič Kosi** and *Andrej Čokl*

Introduction

The use of biological control as a valuable tactic in pest management programs requires conservation and utilization of natural enemies to prevent pest outbreaks and provide control over economically damaging populations of pest species (Todd et al. 1994).

Predatory stink bugs of the subfamily Asopinae (Pentatomidae) are considered to be beneficial, as they prey on a variety of insect pest species (De Clercq 2000), and many of them have the potential to be used in biological control and integrated pest management programmes. Both nymphs and adults attack different pest species, and suck the body fluids from their prey with a needle-like beak (De Clercq et al. 2014) (Fig. 3.1).

Of the approximately 300 known asopine species, only a few have been extensively studied. These studies mainly focusing on the different aspects of life-history and geographical distribution of Asopinae, their food preferences, rearing conditions and mass production, colonization potential, fecundity and fertility, and their potential to control economically important insect pests (De Clercq 2000). The most studied species are the predators of the New World, such as *Podisus* spp., *Perillus*

Department of Organisms and Ecosystems Research, National Institute of Biology, Večna pot 111, SI-1000 Ljubljana, Slovenia.
Email: Andrej.cokl@nib.si
* Corresponding author: Alenka.Žunič-Kosi@nib.si

Figure 3.1. Males of *Podisus maculiventris* feeding on a moth caterpillar.

bioculatus (Fabricius), *Brontocoris tabidus* (Signoret), *Stiretrus anchorago* (Fabricius), *Alcaeorrynchus grandis* (Dallas), and some representatives of the Old Word species *Picromerus bidens* (L.), *Eocanthecona furcellata* (Wolff), and *Arma* sp. (De Clercq 2000, De Clercq et al. 2014).

A successful biological control agent should exhibit positive rapid density response and a high reproductive rate. Studies of reproductive behaviour are therefore an important aspect of evaluating the success of asopines as biocontrol agents. In phytophagous stink bugs, chemical signals, e.g., attractant pheromone (Aldrich 1995, 1998), and substrate-borne (vibrational) signals play an important role in reproductive behaviours; calling, courtship, and rival signals are involved in gender and species recognition, in the finding, selection and stimulation of a potential mate (Čokl and Virant Doberlet 2003, Gogala 2006). Recently, a few studies have also shown the importance of vibrational signalling in the reproductive behaviour of asopines (Gogala 2006, Žunič et al. 2008, Čokl et al. 2009, 2011, Shestakov 2008, Laumann et al. 2013, Čokl et al. 2014). In addition, some of the species can use vibrational signals as information to locate prey (Coppel and Jones 1962, Mukerji and LeRoux 1965, Tostowaryk 1971, Pfannenstiel et al. 1995). However, the amount of knowledge regarding this thematic in the literature is still insufficient, so it is important to address the limitations in the understanding of this important research area in the asopines and their role in augmentative biological control of insect pests.

In view of this, we will first provide an overview of the current knowledge on intraspecific communication mediated by vibrational signals in asopines, as crucial component of mating behaviour that plays an important role in gender and species recognition, as well as in finding a potential mate. We will then discuss the use of vibrational signals in interspecific relationships, and finally, address the impacts of insecticides on insect behaviour, which may influence population dynamics of target and non-target insect species, and indirectly interfere with sustainable pest management programmes.

The Role of Vibrations in Mating Behaviour

There are several important issues that need to be addressed in the mate attraction process, which are essential for successful reproduction; species and gender identification, sexual receptivity, and localization of a potential mate (Bradbury and Vehrencamp 1998).

Although there are quite a few studies on asopines as biological control agents, their reproductive behaviour has rarely been studied. Unlike many other groups of economically important predator or parasitoid species, the reproductive behaviour of which has been exploited and used in integrated management tactics, the knowledge related to mating behaviour and to the role of vibrational mating signals is scarce in case of asopines. Males of many asopine species were found to produce attractant pheromones (Aldrich 1998); on the contrary, vibrational signals have been identified for only a few species so far. Vibrational signalling has been studied to some extent in *Podisus maculiventris* (Say) (Žunič et al. 2008, Čokl et al. 2009), *Podisus nigrispinus* (Dallas) (Laumann et al. 2013), *P. bidens* (Gogala 2006, Shestakov 2008, Čokl et al. 2011), *Troilus luridus* (Fabricius) (Shestakov 2008), *Zicrona caerulea* (L.) (Shestakov 2008) and *Arma custos* (Fabricius) (Shestakov 2008) (Table 3.1).

A review of the reproductive behaviour of the asopines investigated to date reveals several types of intraspecific communication strategies, a wide variety of vibrational repertoires and several different mechanisms of signal production (Fig. 3.2). In addition, the studies have highlighted several similarities as well as differences between the roles of vibrational signalling in predatory stink bugs and the well-known phytophagous pentatomids.

Many heteropteran species like those of the family Cydnidae, communicate with vibrational signals produced by stridulation and abdominal vibration (Gogala 2006). In comparison to them, the vibrational repertoire of predatory stink bug species is more extensive and the production mechanisms are more diverse: signals are produced by vibration of the abdomen, tremulation of the whole body, and percussion with the legs (Fig. 3.2) (Žunič et al. 2008, Shestakov 2008, Čokl et al. 2011, Laumann et al. 2013). Different types of signals are emitted in different behavioural contexts and are characterized by different spectral and temporal parameters.

Table 3.1. Temporal and spectral characteristics of the vibrational signals produced by asopine males and females during mating behaviour.

Species	Signal	Production mechanism	Pulse duration (ms)/repetition time (ms)/frequency (Hz)	Behavioural context (function)
Podisus maculiventris (Žunič et al. 2008)	MS-V	Abdomen vibration	190 ± 30/449 ± 78/90–140	First song, calling; triggering female searching behaviour
	MS-T	Tremulation of a body	96–138/286–347/8–22	Close-range, courtship; possibly rivalry
	MS-comp	Tremulation of a body + abdomen vibration	260 ± 32/437 ± 57/107–152	Calling; close-range, courtship
	MS_{PERC}	Striking first leg on the substrate	100–134/464–510/97	Not related to a specific behavioural context
Podisus nigrispinus (Laumann et al. 2013)	FS1	Abdomen vibration	191–291/471 ± 213/69–87	First song, calling; triggering male vibratory response; copulatory; post-copulatory
	MS1	Abdomen vibration	64–100/508 ± 101/90–145	As a response to female song; calling
	MS2	Abdomen vibration	75–146/107–180/108–145	As a response to female song; copulatory; post-copulatory
	TREM	Tremulation of a body in male, female	84 ± 31/49–1129/137–232	Spontaneous, calling; rejection
Picromerus bidens (Čokl et al. 2011) (Shestakov 2008)	FS	Abdomen vibration	75–104/91–120/90–150	First song; triggering male vibratory response
	FS_{TREM}	Tremulation of a body	81–123/irregular/broad bended	During abdomen vibrations; close-range
	MS1	Abdomen vibration	63–209 ms/93–549 ms/75–250	As a response to female song

	MS2	Abdomen vibration	180–780/235 ± 30/160–200	As a response to female song
	MS3	Abdomen vibration	142–234/175–283/107–207	As a response to female song
	MS4	Abdomen vibration	90–258/206–511/104–211	Rivalry, close-range male-male interaction
	MS_{TREM}	Tremulation of a body	81–123/irregular/broad bended	During vibrational signals; close-range
Troilus luridus (Shestakov 2008)	MS	Abdomen vibration	221/118–256/85–140$^{1st\ harm}$; 185–240$^{2nd\ harm}$	Alone; in the presence of a female; triggering female directional movement
Zicrona caerulea (Shestakov 2008)	MS	Abdomen vibration	1980/irregular/90–130$^{1st\ harm}$; 200–280$^{2nd\ harm}$; 320–360$^{3rd\ harm}$	Alone; in the presence of a female; calling
Arma custos (Shestakov 2008)	MS	Abdomen vibration	223–142/irregular/50–200	Calling

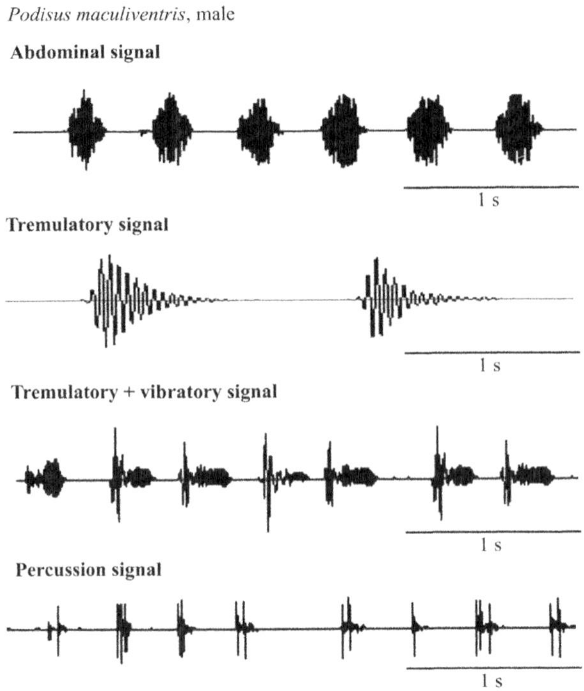

Figure 3.2. Oscillograms of the signals produced by different mechanisms (abdomen vibration, tremulation of the whole body, composite vibrations (tremulation of the body + vibration of the abdomen), and percussion with front legs) in *Podisus maculiventris* males.

Signals produced by abdomen vibration

Communication signals produced by vibrating the abdomen are widely used in Pentatominae, and also in asopines; signals produced by dorso-ventral movements of the abdomen, without touching the substrate, have been recorded in all the asopine species analyzed so far (Žunič et al. 2008, Shestakov 2008, Čokl et al. 2011, Laumann et al. 2013). An exception, with respect to the mode of production, was observed in *T. luridus*, where males produced abdominal vibrations with their bodies pressed to the substrate rather than raised on the legs (Shestakov 2008). In each of the investigated species, abdominal vibrations were emitted as the initial signal in the vibrational repertoire. Signals were produced either spontaneously, with no prior contact or presence of the conspecifics, or/and as a response to the calling mates or rivals. Thus, abdominal vibrations seem to have a calling function in the asopines (Table 3.1), which is also common in phytophagous pentatomids. In *P. bidens* and *P. nigrispinus*, the females emitted calling songs (Fig. 3.3) spontaneously. This triggered emission of the males' vibrational response (Čokl et al. 2009, Laumann et al. 2013) and duet formation, which

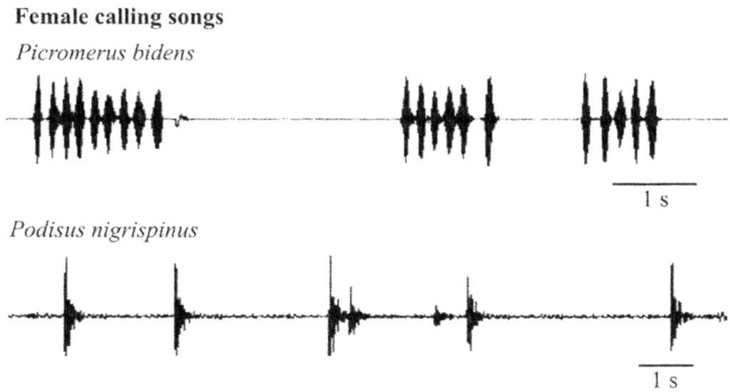

Figure 3.3. Example of female signals produced by abdomen vibrations of *Picromerus bidens* (above) and *Podisus nigrispinus* (below).

was, in turn, important for inducing the searching behaviour of one or both sexes. *P. bidens* females produced songs with characteristics similar to those described for other pentatomids like for example *Nezara viridula* (L.) (Čokl et al. 2000), with species-specific duration and regular repetition rate of pulse trains emitted from the same location (Čokl et al. 2011). In contrast, *P. nigrispinus* females produced more irregularly structured signals, with their temporal and spectral characteristics largely varying between individuals (Laumann et al. 2013). In *P. maculiventris* (Žunič et al. 2008), *T. luridus*, *Z. caeruela*, and *A. custos* (Shestakov 2008) only males were observed to emit calling songs, while females did not show any vibrational response at all. So far the lack of a female signal has been considered as an exception in the vibrational repertoire in the entire family Pentatomidae. Males of many asopine species produce species-specific attractant pheromones, which are involved in aggregating males and females, as well as nymphs (Aldrich 1998). Thus, the irregular and less stereotypical female calling song in *P. nigrispinus*, and the absence of female vibrational activity in the other four investigated species might indicate that in asopines, vibrations as a part of the mate finding process, are less important than in case of other pentatomids, and that pheromones are essential for aggregation of conspecifics, as discussed by Laumann and co-workers (2013). On the other hand, the silent female phenomenon may also be related to the female reproduction strategy. Production of vibration signals is calorically expensive (e.g., Kuhelj et al. 2015) and may influence female fitness. In addition, vibrationally orienting predation and parasitism may substantially reduce females' life span and thus their reproductive opportunities.

The males vibrate their abdomens and produce one or more species-specific type of songs (Table 3.1). *P. bidens* and *P. nigrispinus* males produced abdominal vibrations (Fig. 3.4) as a response to short female calling signals. Their initial call was short and simple, which was readily transformed

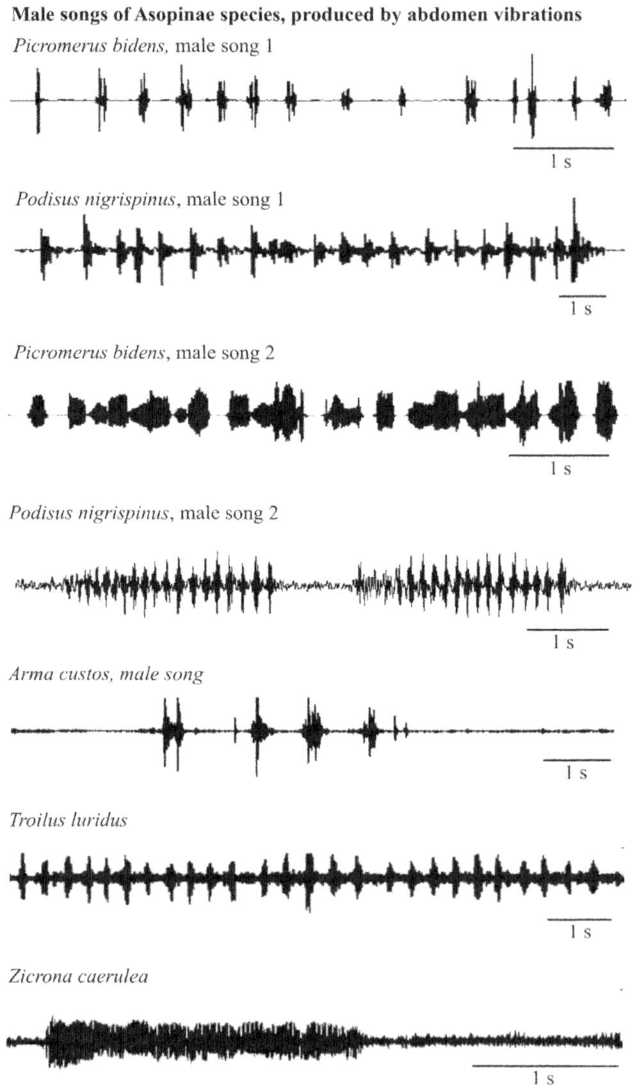

Figure 3.4. Oscillograms of male songs produced by abdominal vibrations in 5 asopine species (*Arma custos, Troilus luridus* and *Zicrona caerulea* from Shestakov 2008, with permission).

into a longer or more complex song type (Čokl et al. 2011, Laumann et al. 2013). These consecutive songs (emitted also during courtship) were the prerequisites for copulation and thus seem to be important in mate recognition, female choice, and sexual stimulation. Instead, *A. custos, T. luridus, Z. caerulea* and *P. maculiventris* males use a less complex repertoire of vibrational signals with a single species-specific type of calling song

(Shestakov 2008, Žunič et al. 2008). Interestingly, asopine males also emitted signals during copulation, indicating that vibrations may also be important in copulatory behaviour, perhaps inducing prolonged copulation and thus more successful sperm transfer or may be connected to mate-guarding, as seen in water striders (Wilcox and Stefano 1991). Furthermore, males of *P. bidens* also produced signals in male-male interactions, during rivalry (Čokl et al. 2011).

Abdominal vibrations of Asopines are spectrally highly variable between individuals and have complex, broadband spectra, with different dominant frequency values even within the same song type (Table 3.1). On the other hand, in phytophagous stink bugs, signals are characterized by low interindividual variability and have uniform frequency properties: narrow band spectra, with a stable dominant peak (not exceeding 200 Hz), different number of higher harmonics (not exceeding 1000 Hz), and with or without the frequency modulated units. As such, these signals are well-tuned to the resonant properties of herbaceous host plants, which act as low-pass filters and due to low attenuation efficiently transmit signals of frequencies below 500 Hz (Čokl 2008). The broadband characteristics of the Asopinae signals might indicate that these signals are used in short-distance rather than in long-distance communication, since broadband signals were shown to be transmitted through plants with higher damping (Čokl et al. 2006). In contrast, the proportion between different frequency components in broadband signals might subserve the function of distance estimation of the vibrations' source for the receiver (Virant-Doberlet et al. 2006).

Tremulation signals

Within the family Pentatomidae, tremulation (vibration of the whole body) has been recorded for the first time in three asopine species: *P. maculiventris* (Fig. 3.2), *P. bidens* and *P. nigrispinus* (Žunič et al. 2008, Čokl et al. 2011, Laumann et al. 2013). Adult bugs produced tremulatory signals by oscillating their bodies parallel to the ground. In *P. maculiventris* only males tremulated their bodies, while in the other two species tremulatory signals were produced by both males and females. In *P. maculiventris*, tremulations alone or in combination with abdominal vibrations were associated with the later stage of courtship and were the prerequisites for successful copulation. On the other hand, in *P. nigrispinus*, females emitted tremulatory signals when rejecting the male trying to copulate. In all three species, tremulatory signals were produced irregularly with variable frequency characteristics within and between species. For example, low frequency (dominant peaks between 8 to 22 Hz) signals have been recorded in *P. maculiventris* (Žunič et al. 2008), while tremulatory signals of broadband spectra, with dominant peaks (below 100 and up to 5 kHz) and high variation between individuals, were recorded in *P. bidens* and *P. nigrispinus* males (Čokl et al. 2011, Laumann et al. 2013) (Table 3.1). Spectral characteristics of the tremulatory signals also varied according to the different substrates (i.e., loudspeaker and host plant) (Čokl et al. 2008, Laumann et al. 2013). It has also been shown that tremulatory signals are more attenuated than the narrowband abdominal vibrations during transmission through plants (Čokl et al. 2009). It seems that the asopine tremulatory signals are primarily involved in the close-range communication between conspecifics (potential mates and/or rivals). Tremulations possibly convey information on fitness of the potential mate, such as its physical condition, age, status and sexual motivation, which in turn reveals their function as courtship signals. For example, previous analyses of vibrational signals engaged in mate advertisement, have confirmed the relationship between body size and spectral characteristics of signals in more than 50 insect species (Cocroft and DeLuca 2006). High interindividual variability was also found in olfactory signals in some heteropteran species. For example in *N. viridula, Piezodorus hybneri* (Gmelin), *Euschistus heros* (Fabricius) and *Riptortus pedestris* (Fabricius), pheromone composition and/or pheromone quantity greatly varied among individuals according to their age and physical condition (Ryan et al. 1995, Endo et al. 2007, Zhang et al. 2003, Mitzuani et al. 2007). In *P. maculiventris* tremulations have also been described during male-male agonistic interactions, indicating that these signals may convey information on the physical condition and vitality of the rivals. Further investigations are needed to confirm the biological function of tremulations in the asopines and their efficiency in conveying information through different substrates.

Drumming signals

Drumming or percussion signals produced by striking the substrate with some part of the body have been described in many arthropods (Hill 2008), but rarely in Pentatomidae. Specific percussion movements have been observed in predatory *P. bidens* males that lift one of their front legs, drop it to the substrate and pull it back again (Gogala 2006). The biological function of percussion signalling is not known. For instance, drumming signals or cues produced by *P. maculiventris* males (Fig. 3.2) have not been connected to any particular behavioural context and have not elicited any specific response in conspecifics (Žunič et al. 2008). On the other hand, these signals may be used in vibrational sounding, possibly for testing transmission properties of the substrate. For example, parasitoid female wasps were shown to use this form of echolocation as a means of host detection (Wäckers et al. 1998).

Courtship behaviour

When asopine males and females are in close proximity, females stop signalling and the pentatomide stereotyped sequence of courtship behaviour follows (i.e., approaching, clasping, courting, copulating, and resting). The courtship behaviour is initiated by the males, who approach and anntenate the females, mount, and rotate their body to achieve copulatory position and copulate. When in copula, the pair is in a line, tail to tail and the male crawls backwards to follow the female. For example, *P. maculiventris* and *P. nigrispinus* males, prior to copulation, always vigorously tremulate their bodies, or produce abdomen vibration signals combined with tremulations (Žunič et al. 2008, Laumann et al. 2013). Females do not show any specific response to the courting males, except in the case of male rejection prior to copulation when they may emit tremulatory signals and move away from the courting male.

The studies on mating behaviour in predatory bugs conducted so far have revealed great opportunities and avenues for future research in this fascinating group of stink bugs. Vibrational signalling as a component of the multimodal and complex mating behaviour of asopines appears to be essential in species and mate recognition, important as a sexual motivation factor, and is a prerequisite for pair formation and successful copulation. Asopinae species are general predators, and as such they are associated with woody and herbaceous plants across a high diversity of habitats such as grasslands, forests, agricultural areas and others (De Clercq 2000, Mahdian et al. 2008). The asopines appear to insure efficient transmission of signals through various substrates and thereby efficient intraspecific vibrational communication by using several vibration producing mechanisms that largely extend their frequency communication window.

Current knowledge on the sexual communication of predatory bugs raises several questions that need to be investigated further, in order to fully understand the production mechanisms of different signal types, in order to elucidate the vibrational repertoire, the behavioural context in concert with the temporal window of vibrational emissions, and the biological function of the specific song. Further research on the mating behaviour of more species together with comparative studies would clarify the specific roles and the functional interplay between visual, chemical and vibrational signalling, all of which determine mating strategies and have a significant impact on reproductive success. We also know very little about the possible role of vibrations in the aggregative behaviour of not only adults but nymphs as well, specifically in species that do not emit chemical attractants. This knowledge is important for the development of traps that incorporate chemical, visual, and vibrational signals. Such traps may lure a greater number of predatory bugs to the target area, manipulate their behaviour, enable monitoring of adult and immature populations, and provide information on spatial and temporal distribution of predator and prey populations. For instance, in laboratory conditions, *P. maculiventris* females stimulated by male vibrational signals moved and located the source of vibration with well-expressed vibrational directionality (Žunič et al. 2008). Combined traps (based on chemical and vibrational signals) are already part of the ongoing studies in the management of phytophagous stink bugs (Čokl and Millar 2009). In addition, a few studies are already showing that manipulating mating behaviour, based on vibrational signals represents an effective method to control insect pests with possibility of being used in the field (e.g., Eriksson et al. 2012, Polajnar et al. 2014). Understanding the mating system of asopine species is crucial for the understanding of their life-history, may enable the progress in rearing and mass-production of predatory stink bugs, and improve their use in controlling pest populations.

Vibrational Signalling in Interspecific Interactions

The reproductive success of predators depends greatly on their ability to find prey. Most asopine species are polyphagous and generalist predators. They prefer to prey on soft-bodied and slow-moving forms of Coleoptera, Lepidoptera (Fig. 3.1), Hymenoptera, Heteroptera and Diptera, but attack eggs as well (McPherson and McPherson 2000, Schaefer 1996, De Clercq 2000). Only a small number of predatory bugs, which have high potential as biological control agents, have received considerable attention from the point of view of the predator-prey relationship. Many of these species detect their prey either by chance or use visual, tactile and olfactory cues for prey detection, recognition and localization. Prey detection was demonstrated by different authors (e.g., Dickens 1999) and has been extensively reviewed by De Clercq (2000). On the other hand, prey detection is commonly mediated by vibrational signals in spiders (Barth 1998, Uetz and Roberts 2002, Virant-Doberlet et al. 2011). It was also demonstrated that the egg parasitoid *Telenomus podisi* (Ashmead) responds with orientated movements towards the female vibrational signal of the phytophagous stink bug *E. heros* (Laumann et al. 2007). Asopines belong to a group of numerous arthropod predators that also feed on phytophagous stink bugs (De Clercq 2000, De Clercq et al. 2002, Vandekerkhove and De Clercq 2004, Saini 1994). It is therefore not surprising that *P. maculiventris* females exhibit a searching response to the calling signals of *N. viridula* (Žunič et al. 2008) that is their potential prey (De Clercq et al. 2002). The female calling song of *N. viridula* is a typical Pentatominae calling signal of stable temporal and spectral characteristics (Čokl 2008). It seems likely therefore that gender- and species-specific vibrational signals as a part of the courtship display may be exploited by asopines in order to detect their prey. Furthermore, the predators also use less specific vibrational cues produced by the walking or feeding of the prey on the substrate (Barth 1998, Meyhöfer and Casas 1999, Brownell and van Hemmen 2001). For example asopine species *P. maculiventris* uses the vibrations produced by caterpillars as a cue to locate prey (Pfannenstiel et al. 1995). Apart from these examples of *P. maculiventris'* response to prey vibrations, very little is known about the use of vibrational signals and cues in the predator-prey relationships between asopines and their prey. To better understand these interactions for the specific predatory bug species, further experiments are needed. For instance, it would be important to identify those incidental vibrational cues or signals used during sexual communication that mediate short range prey detection and prey quality assessment in asopine predators. These results could be used when applying asopines as biological control agents in the managed field systems.

Predators influence the size of the prey populations either through direct consumption or by influencing prey behaviour, physiology, development or morphological traits (Werner and Peacor 2003, Preisser et al. 2005). Consequently it is very important to understand how the prey perceive predation risk, which cues they use to detect predators and how they deal with the conflict between foraging and avoiding predation. Potential prey can detect predators by visual, tactile, chemical, vibrational or other cues. Data on the impact of predator cues on prey behaviour in insects are mainly focused on tactile, chemical, airborne, and visual cues (e.g., Plummer and Camhi 1981, Gnatyz and Kämper 1990, Dicke and Grostl 2001, Williams and Wise 2003, Ninkovic et al. 2013). Recently Herman and Thaler (2014) have shown the importance of chemical cues in detecting predatory stink bugs. The larvae of Colorado potato beetle, *Leptinotarsa decemlineata* (Say), perceived predation risk mediated by sex-specific chemical cues and exhibited a clear response by reducing their feeding behaviour. However, only a few researchers have examined the substrate-borne, vibrational signals that can carry information useful in predation risk assessment. For example, Djemai and co-workers (2001) demonstrated a well-expressed sensory match between parasitoid wasp's vibrational cues and the behavioural response of its host. With regard to risk assessment it was also shown that caterpillars discriminate between abiotic vibrations and vibrations produced by conspecifics and predators, and respond differently to different vibrations (Castellanos and Barbosa 2006, Guedes et al. 2012). Still, very little is known about the risk assessment and cues that have non-consumptive effects in larvae and adults of the prey species. Consequently, further experiments are needed in order to better understand the mechanism by which prey use incidental vibrational cues produced by predators during walking or feeding on plants, or their intraspecific communication signals. In addition, more studies are needed to investigate the perception of predator-related cues (of different modalities) as well as anti-predator behavioural responses and other non-consumptive effects on prey. These investigations should be directed not only to adult prey, but also at the different larval stages, in order to evaluate the perceptive abilities and the effect of predator cues on different life stages (Herman and Thaler 2014). It is also very important to evaluate whether cues from shared predators induce general anti-predator responses or a species-specific response, and to investigate whether vibrations carry reliable information about the risk magnitude. Determination of the sensory basis of non-consumptive effects in prey is essential for understanding the ecological consequences of predator-prey interactions and the dynamics specifically between the economically important prey and predator species. This knowledge is important for practical applications in the managed systems of biological control for manipulating the behaviour, physiology and development of prey and for maximizing the predation efficiency of biological control agents, such as the asopines.

Since only a few asopine species have been investigated with regard to predator-prey interactions, further investigations would be of great importance in the development of pest management strategies and regulating the populations of economically important species and their biorational control.

In addition, predators' survival, abundance, biodiversity and their ability to provide essential ecosystem services depend greatly on their living conditions and ecosystem health. Insecticides that are used to control insect pests can also affect non-target predator species by inducing sublethal physiological and/or behavioural effects on them (e.g., Haynes 1988, Stark and Banks 2003, Desneux et al. 2007). Predators may be exposed to sublethal doses of insecticides through different exposure routes, like for example by intercepting spray droplets, contacting insecticide residues on the treated substrates, or indirectly by the food-chain uptake (Al-Deeb et al. 2001, Torres and Ruberson 2004, Walker et al. 2007, Broughton et al. 2013). A number of studies have investigated the sublethal effects of insecticides on various behaviours of insect predators over the past few years (e.g., Desneux et al. 2007, Thornham et al. 2007, De Castro et al. 2013, Martinou et al. 2014). Several studies have also documented the deleterious sublethal effects on some of the most common biological control agents of predatory stink bugs, specifically on their development, longevity, oviposition and fecundity (Mohaghegh et al. 2000, Torres et al. 2002, 2003, Mahdian et al. 2007). In addition, Malaquias and co-workers (2014) showed that sublethal doses of certain insecticides also affected the predatory behaviour of *P. nigrispinus*. These studies showed great species-specificity in the susceptibility of asopines to a specific group of insecticides (e.g., to pyrethroids, neonicotinoids). Their susceptibility to a toxic compound also depends on the age of the insect, on time and on the route of exposure. Nevertheless, the non-lethal effects of insecticides on the asopines´ behaviour have not been sufficiently studied, and many more aspects of the sublethal effects on the behaviour of predatory bugs are yet to be revealed. For instance, further studies would be needed in order to develop and implement sensitive behavioural bioassays that would help to understand and evaluate the potential sublethal impacts of neurotoxic compounds on specific behavioural responses of asopines, such as mobility, orientation, prey finding, feeding behaviour, and intra- and interspecific communication, all of which may lead to disorder in the population dynamics and influence their efficiency as biological control agents (Haynes 1988, Desneux et al. 2007). Identification of the sublethal effects of insecticides on the behaviour of asopine would be of great help in application studies and in implementation of their results in IPM tactics to ensure conservation of biological control agents.

Concluding Remarks

Biological control and manipulation of agroecosystems using natural enemies is among the earliest pest control measures; bamboo poles were used in orchards to facilitate predation of citrus pests by predatory ants many centuries ago in China (Coulson et al. 1982). Asopinae are a cosmopolitan subfamily of predatory bugs that have been widely recognized as important agents of biological control (Panizzi et al. 2000). This chapter lays out the progression of the investigation of this important group of natural enemies and summarizes the information on their sexual communication and interspecific interactions.

Vibrational cues and signals play a crucial role in the life of asopine stink bugs and other insects. Asopines use vibrational signals during intraspecific communication among others also to localize a potential mate or rival. In addition, vibrational cues and signals may mediate predator-prey interactions. As such, vibrational signalling is directly related to the reproductive success of predatory bugs and influences their effectiveness as biological control agents. Asopine stink bugs have a great potential for use in augmentation programmes. Knowledge gained from studies on basic behavioural processes within intra- and interspecific interactions, including vibrational commmunication, will improve and provide specialized expertise on predatory bugs as biological control agents. This would greatly contribute to the development of environmentally- and target-oriented insect pest management tactics to control many key agricultural pest species.

Minimizing the environmental impact of insecticides is one of the fundamental aims of the integrated pest management approaches. Along with measurements of the direct toxicity of insecticides, their sublethal effects on predator behaviour and physiology must be evaluated, in order to fully assess the negative impacts of insecticides, their selectivity, and compatibility with natural enemies. This would provide key information for the development and implementation of holistic biorational control and also aid in reducing the risks associated with insecticide use.

Acknowledgements

We thank the Slovenian Research Agency (Slovenia), the National Council for Scientific and Technological Development (CNPq), the Brazilian Corporation of Agricultural Research (EMBRAPA) and the Research Support Foundation of the Federal District (FAP-DF) (Brazil) for financial support of insect vibratory and chemical communication research.

References

Al-Deeb, M.A., G.E. Wilde and R.A. Higgins. 2001. No effect of *Bacillus thuringiensis* on the predator *Orius insidiosus* (Say) (Hemiptera: Anthocoridae). Environ. Entomol. 30: 625–629.

Aldrich, J.R. 1995. Chemical communication in the true bugs and parasitoid exploitation. pp. 318–363. *In*: Cardé, R. and W. Bell (eds.). Chemical Ecology of Insects. Chapman & Hall, NY, USA.

Aldrich, J.R. 1998. Status of semiochemical research on predatory Heteroptera. pp. 33–48. *In*: Coll, M. and J.R. Ruberson (eds.). Predatory Heteroptera: Their Ecology and Use in Biological Control. Proc. Thomas Say Publ. Entomological Society of America, Lanham, MD, USA.

Barth, F.G. 1998. The vibrational sense of spiders. pp. 228–278. *In*: Hoy, R.R., A.N. Popper and R.R. Fay (eds.). Comparative Hearing: Insects. Springer, New York, NY, USA.

Bradbury, J.W. and S.L. Vehrencamp. 1998. Principles of Animal Communication. Sinauer Associates, Inc., Sunderland, Mass, USA.

Brownell, P.H. and J.L. van Hemmen. 2001. Vibration sensitivity and a computational theory for prey-localizing behavior in sand scorpions. Amer. Zool. 41: 1229–1240.

Broughton, S., J. Harrison and T. Rahman. 2013. Effect of new and old pesticides on *Orius armatus* (Gross)—an Australian predator of western flower thrips, *Frankliniella occidentalis* (Pergande). Pest Manag. Sci. 70: 389–397.

Castellanos, I. and P. Barbosa. 2006. Evaluation of predation risk by a caterpillar using substrate-borne vibrations. Anim. Behav. 72: 461–469.

Cocroft, R.B. and P.A. De Luca. 2006. Size-frequency relationships in insect vibrational signals. pp. 99–110. *In*: Drosopoulos, S. and M.F. Claridge (eds.). Insects Sounds and Communication: Physiology, Behaviour, Ecology and Evolution. CRC Press, Boca Raton, FL, USA.

Coulson, J.R., W. Klaasen, R.J. Cook, E.G. King, H.C. Chiang, K.S. Hagen and W.G. Yendol. 1982. Notes on biological control of pests in China. pp. 1–192. *In*: Biological Control of Pests in China. US Department of Agriculture Washington, DC: USDA-OICD, USA.

Coppel, H.C. and P.A. Jones. 1962. Bionomics of *Podisus* spp. associated with the introduced pine sawfly, *Diprion similis* (Htg.), in Wisconsin. Wis. Acad. Sci. Arts Let. 51: 31–56.

Čokl, A., M. Virant-Doberlet and N. Stritih. 2000. The structure and function of songs emitted by southern green stink bugs from Brazil, Florida, Italy and Slovenia. Physiol. Entomol. 25: 196–205.

Čokl, A. and M. Virant-Doberlet. 2003. Communication with substrate-borne signals in small plant-dwelling insects. Annu. Rev. Entomol. 48: 29–50.

Čokl, A., C. Nardi, J.M.S. Bento, E. Hirose and A.R. Panizzi. 2006. Transmission of stridulatory signals of the burrower bugs, *Scaptocoris castanea* and *Scaptocoris carvalhoi* (Heteroptera: Cydnidae) through the soil and soybean. Physiol. Entomol. 31: 371–381.

Čokl, A. 2008. Stink bug interaction with host plants during communication. J. Insect Physiol. 54: 1113–1124.

Čokl, A., A. Žunič and J.G. Millar. 2009. Transmission of *Podisus maculiventris* tremulatory signals through plants. Cent. Eur. J. Biol. 4: 585–594.

Čokl, A. and J.G. Millar. 2009. Manipulation of insect signaling for monitoring and control of pest insects. pp. 279–316. *In*: Ishaaya, I. and A.R. Horowitz (eds.). Biorational Control of Arthropod Pests, Springer, Dordrecht The Netherlands.

Čokl, A., A. Žunič and M. Virant-Doberlet. 2011. Predatory bug *Picromerus bidens* communicates at different frequency levels. Cent. Eur. J. Biol. 6: 431–439.

Čokl, A., M. Zorović, A. Žunič Kosi, N. Stritih and M. Vrant-Doberlet. 2014. Communication through plants in a narrow frequency window pp. 171–195. *In*: Cocroft, R.B., M. Gogala, P.S.M. Hill and A. Wessel (eds.). Studying Vibrational Communication. Springer, New York, NY, USA.

De Clercq, P. 2000. Predaceous stink bugs (Pentatomide: Asopinae). pp. 737–789. *In*: Schaefer, C.W. and A.R. Panizzi (eds.). Heteroptera of Economic Importance, CRC Press, Boca Raton, FL, USA.

De Clercq, P., K.W. Wyckhuys, H.N. De Oliveira and J.K. Klapwijk. 2002. Predation by *Podisus maculiventris* on different life stages of *Nezara viridula*. Fla. Entomol. 85: 197–202.

De Clercq, P., T. Coudron and E.W. Riddick. 2014. Production of heteropteran predators. pp. 57–100. *In*: Morales-Ramos, J.A., M. Guadalupe Rojas and D.I. Shapiro-Ilan (eds.). Mass Production of Beneficial Organisms: Invertebrates and Entomopathogens. Academic Press, London, UK.

De Castro, A.A., A.S. Corrêa, J.C. Legaspi, R.N. Guedes, J.E. Serrão and J.C. Zanuncio. 2013. Survival and behavior of the insecticide-exposed predators *Podisus nigrispinus* and *Supputius cincticeps* (Heteroptera: Pentatomidae). Chemosphere 93: 1043–50.

Desneux, N., A. Decourtye and J.-M. Delpuech. 2007. The sublethal effects of pesticides on beneficial arthropods. Annu. Rev. Entomol. 52: 81–106.

Dicke, M. and P. Grostal. 2001. Chemical detection of natural enemies by arthropods: an ecological perspective. Ann. Rev. Ecol. Systemat. 1–23.

Dickens, J.C. 1999. Predator-prey interactions: olfactory adaptations of generalist and specialist predators. Agric. For. Entomol. 1: 47–54.

Djemai, I., J. Casas and C. Magal. 2001. Matching host reactions to parasitoid wasp vibrations. Proc. Biol. Sci. 268: 2403–2408.

Endo, N., T. Yasuda, K. Matsukura, T. Wada, S.E. Muto and R. Sasaki. 2007. Possible function of *Piezodorus hybneri* (Heteroptera: Pentatomidae) male pheromone: effects of adult age and diapause on sexual maturity and pheromone production. Appl. Entomol. Zool. 42: 637–641.

Eriksson, A., G. Anfora, A. Lucchi, F. Lanzo, M. Virant-Doberlet and V. Mazzoni. 2012. Exploitation of insect vibrational signals reveals a new method of pest management. PLoS ONE 7(3): e32954.

Gogala, M. 2006. Vibratory signals produced by Heteroptera–Pentatomorpha and Cimicomorpha. pp. 275–295. *In*: Drosopoulos, S. and M.F. Claridge (eds.). Insects Sounds and Communication: Physiology, Behaviour, Ecology and Evolution. CRC Press, Taylor & Francis Group, Boca Raton, FL, USA.

Gnatyz, W. and G. Kämper. 1990. Digger wasps against crickets. II. An airborne signal produced by a running predator. J. Comp. Physiol. A 167: 551–556.

Guedes, R.N.C., S.M. Matheson, B. Frei, M.L. Smith and J.E. Yack. 2012. Vibration detection and discrimination in the masked birch caterpillar (*Drepana arcuata*). J. Comp. Physiol A. 198: 325–335.

Haynes, K.F. 1988. Sublethal effects of neurotoxic insecticides on insect behavior. Annu. Rev. Entomol. 33: 149–168.

Hermann, S.L. and J.S. Thaler. 2014. Prey perception of predation risk: volatile chemical cues mediate non-consumptive effects of a predator on a herbivorous insect. Oecologia. 176: 669–676.

Hill, P.S.M. 2008. Vibrational Communication in Animals. Harvard University Press, Massachusets, USA.

Kuhelj, A., M. de Groot, A. Blejec and M. Virant-Doberlet. 2015. The effect of timing of female vibrational reply on male signalling and searching behaviour in the leafhopper *Aphrodes makarovi*. PLoS ONE 10(10): e0139020.

Laumann, R.A., M.C.B. Moraes, A. Čokl and M. Borges. 2007. Eaves dropping on sexual vibratory signals of stink bugs (Hemiptera: Pentatomidae) by the egg parasitoid *Telenomus podisi*. Anim. Behav. 73: 637–649.
Laumann, R.A., A. Kavčič, M.C.B. Moraes, M. Borges and A. Čokl. 2013. Reproductive behaviour and vibratory communication of the neotropical predatory stink bug *Podisus nigrispinus*. Physiol. Entomol. 378: 71–80.
Mahdian, K., L. Tirry and P. De Clercq. 2007. Functional response of *Picromerus bidens*: effects of host plant. J. Appl. Entomol. 131: 160–164.
Mahdian, K., T. Van Leeuwen, L. Tirry and P. De Clercq. 2008. Susceptibility of the predatory stink bug *Picromerus bidens* to selected insecticides. BioControl 52: 765–74.
McPherson, J.E. and R.M. McPherson. 2000. General introduction to stink bugs. pp. 1–6. *In*: Stink Bugs of Economic Importance in America North of Mexico. CRC Press, Boca Raton, FL, USA.
Malaquias, J.B., F.S. Ramalho, C. Omoto, W.A. Godoy and R.F.E. Silveira. 2014. Imidacloprid affects the functional response of predator *Podisus nigrispinus* (Dallas) (Heteroptera: Pentatomidae) to strains of *Spodoptera frugiperda* (J.E. Smith) on Bt cotton Ecotoxicology 23: 192–200.
Martinou, A.F., N. Seraphides and M.C. Stavrinides. 2014. Lethal and behavioral effects of pesticides on the insect predator *Macrolophus pygmaeus*. Chemosphere 96: 167–173.
Meyhöfer, R. and J. Casas. 1999. Mini review. Vibratory stimuli in host location by parasitic wasps. J. Insect Physiol. 45: 967–971.
Mizutani, N., T. Yasuda, T. Yamaguchi and S. Moriya. 2007. Individual variation in the amounts of pheromone components in the male bean bug, *Riptortus pedestris* (Heteroptera: Alydidae) and its attractiveness to the same species. Appl. Entomol. Zool. 42: 629–636.
Mohaghegh, J., P. De Clercq and L. Tirry. 2000. Toxicity of selected insecticides to the spined soldier bug, *Podisus maculiventris* (Heteroptera: Pentatomidae). Biocontrol Sci. Technol. 10: 33–40.
Mukerji, M.K. and E.J. LeRoux. 1965. Laboratory rearing of a Quebec strain of the pentatomid predator, *Podisus maculiventris* (Say) (Hemiptera: Pentatomidae). Phytoprotection 46: 40–60.
Ninkovic, V., Z. Feng, U. Olsson and J. Pettersson. 2013. Ladybird footprints induce aphid voidance behavior. Biol. Control 65: 63–71.
Panizzi, A.R., J.E. McPherson, D.G. James, M. Javahery and R.M. McPherson. 2000. Stink bugs (Pentatomidae). pp. 421–474. *In*: Schaefer, W.C. and A.R. Panizzi (eds.). Heteroptera of Economic Importance. Boca Raton, USA.
Pfannenstiel, R.S., R.E. Hunt and K.V. Yeargan. 1995. Orientation of a hemipteran predator to vibrations produced by feeding caterpillars. J. Insect Behav. 8: 1–9.
Plummer, M.R. and J.M. Camhi. 1981. Discrimination of sensory signals from noise in the escape system of the cockroach: the role of wind acceleration. J. Comp. Physiol. 142: 347–357.
Polajnar, J., A. Eriksson, A. Lucchi, M. Virant-Doberlet and V. Mazzoni. 2014. Manipulating behaviour with substrate-borne vibrations—potential for insect pest control. Pest Manag. Sci. 71: 15–23.
Preisser, E.L., D.I. Bolnick and M.F. Benard. 2005. Scared to death? The effects of intimidation and consumption in predator-prey interactions. Ecology 86: 501–509.
Ryan, M.A., C.J. Moore and G.H. Walter. 1995. Individual variation in pheromone composition in *Nezara viridula* (Heteroptera: Pentatomidae): how valid is the basis for designating "pheromone strains"? Comp. Biochem. Physiol. B 111: 189–193.
Schaefer, C.W. 1996. Bright bugs and bright beetles: asopine pentatomids (Hemiptera: Heteroptera) and their prey. pp. 18–56. *In*: Alomar, O. and R.N. Wiedenmann (eds.). Zoophytophagous Heteroptera (Thomas Say Publications in Entomology). Entomological Society of America, Lanham, MD, USA.
Saini, E. 1994. Aspectos morfologicos y biologicos de *Podisus connexivus* Bergroth (Heteroptera: Pentatomidae). Rev. Soc. Entomol. 121: 327–330.
Shestakov, L.S. 2008. Studies of vibratory signals in pentatomid bugs (Heteroptera, Asopinae) from European Russia. Entomol. Rev. 88: 20–25.

Stark, J.D. and J.E. Banks. 2003. Population-level effects of pesticides and other toxicants on arthropods. Annu. Rev. Entomol. 48: 505–519.

Thornham, D.G., C. Stamp, K.F.A. Walters, J.J. Mathers, M. Wakefield, A. Blackwell and K.A. Evans. 2007. Feeding responses of adult seven-spotted ladybirds, *Coccinella septempunctata* (Coleoptera, Coccinellidae), to insecticide contaminated prey in laboratory arenas. Biocontrol Sci. Techn. 17: 983–994.

Todd, J.W., R.M. McPherson and D.J. Boethel. 1994. Management tactics for soybean insects. pp. 115–117. *In*: Higley, L.G. and D.J. Boethel (eds.). Handbook of Soybean Insects Pests. Entomological Society of America Publications, Laham, MD, USA.

Torres, J.B., C.S.A. Silva-Torres, M.R. Silva and J.F. Ferreira. 2002. Compatibility of insecticides and acaricides to the predatory stink bug *Podisus nigrispinus* (Dallas) (Heteroptera: Pentatomidae) on cotton. Neotrop. Entomol. 31: 311–317.

Torres, J.B., C.S.A. Silva-Torres and R. Barros. 2003. Relative effects of the insecticide thiamethoxam on the predator *Podisus nigrispinus* and the tobacco whitefly *Bemisia tabaci* in nectaried and nectariless cotton. Pest Manag. Sci. 59: 315–23.

Torres, J.B. and J.R. Ruberson. 2004. Toxicity of thiamethoxam and imidacloprid to *Podisus nigrispinus* (Dallas) (Heteroptera: Pentatomidae) nymphs associated to aphid and whitefly control in cotton. Neotrop. Entomol. 33: 99–106.

Tostowaryk, W. 1971. Life history and behavior of *Podisus modestus* (Hemiptera: Pentatomidae) in boreal forest in Quebec. Can. Entomol. 103: 662–674.

Uetz, G.W. and J.A. Roberts. 2002. Multi-sensory cues and multi-modal communication in spiders: insights from video/audio playback studies. Brain Behav. Evol. 59: 222–230.

Vandekerkhove, B. and P. De Clercq. 2004. Effects of an encapsulated formulation of lambda-cyhalothrin on *Nezara viridula* and its predator *Podisus maculiventris* (Heteroptera: Pentatomidae). Fla. Entomol. 87: 112–118.

Virant-Doberlet, M., A. Čokl and M. Zorović. 2006. Use of substrate vibrations for orientation: from behaviour to physiology. pp. 81–97. *In*: Drosopoulos, S. and M.F. Claridge (eds.). Insects Sounds and Communication: Physiology, Behaviour, Ecology and Evolution. CRC Press, Taylor & Francis Group, Boca Raton, FL, USA.

Virant-Doberlet, M., R.A. King, J. Polajnar and W.O.C. Symondson. 2011. Molecular diagnostics reveal spiders that exploit prey vibrational signals used in sexual communication. Mol. Ecol. 20: 2204–2216.

Wäckers, F.L., E. Mitter and S. Dorn. 1998. Vibrational sounding by the pupal parasitoid *Pimpla* (*Coccygomimus*) *turionellae*: An additional solution to the reliability–detectability problem. Biological Control 11: 141–146.

Walker, M.K., M.A.W. Stufkens and A.R. Wallace. 2007. Indirect non-target effects of insecticides on Tasmanian brown lacewing (*Micromus tasmaniae*) from feeding on lettuce aphid (*Nasonovia ribisnigri*). Biol. Control 43: 31–40.

Werner, E.E. and S.D. Peacor. 2003. A review of trait-mediated indirect interactions. Ecology 84: 1083–1100.

Wilcox, R.S. and J. Di Stefano. 1991. Vibratory signals enhance mate-guarding in a water strider (Hemiptera: Gerridae). J. Insect Behav. 4: 43–50.

Williams, J.L. and D.H. Wise. 2003. Avoidance of wolf spiders (Araneae: Lycosidae) by striped cucumber beetles (Coleoptera: Chrysomelidae): laboratory and field studies. Environ. Entomol. 32: 633–640.

Zhang, A., M. Borges, J.R. Aldrich and M. Camp. 2003. Stimulatory male volatiles for the Neotropical brown stink bug, *Euschistus heros* (F.) (Heteroptera: Pentatomidae). Neotrop. Entomol. 32: 713–717.

Žunič, A., A. Čokl, M. Virant Doberlet and J.M. Millar. 2008. Communication with signals produced by abdominal vibration, tremulation, and percussion in *Podisus maculiventris* (Heteroptera: Pentatomidae). Ann. Entomol. Soc. Am. 101: 1169–1178.

CHAPTER 4

Communication as the Basis for Biorational Control

Andrej Čokl,[1,*] *Maria Carolina Blassioli-Moraes*,[2] *Raul Alberto Laumann*[2] and *Miguel Borges*[2]

Introduction

The successful development of biorational pest control techniques based on interference with communication demands general and comparative knowledge of the insect and target pest species' biology, ecology and behaviour. Insects, with more than a million species, owe their extreme evolutionary success to their ability to quickly adapt to different environmental conditions. Species reproduction depends on copulation, which occurs as a result of communication within the different phases of complex mating behaviour. Mating in social species occurs under different conditions as compared to solitary species. Insect societies require stable inner organization and a recognized hierarchy with a synergy of duties, including tasks during reproduction. Solitary insects, on the other hand, need to first to find a mate in the open field using long distance calling that is generally not necessary in case of social species. Differences between the social and solitary ways of life are reflected in communication. Solitary Pentatominae stink bugs live, feed and mate exclusively on plants that significantly determine the specificity of their mating behaviour and communication. To understand the specificity of information exchange in

[1] Department of Organisms and Ecosystems Research, National Institute of Biology, Večna pot 111, SI-1000 Ljubljana, Slovenia.
[2] Laboratório de Semioquímicos, Embrapa Recursos Genéticos e Biotecnologia, Avda. W5 Norte (Final), 71070-917, Brasilia, DF, Brazil.
Emails: carolina.blassioli@embrapa.br; raul.laumann@embrapa.br; miguel.borges@embrapa.br
* Corresponding author: andrej.cokl@nib.si

Pentatominae, we present in this chapter basic knowledge on the different aspects of insect communication, focusing mainly on the role of chemical and vibratory signals.

Communication

We can describe communication most simply as a complex of processes that enable the sender (actor) and the receiver (reactor) to exchange information incorporated into signals transmitted through different transmission media (air, solids, or water). Communication is probably the most ancient phenomenon driving life in all surroundings at the molecular, cell, organ, organism and society levels.

The first living organisms were required to develop and maintain controlled homeostasis by sensing and reacting to the changes in their physical surroundings. This likely triggered the development of broad spectrum receptors that through evolution have specialized into highly sensitive unimodal sensory organs. An increasing number of organisms required communication with others in order to establish and hold societies together at the relevant hierarchical level, and in case of solitary species, to meet a mate in the field. Life in water, air or on solid substrates led to the evolution of signal production and sensory organs tuned to the properties of the transmission medium. Mating in surroundings occupied by different species demanded mechanisms to prevent hybridization. In this respect, the evolution of multimodal communication was directed towards differentiation and specialization of signals tuned to the properties of the medium and with species-specific characteristics that were less modified during transmission. Among many other tasks, communication solves the problems concerning species, gender or society membership identity; moods such as dominance, fear or aggression; intention of activity such as approach, fight or flight; and determining the location of a mate or food (Bradbury and Vehrencamp 2011). We can differentiate between three kinds of semantic information in animal communication (Krebs and Dawkins 1984). Intention is transmitted by the sender as information about what to do next in a strategic way, as a quality assessment connected with persuading the receiver to pay attention to the sender's characteristics, and finally, signalling systems can convey information about the environment.

We can largely describe the elements of communication following the work of Sebeok (1962) and Barnard (2004).

The *sender* is an individual who emits a signal, and the *receiver* is an individual whose probability of behaving in a particular way is altered by this signal. The *channel* is described as the medium through which the signal is transmitted. The rate of information transmission through air, water or substrate depends on the channel's bandwidth and signal-to-noise ratio.

The bandwidth is determined by the channel's range of events transmitted per unit time that can be detected in a meaningful way. The transmission capacity of a channel is low if it cannot switch rapidly from one symbol to another. The *signal* may be generally described as an event such as a gesture or position of the body, sound, vibration or touch, that shows the concerned parties that something exists or is about to happen. In true communication, this excludes incidental transmission or transmission of misinformation if the signal is a symbol for which different signals could be substituted if the sender and receiver agree in the substitution (Dusenberry 1992). Sebeok (1962) describes the signal as the behaviour (e.g., posture, display, and vocalisation) transmitted by the sender. From a physiological standpoint, the signal may also be determined to be a *stimulus*. The meaning of the signal is connected to the pattern that has been shaped by natural selection to enhance its information-transmitting function, in contrast to *cues* that have not been shaped by natural selection (Seeley 1989). The *context* is the setting in which the signal is transmitted and received. The *noise* is background activity that is irrelevant to the transmitted signal and can also be described as unpredictable patterns that are not part of the signal. The *code* is described as the complete set of possible signals and contexts, or in more general terms, as a system of symbols that give a certain arbitrary meaning. Finally, Smith (1965) describes the *message* as what the signal encodes about the sender and *meaning* as what the receiver construes from the signal. Signals that carry only one message (and function in terms of presence or absence) are described as discrete or digital and those with different intensities and temporal and frequency characteristics are called graded or analogue (Sebeok 1962). Signals may be connected to just one specific (single) message or to different messages in different contexts, emitted at different times and to different individuals (Barnard 2004). Complex signals may convey the same information through different channels.

Animals can use complex signals with specific information that trigger several or just one stereotyped response. For example, insect directional movement towards a calling mate may be mediated by visual, olfactory, airborne and/or substrate-borne signals that all carry the same information, namely, the position of the sender in three-dimensional space. The choice of the communication channel (modality) is determined by its transmission properties and environmental conditions. The efficiency of information exchange and the communication distance fundamentally depend on the tuning between signal and transmission medium characteristics.

Evolution favours ritualization and stereotypy of signals and displays in order to decrease the ambiguity of information (Cullen 1966). The signaller emits different signals, each with a definite meaning, with the intention of avoiding the risk of confusing the receiver. In this context, communication may be viewed as an honest partnership between the signaller and the receiver, in that both benefit in the process of information exchange. Another aspect focuses attention on the cost and benefit relationship that inevitably causes the sender to manipulate the receiver. Because the sender plays an active role in the entire process, it is in the sender's own best interest to increase the benefits against the costs. Manipulation of the receiver is also accomplished by decreasing the amount of useful information with the intention of hiding the sender's weak points or by exaggerating in order to force the receiver to respond.

Guilford and Dawkins (1991) differentiate between the strategic and tactical designs of the signal. The strategic component of a signal is related to the reliability of the signal (the handicap principle) (Bradbury and Vehrencamp 2011) and the tactical design is related to the efficiency of transmission through the medium, together with the level of signal understanding or interpretation. The tactical design is determined by the transmission properties of the environment and the medium through which the signals are conveyed. Substrate, air and water differently distort the transmitted signal and are differently sensitive to noise. Generally, air is a relatively inert transmission medium that enables communication by airborne, chemical and visual signals at longer distances while substrates such as plants limit communication due to their architecture and size to shorter distances and significantly change a signal's amplitude, frequency and temporal characteristics. Communication distance also depends on the energy required for signal production, on signal attenuation by transmission through different media and environments, and finally, on the sensitivity of the receiver's sensory apparatus. The characteristic sensitivity to noise is another important criterion in the choice of transmission channel. Depending on the goal of communication, animals choose between different *specificity* levels of their signals. *Localization* of the signal emitter demands the use of signals with the highest level of directionality information, and the efficiency of species or gender *recognition* of the signaller depends, among other factors, on the ability to incorporate the information into signal characteristics less influenced by the medium and environmental noise. The *range* of communication often determines the size of an animal's territory.

Abiotic and biotic environmental noise significantly influences behaviour and communication, necessitating the evolution of different mechanisms in order to increase the contrast between these signals (Sueur 1992). The emitter may change frequency or some other characteristic of the signal (Aubin 1994, Cocroft and Rodriguez 2005, Čokl et al. 2015), switch to a different sensory channel (Gerhardt 1983) or simply move to a quieter place.

In spite of all the benefits, honesty and general necessity of communication, signalling also has several disadvantages. Calling mates attract predators and parasitoids by their emissions, as exemplified by male field crickets, *Gryllus integer* (Scudder), that attract parasitic tachinid flies (Cade 1979) by their stridulatory calls. The egg parasitoids *Telenomus podisi* (Ashmead) eavesdrop on the sexual vibratory signals of stink bugs (Laumann et al. 2007, 2011a). Interactions mediated by cues between herbivorous and natural enemies are widely explored in biological control and are further explained in Chapter 11 of this book.

Stink Bug Mating Behaviour and Communication—A Comparative Approach

Mating behaviour

The reproductive success of insects depends on a sequence of long- and short-range exchanges of information, involving chemical, visual and mechanical signals. The first studies on mating behaviour in solitary Pentatomidae revealed in *Chlorochroa ligata* (Say), *Cosmopepla bimaculata* (Thomas) and *Murgantia histrionica* (Hahn), a sequence of steps comprising their short-range courtship behaviour without addressing the communication signals (Fish and Alcock 1973, Lanigan and Barrows 1977). However, these studies have shown male directional movement towards females, indicating the presence of attracting chemical and/or mechanical signals emitted by the female. The first complete description of mating behaviour within Pentatomidae was carried out for *Nezara viridula* (L.) by Borges et al. (1987), showing two main stereotyped sequences of behaviours: long-range mate location and short-range courtship. The short-range courtship behavioural patterns of *N. viridula* are similar to those shown by Fish and Alcock (1973). Borges et al. (1987) described mutual antennation of the mate's body as the first step, followed by the male butting the female's abdomen with its head, elevation of the receptive female's abdomen, and the male pivoting and finally inserting its aedaegus into the female genital opening (Fig. 4.1). Similar short-range courtship behaviour has been described in many other pentatomine stink bug species, such as recently in *Dichelops melacanthus* (Dallas) (Blassioli-Moraes et al. 2014).

Borges et al. (1987) first described the long-range attraction of mates mediated by chemical signals, showing that female movement was oriented towards live males or towards air-entrainment extracts containing male sex pheromone compounds. In both situations, the females responded by elevating their antennae, moving upwind towards the odour source and fanning their wings. Differences between the two tests were observed at close range to the odour source: the typical short-range female behaviour connected with courtship was displayed only when they reached the living male, and Borges et al. (1987) suggested that signals of other modalities also have to be involved in order to induce female short-range behaviour. Pheromone-mediated long-range communication has generally been attributed to most stink bug species, although in some, such as

Figure 4.1. Typical short-range courtship patterns of stink bugs (Pentatominae). a = antennation of the female and/or female antennation of the male, b = the male antennating and butting the female, c = the male butting the female' abdomen, followed by abdominal elevation of the receptive female, d = the male pivoting, exposing his aedeagus, e = mating position (Photographs: Douglas Maccagnanan, behavioural sequences from the description in Borges et al. 1987).

D. melacanthus (Blassioli-Moraes et al. 2014), male-emitted pheromones have not yet been identified.

Early suggestions regarding the role of signals of modalities other than male pheromones (Borges et al. 1987) in communication on a common substrate (plant) have been confirmed in all examined Pentatominae stink bug species so far (Chapter 6). Vibratory communication signals were described more than 40 years ago in *N. viridula* (Čokl et al. 1972) and subsequent research in all investigated Pentatominae species has demonstrated the decisive role of substrate-borne vibratory communication along the plant. At the present state of knowledge, we can differentiate between three distance-dependent phases of mating behaviour in plant-dwelling Pentatominae species. Long-range communication in the field is mediated by chemical signals (see below), medium-range communication on the plant proceeds via the substrate-borne component of vibratory emissions (see below) and previously described short-range communication is conducted by the use of visual, chemical and contact mechanical signals (Fig. 4.2). Calling at the long- and medium-range scales provides information on location and species identity; during courtship at short distances, recognition becomes more precise by visual and chemical contact and mechanical signals. Communication during mating behaviour also includes signalling connected with rivalry, female rejection of a male attempting to copulate and emission of signals by mates in copula (Chapter 6).

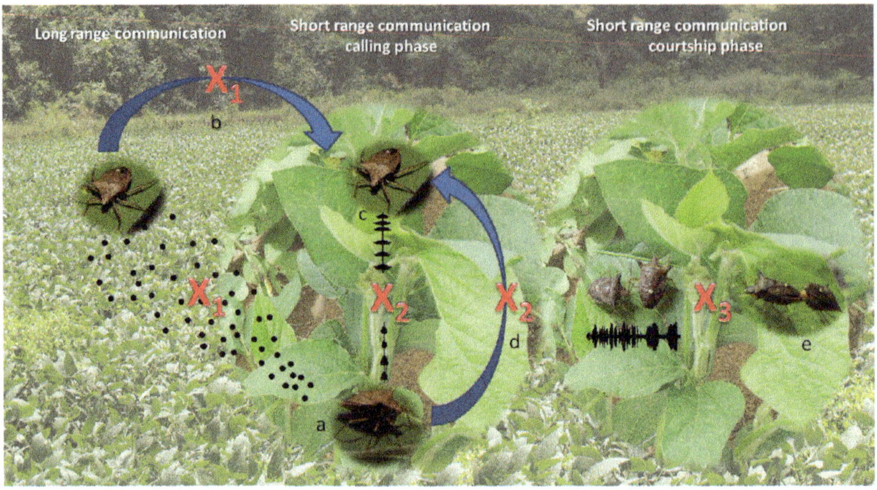

Figure 4.2. Typical long- and short-range steps of sexual communication in stink bugs and possibilities to interfere/disrupt them for biorational management development. a = male pheromone emission, b = female directional movement towards the pheromone source, c = while on the same plant, the female starts to emit calling vibratory signals with male responses, d = directional movements of the male to follow the female calling signals, e = physical contact and multimodal communication with courtship vibratory and visual signals and short-range chemicals, pairs mate. X_1 = interference in chemical communication, preventing females from reaching males' plants, X_2 = interruption of the calling phase, preventing directional movement of the males and inhibiting pairs formation, X_3 = possibility of disrupting courtship using an alarm or advertising signals (Stink bugs photographs Cecilia Vieira and Douglas Maccagnan, landscape and plant photographs Raúl Laumann).

Communication with chemical signals

Insects communicate via different chemical signals that are used in a sequence of steps to provide information in order to locate hosts, find mates and food and to repel enemies, among other reasons. Like acoustic and visual communication, chemical signals occur in a narrow emission and reception band (Howse 1998). Chemical communication has a very low plasticity as compared with visual and acoustic communication because it cannot be modulated to avoid noise or overlapping by other chemical signals. However, chemical communication signals are species-specific and provide a reliable cue. Pheromones are used in intraspecific communication. The main examples are sex and sex aggregation pheromones and defensive compounds (alarm pheromones), as well as allelochemicals, which are used in interspecific communication. Plant volatiles are important in locating host habitat or in avoiding predators and parasitoids. In general, chemical communication is a very reliable way to exchange information in Pentatomidae. Sex pheromones in stink bugs are produced by males and are species-specific (Chapter 5). In contrast to vibratory signals, chemical emissions do not overlap and specific communication channels are used. Each molecule or blend of molecules carries different kinds of information (Howse 1998). Communication using chemical signals has several advantages: volatiles are transmitted through the air and are active in the dark and at long distances after crossing several obstacles between the receiver and the emitter. The efficiency of chemical signal transportation depends on air currents, humidity, temperature, UV incidence, and the types of obstacles along its way. For example, chemicals can be absorbed by vegetation used as a barrier for wind and UV rays may interfere with communication by causing isomerisation or degradation of the chemical molecules. Biosynthesis of chemical signals prevents their modulation (Morgan 2010), and in case of the presence of molecules emitted by conspecifics or heterospecifics, signal-to-noise ratio cannot be increased by the production of different molecules. Any temporal pattern created during pheromone emission is lost during transmission independently if it proceeds by diffusion or by current flow (Dusenberry 1992). However, pheromones are generally redundant signals (Howse 1998), meaning that the receiver can process the information even when the composition of the pheromone is not characteristic of the species, similar to how we can read and interpret text even when the letters are placed in the wrong order, as in: agreggaiotn pehoromne.

Several species of Lepidoptera respond to chemicals that have a chemical structure very similar to that of their own pheromones. Such examples have not yet been found in Pentatomidae, however, field studies using the sex pheromones of *Euschistus heros* (Fabr.) in different areas of Brazil showed that pheromone-baited traps attracted not just *E. heros* females but also *Piezodorus guildinii* (Westwood), *Thyanta perditor* (Fabr.), *Edessa meditabunda* (Fabr.) and *D. melacanthus* (Borges et al. 1998, 2011). Similarly, traps baited with the sex pheromones of *T. perditor* are able to attract *E. heros*, *E. meditabunda*, *P. guildinii* and *N. viridula* (Laumann et al. 2011b). The sex pheromone molecules of all of these species do not show any chemical similarity (see details in Chapter 5), therefore suggesting that these insects have learnt to interpret signals from other species in order to locate host plants. Use of the sex pheromones of *Riptortus clavatus* (Thunberg) (Alydidae) by *Piezodorus hybneri* (Gmelin) (Pentatomidae) was also reported in Japan (Endo et al. 2006), with the authors of the study suggesting that the latter species uses the pheromone of *R. clavatus* as a kairomone to search for food plants. Aldrich et al. (2007) offered an alternative hypothesis to cross-attraction, suggesting that it could be related to a defence mechanism of stink bug aggregations. For these interactions, Pentatomidae species most likely have odorant binding proteins (OBPs) that recognize all of these signals.

Recently, Farias et al. (2015) reported the first study identifying the odorant binding proteins of Pentatomidae. This study used RNA-Seq to analyse transcriptomes for OBPs from three stink bug pest species

(*E. heros*, *Chinavia ubica* (Rolston), and *D. melacanthus*), as well as from their egg parasitoid, *T. podisi*. In general, the deduced amino acid sequences of the full-length OBPs had low intraspecific similarity, but a very high similarity between two pairs of OBPs from *E. heros* and *C. ubica* (76.4 and 84.0%) and between two pairs of OBPs from the parasitoid and its preferred host *Euschistus heros* (82.4 and 88.5%) was found. The similar pairs of OBPs from *E. heros* and *C. ubica* may suggest that they retain the same biological function to bind a ligand identified in both species, and the similar OBPs in *T. podisi* and *E. heros* support the laboratory and field experiments showing that *T. podisi* uses the sex pheromone of *E. heros* as a cue to locate its host (Borges et al. 1998, Silva et al. 2006).

The high efficiency and selectivity of intraspecific chemical communication has been reported in several studies conducted under laboratory conditions (Blassioli-Moraes et al. 2008, Millar 2005, Borges et al. 2006, 2007). However, the results obtained under field conditions have shown plasticity in response to a variety of chemical signals acting as sex or sex aggregation pheromones (Borges et al. 1998, 2011). To understand these mechanisms, one needs to study the process of odour molecule recognition at the molecular level in stink bug olfactory receptor cells and how odour molecules are used to extract information in a complex ecological context. These studies will provide information on the species-specificity of chemical communication and how to use chemical signals (semiochemicals) such as pheromones and parapheromones to manipulate insect behaviour.

The average life span of 30 to 60 days for adult stink bugs, as determined under laboratory conditions (Silva et al. 2004, Michereff et al. 2011), is longer than that of moths. Consequently we may expect not only intra- but also interspecific chemical communication in the field. Studies have shown that male and female stink bugs are able to recognize constitutive volatiles from plants and that they avoid plants having herbivory of conspecifics (Michereff et al. 2011). Such behaviour may reduce competition and attack by natural enemies, as it has been shown that damage to plants triggers the release of volatile compounds that attract egg parasitoids (Colazza et al. 2004, Blassioli-Moraes et al. 2005, 2008). Despite all of the information described here and in Chapter 5, there are still very few studies that have addressed the process of perception of odour in Pentatomidae, how these chemicals are released in nature, and how the odour plume functions in air transmission to reach its receivers.

Communication with vibratory signals

Vibration is a basic phenomenon in almost any physical system. It is commonly described as the periodic back-and-forth motion of particles of an elastic body or medium as a consequence of displacement from equilibrium with the tendency to restore it. Most mechanisms that enable life on our planet are based on vibration, from molecules to organisms and behaviour. Despite intensive research on the influence of different vibration-induced events, there are still many animal reactions to vibration that remain less understood. For example, Anderson (1978), Kenneally (2005) and Hill (2008) describe several examples of aberrant animal behaviour and sensitivity related to vibration phenomena connected with earthquakes and tsunamis. Elephants and tigers have been shown to escape to safe places prior to tsunami arrival, thereby saving their lives, whereas humans do not.

The evolution of communication has guided adaptation in both the sender and receiver. Signallers have tuned the characteristics of their emissions to the properties of the transmission medium and the receiver's sensory apparatus has become specialized to detect unimodal complex communication signals transmitted through different media with high sensitivity; the evolution of underlying neuronal networks followed to process, filter and integrate information at higher levels. Plants represent the most natural insect environment, and communication in such surroundings demands specific adaptations (Chapter 7). Plant-dwelling insects have chosen from among different modalities of mechanical signals (Greenfield 2002), predominantly communication using substrate-borne vibratory signals (Cocroft and Rodriguez 2005). The authors estimate that 80% of insect families (over 195,000 taxa) communicate by substrate-borne vibrations, either exclusively or in concert with other signals. The exclusive use of substrate-borne vibrational communication was documented in approximately 150,000 species (74% of the families). Since the time of this study, the number of species recorded with prevalent vibrational signalling has been increased well above the estimated values.

Since the pioneering work of Michelsen and co-authors (1982), the mechanical transmission properties of plants have been described in different books (Barth 2002, Michelsen 2014) and review articles (Čokl and Virant-Doberlet 2003, Čokl 2008). Bending waves are propagated with frequency-dependent velocity in the range below 5 kHz and independently of this factor at higher frequencies, the low-pass filtering properties of plants represent an advantage by allowing communication with lower frequency signals, the amplitude of plant-transmitted signals does not decrease linearly with distance, the temporal characteristics of a signal change due to reflections at the plant-air phase border and plant resonance characteristics modify the input signal spectra (Chapter 7). Vibration transmission properties of soil and sand have also been recently described and discussed by Michelsen (2014) and Devetak (2014).

Insects produce vibratory communication signals via a variety of different mechanisms, including drumming (percussion), stridulation, vibration of the whole body (tremulation) or its parts and by the buckling of tymbals. Each vibration-producing mechanism determines the basic frequency characteristics of the emitted signal.

The main vibration-producing mechanism in stink bugs is the vibration of the abdomen against the thorax without touching the substrate (Chapter 6). Similar mechanisms have been described in *Cupiennius* spiders (Barth 2002), female planthoppers of the family Delphacidae (Claridge 1985), male whiteflies (Kanmiya 2006) as well as *Apis mellifera* (L.) during the waggling of their abdomen (Hrncir et al. 2006). Such mechanisms allow insects with a small body mass to apply sufficient force on green rather than woody plants (Morris 1980). High amplitude signals produced by the vibration of the whole body (tremulation) and wing fluttering (buzzing) have been recently observed and described in Pentatominae (Kavčič et al. 2013).

Drumming occurs in many insect groups (Hill 2008). Megalopteran species of the genus *Sialis* produce vibratory signals by the percussion of their abdomen and wings (Ruprecht 1975), drumming with antennae has been described in wasp queens (Pratte and Jeanne 1984) and *Meconema thalassinum* (De Geer) bush crickets drum by tapping one leg on the substrate (Sismondo 1980). Percussion signals have also been described in Asopinae (Chapter 3) and Pentatominae (Chapter 6). Stridulation produces signals with airborne and substrate-borne components via friction of the different body parts against each other. Gogala (2006) reviewed stridulation in Heteroptera infraorders Pentatomorpha and Cimicomorpha, describing in Cydnidae an interesting example of signal emission by two different mechanisms: stridulation and abdomen vibration. Stridulation in Orthoptera is the main sound-producing mechanism with both airborne and substrate-borne components. The use and combination of both are discussed in a review by Rössler and co-authors (2006). Orthopteran stridulatory signals are combined in some species with vibratory signals produced by drumming (Keuper et al. 1985) or tremulation (Morris 1980). Stridulation has not yet been described in any Pentatomidae stink bug species.

Signals produced by tymbal or tymbal-like structures are described in cicadas (for a review, see Hill 2008). Planthoppers, leafhoppers, true and moss bugs communicate by substrate-borne vibrations produced by tymbal-like organs (Wessel et al. 2014). The tymbal-like role of the fused first and second abdominal tergits in the production of vibratory signals in Pentatominae has not yet been confirmed.

Mechanoreceptors are the most ancient sensory organs (Kung 2005). Each cell outside the nervous system is sensitive and responds to vibrational stimuli, and almost any mechanoreceptor on and in the body acts as a vibration receptor. Communication with the high diversity of insect airborne, substrate- and water-borne vibratory signals demanded the evolution of different types of mechanoreceptors with a high ability to detect and code specific parameters of adequate stimuli in a broad sensitivity range, and these systems are based on various morphological adaptations. The vibratory sensory system of insects is composed mostly of sensitive subgenual and other leg chordotonal organs supported by mechanoreceptors that are primarily involved in detecting other mechanical modalities. These include Johnston's organ and hair sensilla, which are both sensitive to fluid flow around the body, or campaniform sensilla, which chiefly detect cuticle strain.

Recently, Lakes-Harlan and Strauss (2014) summarized data on the functional morphology and evolutionary diversity of vibration receptors in insects. The authors described different types of sensilla according to their mechanoreceptive function in detecting vibration. The campaniform sensilla found on the body surface act primarily as proprioreceptors responding to cuticular strain. Kühne (1982) demonstrated that campaniform sensilla in locusts and bushcrickets respond to leg vibration in a phase-locked manner, with the highest sensitivity at approximately 50 Hz and the frequency range extending to a few hundred Hz. Filiform hairs acting as sensitive touch receptors also respond directly or indirectly to the vibrations produced by air particle movement caused by the activity of other organisms or the vibration of objects in their vicinity. For example, Goodwyn and co-workers (2009) demonstrated that trichobotria act as vibration receptors in water striders.

The most sensitive insect vibratory receptor organs are located in the legs. Scolopidial organs that span between the segments of a joint and control its position and angular velocity or are attached at different points of the same segment detect the vibrations transmitted over the cuticle, haemolymph of the blood channel and/or air in the trachea with high sensitivity. Shaw (1994), for example, demonstrated that airborne sound elicits subgenual organ responses in cockroaches with no tympana. The joint chordotonal organs present in all insects are basically proprioreceptors, controlling joints between the femur and tibia (femoral chordotonal organ), tibia and tarsus (tibial chordotonal organ) and between different parts of the tarsus (tarsal chordotonal organ). It has been demonstrated that these organs are in many species sensitive to vibrations of the substrate. The proximal scoloparium of the femoral chordotonal organs of locusts, for example, detects vibrations in the frequency range below 300 Hz (Field and Pflüger 1989) and in the green lacewing, *Chrysoperla carnea* (Stephens) (Neuroptera), the femoral chordotonal responds best to vibrations of approximately 1000 Hz (Devetak and Amon 1997). Water striders orient to their prey by detecting vibrations of the water surface with tarsal chordotonal organs (Murphey 1971, Wiese 1972, Goodwyn et al. 2009).

Subgenual organs are composed of a group of scolopidia situated in the proximal part of the tibiae. In different species the subgenual organ is part of a complex system together with other tibial chordotonal organs such as the distal organ, intermediate organ, Nebenorgan, accessory organ and crista acoustica (for a review see Lakes-Harlan and Strauss 2014). The anatomy of subgenual organs and the number of scolopidia vary significantly in different insect groups (Lakes-Harlan and Strauss 2014): the subgenual organ of parasitoid wasps is composed of 300 to 400 scolopidia (Vilhelmsen et al. 2008) and just 2 have been described in that of the green stink bug, *N. viridula* (Michel et al. 1982). Subgenual organs express extremely high sensitivity to different frequency ranges: the cockroach *Periplaneta americana* (L.) vibratory threshold lies at displacements between 0.25 and 5×10^{-9} m at frequencies of approximately 1.6 kHz (Shaw 1994), the acceleration threshold in lacewings ranges around 0.02 m/s^2 at frequencies of approximately 1 kHz (Devetak and Amon 1997) and the high frequency vibratory receptor cell of the *N. viridula* subgenual organ responds to frequencies between 0.6 and 1 kHz with a threshold acceleration of approximately 0.01 m/s^2 (Čokl 1983). Different comparative aspects of the morphology and function of the subgenual and other leg vibratory receptor organs were recently reviewed and discussed in detail by Lakes-Harlan and Strauss (2014).

Opportunities for Stink Bug Biorational Control Management

Stink bugs are currently one of the most important agricultural pests having a major worldwide impact on agriculture. Interference with communication offers an environmentally friendly strategy for use in biorational pest control management. Its development and application require detailed knowledge of the target pest species' general biology, behaviour and landscape ecology obtained under natural conditions in the field. Stereotypical stink bug behaviour and communication during the reproductive phase of life offer such an opportunity (Čokl and Millar 2009). The most suitable step for interfering in communication is during the calling phase, which occurs at long ranges by chemical signals and at medium ranges on the plant by vibratory signals (Fig. 4.2).

The oriented movements towards the source of pheromones or vibratory calling signals attract insects to specific areas, offering opportunities for monitoring and/or application of control measures that can be conducted efficiently in more restricted areas (see details in Chapters 5, 11). Until now, the most successful and secure way to disrupt mating and decrease population levels has been shown to be at the pheromone communication level (Witzgall et al. 2010) because insects cannot efficiently adapt and modulate the emission of chemical signals to avoid interference. In high-density populations, mates can find each other by chance, and interference with vibratory communication may, in this case, decrease mating success (Chapter 11). Interference with the courtship phase of mating behaviour is more complex and demands a basic knowledge of the synergy among the signals of different modalities. Alarm chemical signals and vibratory repellent or even rival signals provide potential tools to remove target populations of pest species from protected areas.

Acknowledgments

We thank the Slovene Research Agency, the National Council for Scientific and Technological Development (CNPq), the Brazilian Corporation of Agricultural Research (EMBRAPA) and the Research Support Foundation of the Federal District (FAP-DF).

References

Aldrich, J.R., A. Khrimian and M.J. Camp. 2007. Methyl 2,4,6-decatrienoates attract stink bugs (Hemiptera: Heteroptera: Pentatomidae) and tachinid parasitoids. J. Chem. Ecol. 33: 801–815.

Andresen, C.J. 1973. Animals, earthquakes, and eruptions. Field Mus. Nat. Hist. Bull. 44: 9–11.

Aubin, T. 1994. Adaptation for acoustic information through the environment: importance of duration parameters in the distress calls of the starling (*Sturnus vulgaris*). Rev. d' Ecol. 49: 405–415.

Barnard, C.J. 2004. Animal Behaviour: Mechanisms, Development, Function and Evolution. Pearson Education Limited, Harlow, UK.

Barth, F.G. 2002. A Spider's World: Senses and Behaviour. Springer, Berlin.

Blassioli-Moraes, M.C., R.A. Laumann, A. Čokl and M. Borges. 2005. Vibrational signals of four Neotropical stink bugs species. Phys. Entomol. 30: 175–188.

Blassioli-Moraes, M.C.B., M. Pareja, R.A. Laumann and M. Borges. 2008. The chemical volatiles (semiochemicals) produced by neotropical stink bugs (Hemiptera: Pentatomidae). Neotrop. Entomol. 37: 489–505.

Blassioli-Moraes, M.C., D.M. Magalhães, A. Čokl, R.A. Laumann, J.P. da Silva, C.C.A. Silva et al. 2014. Vibrational communication and mating behaviour of *Dichelops melacanthus* (Hemiptera: Pentatomidae) recorded from loudspeaker membranes and plants. Phys. Entomol. 39: 1–11.

Borges, M., P.C. Jepson and P.E. Howse. 1987. Long-range mate location and close range courtship behavior of the green stink bug, *Nezara viridula* and its mediation by sex pheromones. Entomol. Exp. Appl. 44: 205–212.

Borges, M., F.G.V. Schmidt, E.R. Sujii, M.A. Medeiros, K. Mori, P.H.G. Zarbin et al. 1998. Field responses of stink bugs to the natural and synthetic pheromone of the neotropical brown stink bug, *Euschistus heros* (Heteroptera: Pentatomidae). Phys. Entomol. 23: 202–207.

Borges, M., M. Birkett, J.R. Aldrich, J.E. Oliver, M. Chiba, Y. Murata et al. 2006. Sex attractant pheromone from the rice stalk stink bug, *Tibraca limbaliventris* Stal. J. Chem. Ecol. 32: 2749–2761.

Borges, M., J.G. Millar, R.A. Laumann and M.C.B. Moraes. 2007. A male-produced sex pheromone from the neotropical redbanded stink bug, *Piezodorus guildinii* (W.). J. Chem. Ecol. 33: 1235–1248.

Borges, M., M.C.B. Moraes, M.F. Peixoto, C.S.S. Pires, E.R. Sujii and R.A. Laumann. 2011. Monitoring the Neotropical brown stink bug *Euschistus heros* (F.) (Hemiptera: Pentatomidae) with pheromone-baited traps in soybean fields. J. Appl. Entomol. 135: 68–80.

Bradbury, J.W. and S.L. Vehrencamp. 2011. Principles of Animal Communication. Sunderland: Sinauer Associates, MA.

Cade, W.H. 1979. The evolution of alternative male reproductive strategies in field crickets. pp. 343–379. *In*: Blum, M. and N.A. Blum (eds.). Sexual Selection and Reproductive Competition in Insects. Academic Press, London.

Claridge, M.F. 1985. Acoustic signals in Homoptera: behaviour, taxonomy, and evolution. Ann. Rev. Entomol. 30: 297–317.

Cocroft, R.B. and R.L. Rodriguez. 2005. The behavioral ecology of insect vibrational communication. Bioscience 55: 323–334.

Colazza, S., A. Fucarino, E. Peri, G. Salerno, E. Conti and F. Bin. 2004. Insect oviposition induces volatile emission in herbaceous plants that attracts egg parasitoids. J. Exp. Biol. 207: 47–53.

Cullen, J.M. 1966. Ritualization of animal activities in relation to phylogeny, speciation and ecology: reduction of ambiguity through ritualization. Phil. Trans. Roy. Soc. London B 241: 363–374.

Čokl, A., M. Gogala and M. Jež. 1972. The analysis of sound signals in the stink bug species *Nezara viridula*. Biol. Vest. 20: 47–53.

Čokl, A. 1983. Functional properties of vibroreceptors in the legs of *Nezara viridula* (L.) (Heteroptera, Pentatomidae). J. Comp. Physiol. A 150: 261–269.

Čokl, A. and M. Virant-Doberlet. 2003. Communication with substrate-borne signals in small plant dwelling insects. Ann. Rev. Entomol. 48: 29–50.

Čokl, A. 2008. Stink bug interaction with host plants during communication. J. Insect Physiol. 54: 1113–1124.

Čokl, A. and J.G. Millar. 2009. Manipulation of insect signalling for monitoring and control pests insects. pp. 279–316. *In*: Ishaaya, I. and A.R. Horowitz (eds.). Biorational Control of Arthropod Pests. Springer, Dordrecht Heidelberg London New York.

Čokl, A., R.A. Laumann, A. Žunič-Kosi, M.C. Blassioli-Moraes, M. Virant-Doberlet and M. Borges. 2015. Interference of overlapping insect vibratory communication signals: An Euschistus heros model. PloS one 10(6): 1–16.

Devetak, D. and T. Amon. 1997. Substrate vibration sensitivity of the leg scolopidial organs in the green lacewing *Chrysoperla carnea*. J. Ins. Physiol. 43: 433–437.

Devetak, D. 2014. Sand-borne vibrations in prey detection and orientation of antlions. pp. 319–330. *In*: Cocroft, R.B., M. Gogala, P.S.M. Hill and A. Wessel [eds.]. Studying Vibrational Communication. Springer, Heidelberg, New York, Dordrecht, London.

Dusenberry, D.B. 1992. Sensory Ecology: How Organisms Acquire and Respond to Information. W.H. Freeman and Co. New York, U.S.A.

Endo, N., T. Wada, Y. Nishiba and R. Sasaki. 2006. Interspecific pheromone cross-attraction among soybean bugs (Heteroptera): Does *Piezodorus hybneri* (Pentatomidae) utilize the pheromone of *Riptortus clavatus* (Alydidae) as a kairomone? J. Chem. Ecol. 32: 1605–1612.

Farias, L.R., P.H.C. Schimmelpfeng, R.C. Togawa, M.M.C. Costa, P. Grynberg, N.F. Martins et al. 2015. Transcriptome-based identification of highly similar odorant-binding proteins among neotropical stink bugs and their egg parasitoid. PLoS ONE 10: e0132286. doi:10.1371/journal.pone.0132286.

Field, L. and H.J. Pflüger. 1989. The femoral chordotonal organ: a bifunctional orthopteran (*Locusta migratoria*) sensory organ? Comp. Biochem. Physiol. A 93: 729–743.

Field, L. and T. Matheson. 1998. Chordotonal organs in insects. Adv. Insect Physiol. 27: 1–228.

Fish, J. and J. Alcock. 1973. The behavior of *Chorochroa ligata* (Say) and *Cosmopepla bimaculata* (Thomas), (Hemiptera: Pentatomidae). Entomol. News 84: 260–268.

Gerhardt, H.C. 1983. Communication and the environment. pp. 82–133. *In*: Halliday, T.R. and P.J.B. Slater (eds.). Animal Behaviour: Volume 2 Communication. Blackwell, Oxford, UK.

Gogala, M. 2006. Vibratory signals produced by Heteroptera-Pentatomorpha and Cimicomorpha. pp. 275–296. *In*: Drosopoulos, S. and M.F. Claridge (eds.). Insect Sounds and Communication: Physiology, Behaviour, Ecology and Evolution. CRC Taylor & Francis Group, Boca Raton, London, New York.

Goodwyn, P.P., A. Wada-Katsumata and K. Okada. 2009. Morphology and neurophysiology of tarsal vibration receptors in the water strider *Aquarius paludum* (Heteroptera: Gerridae). J. Insect Physiol. 55: 855–861.

Greenfield, M.D. 2002. Signallers and Receivers: Mechanisms and Evolution of Arthropod Communication. Oxford University Press, New York.

Guilford, T. and M.S. Dawkins. 1991. Receiver psychology and the evolution of animal signals. An. Behav. 42: 1–14.

Hill, P.S.M. 2008. Vibration Communication in Animals. Harvard University Press, Cambridge, Massachusetts, London.
Howse, P.E. 1998. Pheromones and behaviour–insect semiochemicals and communication. pp. 6–37. *In*: Howse, P., I. Stevens and O. Jones (eds.). Insect Pheromones and their Use in Pest Management. Chapman & Hall, London, UK.
Hrncir, M., V.M. Schmidt, D.L.P. Shorkopf, S. Jarau, R. Zucchi and F.G. Barth. 2006. Vibrating the food receivers: A direct way of signal transmission in stingless bees (*Melipona seminigra*). J. Comp. Physiol. 192: 879–887.
Kanmiya, K. 2006. Communication by vibratory signals in Diptera. pp. 381–396. *In*: Drosopoulos, S. and M.F. Claridge (eds.). Insect Sounds and Communication: Physiology, Behaviour, Ecology and Evolution. CRC Taylor & Francis Group. Boca Raton, London, New York.
Kavčič, A., A. Čokl, R.A. Laumann, M.B. Blassioli-Moraes and M. Borges. 2013. Tremulatory and abdomen vibration signals enable communication through air in the stink bug *Eushistus heros*. PloS One 8: 1–10.
Kenneally, C. 2005. Do they know something we don't? Animals' senses may have helped them to survive the tsunami. The Boston Globe, January 11, C1, C4.
Keuper, A., C.W. Otto and A. Schatral. 1985. Airborne sound and vibration signals of bushcrickets and locusts. Their importance for the behaviour in the biotope. pp. 135–142. *In*: Kalmring, K. and N. Elsner (eds.). Acoustical and Vibrational Communication in Insects. Paul Parey, Berlin, Germany.
Krebs, J.R. and R. Dawkins. 1984. Animal signals: mind reading and manipulation. pp. 380–402. *In*: Krebs, J.R. and R. Dawkins (eds.). Behavioural Ecology: An Evolutionary Approach. Blackwell, Oxford, UK.
Kühne, R. 1982. Neurophysiology of the vibration sense in locusts and bushcrickets: Response characteristics of single receptor units. J. Ins. Physiol. 28: 155–163.
Kung, C. 2005. A possible unifying principle for mechanosensation. Nature 436: 647–654.
Lakes-Harlan, R. and J. Strauss. 2014. Functional morphology and evolutionary diversity of vibration receptors in insects. pp. 277–302. *In*: Cocroft, R., M. Gogala, P.S.M. Hill and A. Wessel (eds.). Studying Vibrational Communication. Springer, Heidelberg, New York, Dordrecht, London.
Lanigan, P.J. and E.M. Barrows. 1977. Sexual behaviour of *Murgantia histrionica* (Hemiptera: Pentatomidae). Psyche. June: 19–197.
Laumann, R.L., M.C. Blassioli-Moraes, A. Čokl and M. Borges. 2007. Eavesdropping on sexual vibratory signals of stink bugs (Hemiptera: Pentatomidae) by the egg parasitoid *Telenomus podisi*. Anim. Behav. 73: 637–649.
Laumann, R.A., A. Čokl, A.P.S. Lopes, J.B.C. Fereira, M.C. Blassioli-Moraes and M. Borges. 2011a. Silent singers are not safe: selective response of a parasitoid to substrate-borne vibratory signals of stink bugs. Anim. Behav. 82: 1175–1183.
Laumann, R.A., M.C.B. Moraes, A. Khrimian and M. Borges. 2011b. Field capture of *Thyanta perditor* with pheromone-baited traps. Pesq. Agrop. Bras. 46: 113–119.
Michel, K., T. Amon and A. Čokl. 1982. The morphology of the leg scolopidial organs in *Nezara viridula* (L.) (Heteroptera, Pentatomidae). Rev. Can. Biol. Exp. 42: 139–150.
Michelsen, A., F. Fink, M. Gogala and D. Traue. 1982. Plants as transmission channels for insect vibrational signals. Behav. Ecol. Sociobiol. 11: 269–281.
Michelsen, A. 2014. Physical aspects of vibrational communication. pp. 199–213. *In*: Cocroft, R., M. Gogala, P.S.M. Hill and A. Wessel (eds.). Studying Vibrational Communication. Springer, Heidelberg, New York, Dordrecht, London.
Michereff, M.F.F., R.A. Laumann, M. Borges, M. Michereff Filho, I.R. Diniz, A. Faria Neto et al. 2011. Volatiles mediating plant-herbivory-natural enemy interaction in resistant and susceptible soybean cultivars. J. Chem. Ecol. 37: 273–285.
Millar, J.G. 2005. Pheromones of true bugs. Topics Curr. Chem. 240: 37–84.
Morgan, D.E. 2010. Biosynthesis in Insects. RSC Publishing, Cambridge, UK.

Morris, G.K. 1980. Calling display and mating behaviour of *Copiphora rhinoceros* Pictet (Orthoptera: Tettigoniidae). Anim. Behav. 28: 42–51.
Murphey, R. 1971. Motor control of orientation to prey by the water strider *Gerris remigis*. Z. vergl. Physiol. 72: 150–167.
Pratte, M. and R.L. Jeanne. 1984. Antennal drumming behaviour in *Polistes* wasps (Hymenoptera: Vespidae). Zeit. Tierpsychol. 66: 177–188.
Rössler, W., M. Jatho and K. Kalmring. 2006. The auditory-vibratory sensory system in bushcrickets. pp. 35–69. *In*: Drosopoulos, S. and M.F. Claridge (eds.). Insect Sounds and Communication: Physiology, Behaviour, Ecology and Evolution. CRC Press, Taylor and Francis Group. Boca Raton, London, New York.
Ruprecht, R. 1975. Kommunikation von *Sialis* (Megaloptera) durch Vibrationsignale. J. Insect Physiol. 21: 340–341.
Sebeok, T.A. 1962. Coding in the evolution of signalling behaviour. Behav. Sci. 7: 430–442.
Seeley, T. 1989. The honey bee colony as a super-organism. Am. Scientist 77: 546–553.
Shaw, S.R. 1994. Re-evaluation of the absolute threshold and response mode of the most sensitive known "vibration" detector, the cockroach's subgenual organ: a cochlea-like displacement threshold and a direct response to sound. J. Neurobiol. 21: 311–322.
Silva, C.C., D.M. Cordeiro, R.A. Laumann, M.C. Blassioli-Moraes, J.A. Barrigossi and M. Borges. 2004. Ciclo de vida e metodologia de criação de *Tibraca limbativentris* para estudos de Ecologia Química. Boletim de Pesquisa e Desenvolvimento (Embrapa) 78: 1–19.
Silva, C.C., M.C.B. Moraes, R.A. Laumann and M. Borges. 2006. Sensory response of the egg parasitoid *Telenomus podisi* to stimuli from the bug *Euschistus heros*. Pesq. Agrop. Bras. 41: 1093–1098.
Sismondo, E. 1980. Physical characteristics of drumming of *Meconema thalassinum*. J. Insect Physiol. 26: 209–212.
Smith, W.J. 1965. Message, meaning and context in ethology. Amer. Naturalist 99: 405–409.
Sueur, J. 1992. Cicada acoustic communication: potential sound partitioning in a multispecies community from Mexico (Hemiptera: Cicadomorpha: Cicadidae). Biol. J. Linnean Soc. 75: 379–394.
Vilhelmsen, L., G.F. Turrisib and R.G. Beutelc. 2008. Distal leg morphology, subgenual organs and host detection in Stephanidae (Insecta, Hymenoptera). J. Nat. Hist. 42: 1649–1663.
Wessel, A., R. Mühlethaler, V. Hartung, V. Kuštor and M. Gogala. 2014. The tymbal: Evolution of a complex vibration-producing organ in the Tymbalia (Hemiptera excl. Sternorrhyncha). pp. 395–444. *In*: Cocroft, R.B., M. Gogala, P.S.M. Hill and A. Wessel (eds.). Studying Vibrational Communication. Springer, Heidelberg New York Dordrecht London.
Wiese, K. 1972. Die mechanorezeptive Beuteortingssystem von *Notonecta*. I. Die Funktion des tarsalen Scolopidialsorgan. J. Comp. Physiol. 78: 83–102.
Witzgall, P., P. Kirsch and A. Cork. 2010. Sex pheromones and their impact on pest management. J. Ch. Ecol. 36: 80–100.

CHAPTER 5

The Semiochemistry of Pentatomidae

*Miguel Borges** and *Maria Carolina Blassioli-Moraes*

Introduction

Stink bugs, similar to many other insects, use multimodal communication, comprising chemical, visual, and vibrational signals. Chemical communication is probably the most important signal used by insects at long distances and at medium to short distancesto locate hosts and find mates and food. Chemical communication in stink bugs has been widely studied, and there are some good reviews on this topic (Millar 2005, McBrien and Millar 1999, and more recently, Weber et al. 2016, in press). The stink bugs' semiochemistry comprises different signals, like sex, aggregation, and alarm pheromones, allomones, kairomones, and synomones. These signals are produced by different structures in the insects, like dimorphic cells, cuticular glands, and metathoracic glands (MTGs) in adults and dorsal abdominal glands (DAGs) in nymphs (Aldrich et al. 1978, Aldrich 1988, Borges and Aldrich 1992, Aldrich et al. 1994, Borges and Aldrich 1994).

The stink bugs (Pentatomidae) constitute the third-largest Heteropteran family and are distributed into eight suborders: Asopinae, Cytocorinae, Discocephalinae, Edessinae, Pentatominae, Phyllocephalinae, Podopinae, and Serbaninae (Panizzi et al. 2000). Except for the Asopinae family, which shall be addressed in Chapter 9, all of the Pentatomidae suborders will be covered in this chapter. The present review will summarise the new advances towards understanding the chemical communication within and between stink bug species and the chemical communication between stink bugs and their natural enemies, including their interactions with plants.

Embrapa Recursos Genéticos e Biotecnologia, Parque Estação Biológica - W5 Norte (final), CEP: 70770-017, Brasília, Brazil.
Email: carolina.blassioli@embrapa.br
* Corresponding author: miguel.borges@embrapa.br

The comprehension of semiochemicals from stink bugs is in its infancy. For some species, such as *Euschistus heros* Fabricius (Aldrich et al. 1994, Borges and Aldrich 1994, Borges et al. 1998, 2011, Silva et al. 2014) and *Halyomoprha halys* (Stål) (Khrimian et al. 2014a, Weber et al. 2014a, Leskey et al. 2012a,b, 2015a, b) a complete study was conducted, with detailed bioassays, chemical elucidation, field trials, and commercial development. For the majority of the stink bug species, however, the studies stop or have been stopped at the initial hint that either a sex or aggregation pheromone exists. There are a few commercial stink bug pheromones available in the market: the compound methyl (2*E*,4*Z*)-2,4-decadienoate, a pheromone of the different species of Nearctic *Euschistus* spp., the compound methyl (2*E*,4*E*,6*Z*)-2,4,6 decatrienoate, the pheromone of *Plautia stali* Scott, the compound methyl-2,6,10-trimethyltridecanoate, the pheromone produced by the Neotropical *E. heros* and the compound 10,11-epoxy-1-bisabolen-3-ol, the pheromone of *Halyomoprha halys*. The first two compounds can be obtained as pure compounds from Bedoukian. All the pheromones mentioned above have been commercialized by different companies and can be easily found on the Internet.

Stink Bug Semiochemicals

Pentatomids produce a wide variety of chemical compounds that can potentially be used to manage these insects (Aldrich 1988, Borges et al. 1998, 2011). Among these compounds are pheromones, which can be classified as either sexual, alarm, or aggregation pheromones (Aldrich 1988) (Tables 5.1 and 5.2). Until now, all the sex and aggregation pheromones reported for stink bugs have been produced by males, and most of them act as sex pheromones; only for *Piezodorus hybneri* (Westwood) does the pheromone appear to be bifunctional, acting either as a sex or aggregation pheromone, depending on the time of the year (Leal et al. 1998). Semiochemicals can be applied for different purposes: population monitoring, mass trapping, sexual confusion, to attract and kill, and for manipulating the behaviour of natural enemies. The stink bug's natural enemies use both sex pheromones and defensive compounds during foraging (Borges et al. 1998, Blassioli-Moraes et al. 2013) in order to locate prey/hosts. Defensive compounds are used for defence against predators, since many are repellent, irritating, or toxic (Eisner et al. 1991).

Pheromones are chemical compounds that mediate interactions between the individuals of the same species, and allelochemicals are chemical compounds that mediate interactions between the individuals of different species. The allelochemicals are divided into three classes: allomones that provide a positive interaction to the emitter (e.g., a defensive compound against a predator); kairomones that provide a positive

Table 5.1. Chemical structure of the sex and sex-aggregation pheromones of stink bugs.

Compounds	Structure	Species
Methyl 2,6,10 trimethyltridecanoate	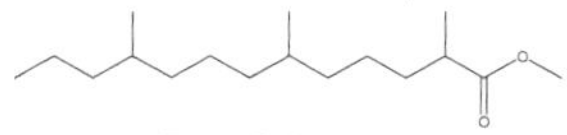	*E. heros, E. obscurus, A. griseus*
Methyl (2*E*,4*Z*)-2,4-decadienoate	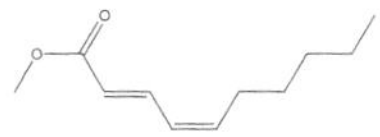	*Euschistus* sp.
Methyl 2,6,10 trimethyldodecanoate	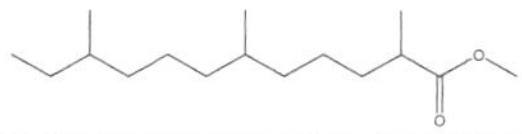	*E. heros*
Methyl (2*E*,4*Z*,6*Z*)-2,4,6-decatrienoate	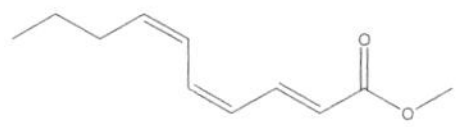	*Thyanta* sp.
Methyl (8*Z*)-8-hexadecenoate		*P. hybneri*
(6*R*,10*S*)-6,10,13-Trimethyltridecan-2-one	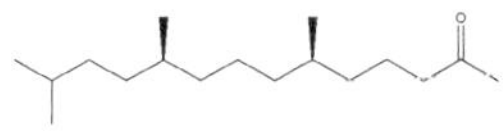	*P. macunaima*
2,4,8-Trimethylpentadecan-1-ol	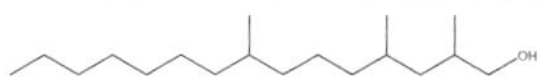	*P. stictica*
Methyl (2*E*,4*E*,6*Z*)-2,4,6-decatrienoate	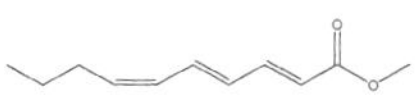	*P. stali*
Methyl 4,8,12-trimethylpentadecanoate	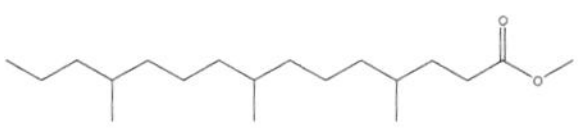	*E. meditabunda*
Methyl 4,8,12-trimethyltetradecanoate	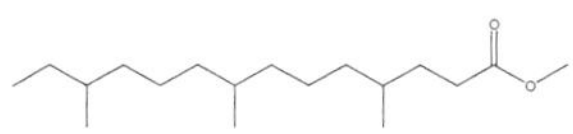	*E. meditabunda*
(2*E*)-2-Octenyl acetate	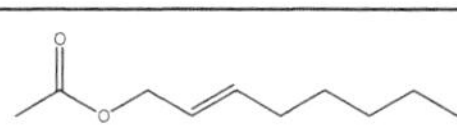	*B. hilaris*
(*R*)-15-Hexadecanolide	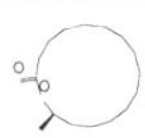	*P. hybneri*
(7*R*)-(+)-β-Sesquiphellandrene	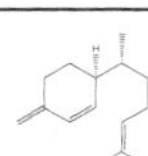	*P. guildinii* *P. hybneri**
(7*S*)-(–)-Zingiberene	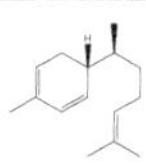	*T. custator acerra*
(7*S*)-(–)-β-Sesquiphellandrene	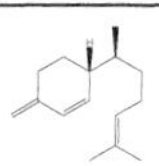	*T. custator acerra*
(7*S*)-(–)-α-Curcumene	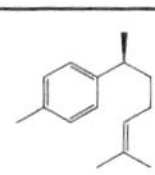	*T. custator acerra*

Table 5.1 contd. ...

...Table 5.1 contd.

Compounds	Structure	Species
cis-(Z)-Bisabolene epoxide *trans-(Z)-Bisabolene epoxide*	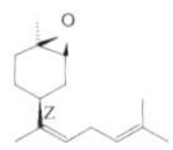 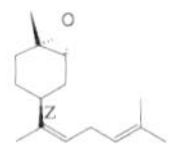	*Chinavia* sp. *Nezara* sp.
Methyl geranate	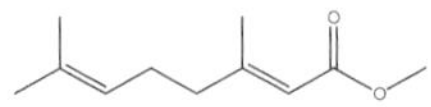	*C. sayi*
Methyl citronellate	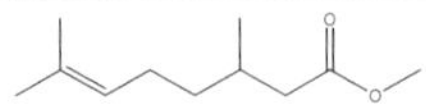	*C. sayi*
Methyl (*E*)-6,2,3-dihydrofarnesoate	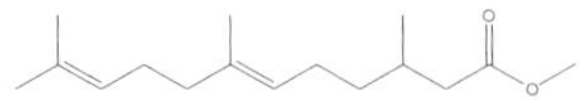	*C. sayi, C. uhleri, C. ligata*
Methyl (2*E*,6*E*)-farnesoate	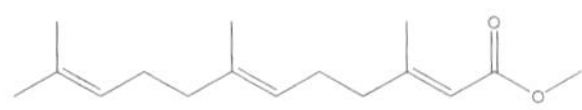	*C. uhleri, C. ligata*
Methyl (*E*)-5-2,6,10-trimethyl-5,9-undecadienoate	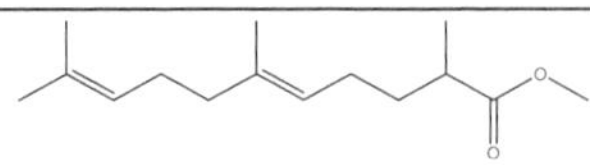	*C. uhleri, C. ligata*
10,11-Epoxy-1-bisabolen-3-ol	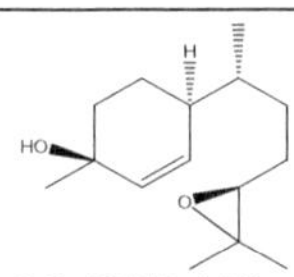 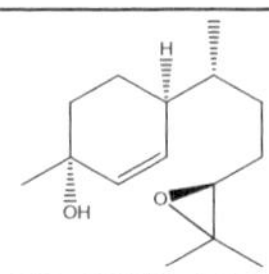 3*S*,6*S*,7*R*,10*S* 3*R*,6*S*,7*R*,10*S*	*H. halys*
10,11-Epoxy-1-bisabolen-3-ol	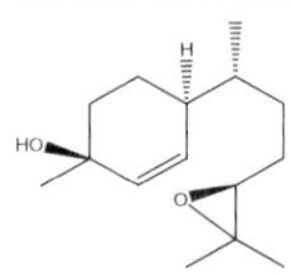 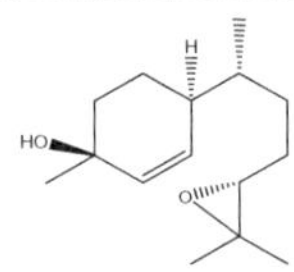 3*S*,6*S*,7*R*,10*S* 3*S*,6*S*,7*R*,10*R*	*M. histriônica*
1,10-Bisaboladien-3-ol	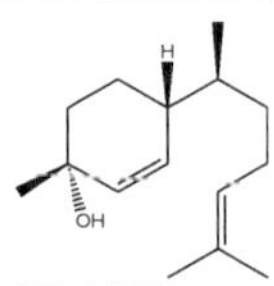 3*R*,6*S*,7*R*	*O. poecilus*
1,10-Bisaboladien-3-ol	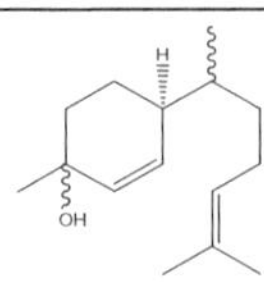 3*RS*,6RS,7*S*	*T. limbativentris, G. spinosa**

**P. hybineri* and *G. spinosa* did not have the absolute configuration determined.

Table 5.2. Chemical structure of the main defensive compounds identified from the MTG and abdominal glands of the Neartic and Neotropical stink bugs.

Compound	Structure
Hydrocarbons	
Undecane	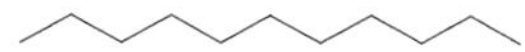
Tridecane	
1-Tridecene	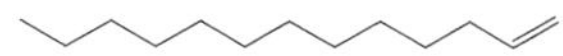
4-Tridecene	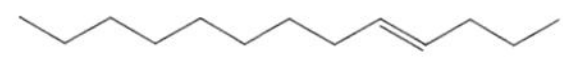
Aldehydes	
Hexanal	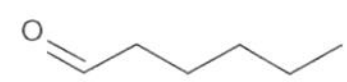
(*E*)-2-Hexenal	
(*Z*)-2-Octenal	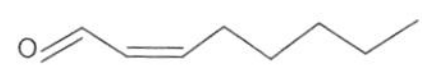
(*E*)-2-Octenal	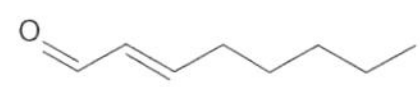
Nonanal	
(*E*)-2-Nonenal	
Decanal	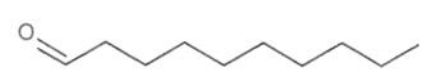
(*Z*)-2-Decenal	
(*E*)-2-Decenal	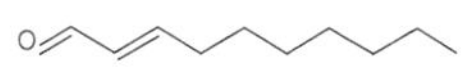
(2*E*,4*Z*)-2,4-Decadienal	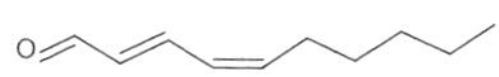
(2*E*,4*E*)-2,4-Decadienal	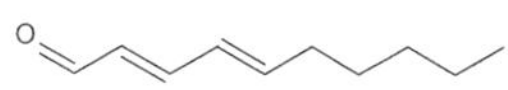
Tridecanal	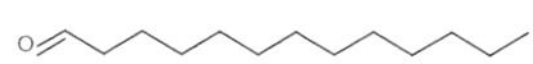
Tetradecanal	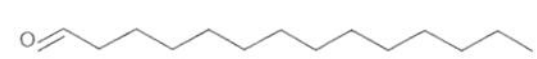
Esters	
Hexyl acetate	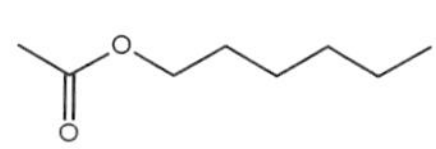
(E)-2-Hexenyl acetate	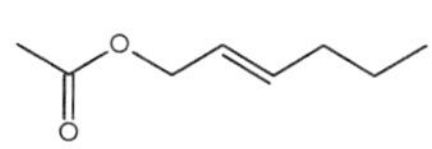
(*E*)-2-Hexenyl butyrate	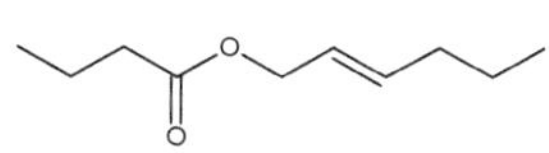
(*E*)-2-Octenyl-acetate	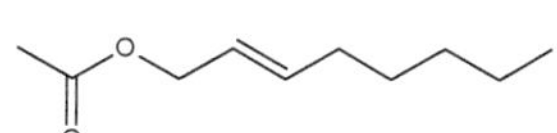

Table 5.2 contd. ...

...Table 5.2 contd.

Compound	Structure
(*E*)-2-Decenyl acetate	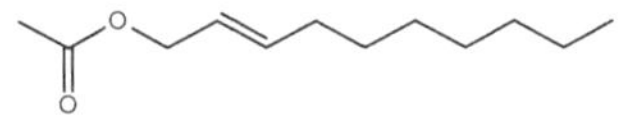
Alcohols	
Hexan-1-ol	
(*E*)-2-Octen-1-ol	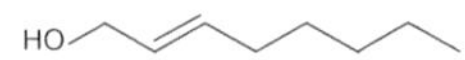
Tridecan-1-ol	
Monterpenoids	
Linalool	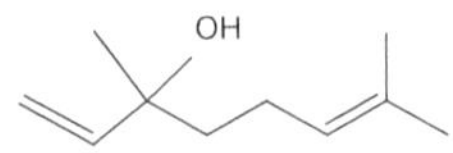
4-Oxo-(E)-2-alkenais	
4-Oxo-(*E*)-2-hexenal	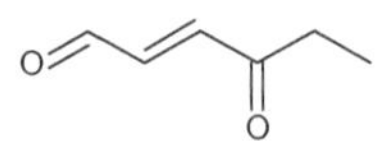
4-Oxo-(*E*)-2-octenal	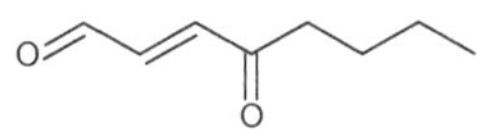
4-Oxo-(*E*)-2-decenal	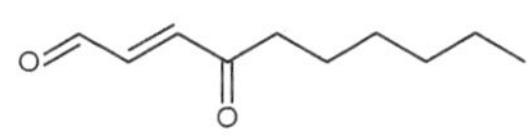
Ketones	
4-Hidroxy-4-methyl-pentanone	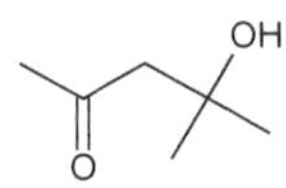
6-Methyl-5-hepten-2-one	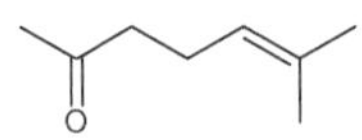

interaction to the receiver (e.g., an odour that is used by a predator to locate its prey); and synomones that provide positive interactions to both the emitter and the receiver (e.g., herbivore-induced plant volatiles that attract an insect that kills the herbivore). The role that a chemical plays in the organism's interactions is context-dependent (Dicke and Sabelis 1988), i.e., the same compound can be involved in one interaction as an allomone and as a kairomone in another. For example, the sex pheromone of *E. heros*, the compound methyl-2,6,10 trimethyltridecanoate, acts as a sex pheromone within the same species, but this compound is used as a kairomone by its main natural enemy, the egg parasitoid *Telenomus podisi* Ashmead (Hymenoptera: Platygastridae) (Borges et al. 1998). In this review, the following terms have been used: "defensive compound" refers to the semiochemicals produced by pentatomids that serve as allomones, or those whose biological importance is still unknown; while "alarm pheromone" refers to the compounds that induce defensive behaviour (mainly dispersal and increased activity) in individuals of the same species. Other terms are; "sex pheromone and sex–aggregation pheromone" which refer to the compounds that are produced by one gender in order to attract the other or by both genders for mating; and "aggregation pheromone" refers to the compounds that are produced by one gender and attract both genders for protection or other behavior, but not for mating proposing. Most of the molecules of the sex and sex-aggregation pheromones produced by stink bugs are species-specific, highly volatile, and chemically unstable. The instability and high volatility of these pheromones suggest that chemical communication occurs at specific distances and that the compounds do not remain in the environment for any longer than is necessary to deliver a message for conspecific communication. This could be a strategy to avoid chemical espionage by natural enemies or host competitors; at first glance, semiochemical instability appears to be disadvantageous, but it is an advantage as it ensures that these chemicals will not last long in the environment.

In order to evaluate whether a chemical compound produced by an insect is a pheromone, one must conduct a series of detailed bioassays; in general, the first step is to identify the emitter of the pheromone. Borges et al. (1987) developed two different olfactometers, which allowed the identification of the probable sex pheromone emitter of *Nezara viridula* L. and all the steps of its mating behaviour. Observing *N. viridula* males and females, Borges and co-workers (1987) described their complete mating behaviour including all the steps that the bugs use in order to recognize mating partners at long and short distances (see Chapter 4). At long distances, *N. viridula* females are attracted to males releasing the sex pheromone. The main methodology used to collect stink bug sex pheromone is air-entrainment volatile collection. Stink bugs do not have sex pheromone-producing glands or special structures to store the sex pheromone. Pavis and Malosse (1986) observed that sexually mature southern green stink bugs, *Nezara* sp. and *Chinavia* sp. produce pheromone from unicellular glands in the ventral abdominal epithelium. Later, Borges (1995) confirmed that males from a Brazilian population of *N. viridula* have pheromone components in their abdominal sternites that attract conspecific females. Male stink bugs, in general, produce very tiny amounts of sex pheromone, nanograms or less, and these compounds are continuously produced and released. Therefore, in order to obtain an experimental amount, it is necessary to conduct the air-entrainment volatile collection of males during optimal hours to accumulate the pheromone for chemical analysis and to conduct bioassays with the crude extracts. On the other hand, the defensive compounds present in the metathoracic scent glands (MTG) in adults and in the dorsal abdominal glands (DAG) in nymphs are produced in higher amounts and are easily extracted from the MTG and DAG using liquid extraction or through air-entrainment.

Pheromones of stink bugs

Agroecus griseus *Dallas [Pentatomidae: Pentatominae: Carpocorini]*

Agroecus griseus is a secondary maize pest in Brazil, but is considered a serious pest when it attacks crops, as it causes irreversible damage to the plants (Gassen 1996). Males of this species produce as sex pheromone, the compound methyl-2,6,10-trimethyltridecanoate (Fávaro et al. 2012) (Table 5.1). The age of the males and the period of pheromone production, have not been investigated and remain to be determined. In order to obtain the sex pheromone, male and female air-entrainment volatile collection was conducted using five individuals of each gender. Methyl-2,6,10-trimethyltridecanoate has eight possible stereoisomers, and the isomer produced by *A. griseus* remains unknown. However, Y-tube olfactometer bioassays conducted in the laboratory, using males and females and a racemic synthetic solution of methyl-2,6,10-trimethyltridecanoate as the odour source, showed that the females were attracted to this racemic mixture, containing the 8 possible isomers (Fávaro et al. 2012). The same compound is also produced as a sex pheromone by two other Pentatomidae stink bugs: *E. heros* and *E. obscurus* (Palisot).

Bagrada hilaris *Burmeister [Pentatomidae: Pentatominae: Strachiini]*

Bagrada hilaris originates from Africa; it has invaded and established itself in the desert region of the southwest U.S. in the last decade. The *Bagrada* sp. is an economically important species that affects cruciferous vegetable crops. Guarino et al. (2008) reported that the females are attracted to the volatiles from air-entrainment male extracts and to the odours emitted from live males; while the males are not attracted to the odours from live females or to female volatiles, from air-entrainment extracts, in Y-tube olfactometer bioassays. The chemical analysis of the male and female air-entrainment extracts did not show gender-specific compounds; quantitative analysis, however, showed that (*E*)-2-octenyl acetate (Tables 5.1 and 5.2) was produced in higher amounts by males, and the authors suggested that this acetate might be the pheromone of this species. Other stink bug males also release this acetate in higher amounts; *Thyanta pallidovirens* (Stål) (McBrien et al. 2002) and *Dichelops melacanthus* (Dallas) as compared to the females of these species.

Chinavia hilaris (Say), *Chinavia ubica* (Rolston), and *Chinavia impicticornis* (Stål) [Pentatomidae: Pentatominae: Nezarini]

The species within the genera *Chinavia* and *Nezara* produce the same compounds as sex pheromones: *trans*-(Z)-bisabolene epoxide (*trans*-Z-EBA) ((Z)-(1′*S*,3′*R*,4′*S*)-(−)-2-(3′,4′-epoxy-4′-methylcyclohexyl)-6-methylhepta-2,5-diene) and the corresponding *cis*-isomer (*cis*-Z-EBA) (Table 5.1) (Baker et al. 1987, Aldrich et al. 1987, 1993, McBrien et al. 2001, Blassioli-Moraes et al. 2012). The species-specificity is guaranteed by the different ratios between the two isomers, *cis*-Z-BAE and *trans*-Z-BAE. It is interesting to note, that in eight of the ten different populations of the *Nezara* spp. studied, the *trans*-Z-isomer is present in higher amounts than the *cis* isomer, and the two other *N. viridula* populations have both isomers in equal quantity. On the other hand, five of the six *Chinavia* species studied have the *cis*-isomer in a higher quantity than the *trans* isomer (Moraes et al. 2008a). There is one exception: the species *C. impicticornis* appears to produce only *trans*-Z-EBA (Blassioli-Moraes et al. 2012). Another interesting point about the chemical communication of these species is that the absolute configuration of the different stereoisomers of *cis* and *trans*-Z-EBA appears to be less important for their mate recognition, as shown by *C. ubica* and *C. impicticornis* (Blassioli-Moraes et al. 2012). In Y-tube olfactometer bioassays using synthetic solutions with different combinations of the stereoisomers, these two species recognized conspecifics using the ratio between the two components and their relative configuration; *cis* and *trans* configurations appear to be the key information for mate recognition. *Chinavia hilaris*, *C. impicticornis*, *C. ubica*, and *N. viridula* produce the two isomers with the same absolute configuration (McBrien et al. 2001, Baker et al. 1987, Blassioli-Moraes et al. 2012). *Chinavia ubica* and *C. impicitcornis* become sexually mature about 7–8 days after the final moult, and release their sex pheromones in a diurnal cycle. *Chinavia ubica* produces 0.23 ± 0.24 μg/24 h of *trans*-(Z)-EBA, and 3.03 ± 4.14 μg/24 h of *cis*-Z-(EBA), whereas the males of *C. impicticornis* produce significantly less amounts of pheromone per day (0.12 ± 0.19 μg/24 h *trans*-(Z)-EBA). Differences in the production of the pheromone do not imply better biological performance; Silva et al. (2015) reported that *C. impicticornis* adults have significantly greater longevity and fecundity than *C. ubica* adults.

Chlorochroa sayi (Stål), *Chlorochroa uhleri* (Stål), and *Chlorochroa ligata* (Say) [Pentatomidae: Pentatominae: Nezarini]

The genus *Chlorochroa* has 25 described species (Buxton 1983), but only three species have had their mating behaviour and semiochemistry explained in more detail: *C. sayi*, *C. ligata*, and *C. uhleri* (Ho and Millar 2001a,b). In *Chlorochroa*, similar to other Pentatomidae, the male is the sex pheromone producer. The three species start to release their sex pheromones a few days after the imaginal moult, which is about 9–12 days in *C. sayi* males and 12–14 days in *C. ligata* and *C. uhleri* (Ho and Millar 2001a,b). *Chlorochroa sayi* produces three compounds as sex pheromones: methyl generate, methyl citronellate, and methyl-(*E*)-6,2,3-dihyrofarnesoate, in a ratio of 100:0.45:1.6 (Table 5.1). *Chlorochroa uhleri* males, in contrast, produce methyl (*R*)-3-(*E*)-6,2,3-dihydrofarnesoate, methyl-(2*E*,6*E*)-farnesoate, and methyl (*E*)-5-2,6-10-trimethyl-5,9-undecadienoate in a ratio of 100:0.9:0.6. *Chlorochroa ligata* produces the same components as *C. uhleri* but in a ratio of 100:0.5:0.4. For these three species, the maximum production of male-specific compounds occurred between 19:00 and 21:00 hr. During peak production, *C. ligata* males produced about 35 ng/bug/hr of the major male-specific compound, *C. uhleri* males produced 8.8 ng/bug/hr, and *C. sayi* males produced 51.6 ng/bug/hr (Ho and Millar 2001a,b).

Edessa meditabunda *(F.) [Pentatomidae: Edessinae]*

Edessa meditabunda is a secondary pest of soybean crops and other grains in Brazil. Two male-specific compounds, methyl 4,8,12-trimethypentadecanoate and methyl 4,8,12-trimethyltetradecanate in a ratio of 98:2 (Table 5.1), were identified as the components of its sex pheromone (Zarbin et al. 2012). The activity of these two components as a sex pheromone was determined through GC-EAD; the female antenna responded only to the major component in the Y-tube olfactometer bioassays. The bioassays with male and female odours showed that the females responded to male odour, but the males and females did not respond to female odour and the males did not respond to male odour. *Edessa meditabunda* produces pheromones continuously during photophase and scotophase, with a higher production during photophase (71%); the authors did not present a quantification of the production of the sex pheromone by males. Bioassays conducted with a synthetic solution containing the major compound and the two components were highly attractive to females (Zarbin et al. 2012). The absolute configurations of the two components remain to be established and the blend needs to be evaluated under field conditions.

Euschistus sp. *[Pentatomidae: Pentatominae: Carpocorini]*

Seven different species of *Euschistus* have had their male-specific pheromone components identified. Methyl (2*E*,4*Z*)-2,4-decadienoate was determined as the major male-specific component in five Nearctic species: *E. conspersus, E. tristigmus, E. politus, E. servus*, and *E. ictericus* (Aldrich et al. 1991) (Table 5.1). On the other hand, for males of the Nearctic stink bug *E. obscurus,* a blend of three components was found to be the sex pheromone: methyl

(2*E*,4*Z*)-2,4-decadienoate, methyl 2,6,10-trimethyltridecanoate and methyl 2,6,10-trimethyldodecanoate (Table 5.1). Field traps using methyl (2*E*,4*Z*)-2,4-decadienoate caught males and females of *E. tristigmus*, *E. politus*, and *E. servus*, indicating that this ester is a sex-aggregation pheromone (Aldrich et al. 1991). Therefore, no *E. obscurus* individuals were trapped in these field experiments. In laboratory bioassays, *E. obscurus* females responded to the male extract containing the three male-specific pheromone components, but the males did not. As opposed to the other Nearctic species, the results of these bioassays indicate that the male-specific compounds of *E. obscurus* work as a sex pheromone for this species. The Neotropical species *E. heros* has a sex pheromone composition that is very similar to that of *E. obscurus*; however, the ratio between the components is different. For *E. heros*, the compounds methyl 2,6,10-trimethyltridecanoate and methyl (2*E*,4*Z*)-2,4-decadienoate are the major components and methyl 2,6,10-trimethyldodecanoate is a minor component (Aldrich et al. 1994, Borges and Aldrich 1994). Zhang et al. (2003) confirmed the ratio of the three components that was proposed by Borges et al. (1998) and Borges and Aldrich (1994), specifically: 53% methyl (2*E*,4*Z*)-2,4-decadienoate, 3% 2,4,6-methyl trimethyldodecanoate, and 44% methyl 2,6,10-trimethyltridecanoate. The biological activity of the three components was confirmed in a laboratory bioassay; methyl 2,6,10-trimethyltridecanoate was the main component that attracted females (Borges et al. 1998) and this compound did not attract males; therefore, as in the case of *E. obscurus*, this compound works as a sex pheromone (Borges et al. 1998, 2011, Silva et al. 2014). Costa et al. (2000) carried out a set of experiments that showed that the racemic mixture of methyl 2,6,10-trimethyltridecanoate was efficiently attractive to females in laboratory bioassays. Methyl-2,6,10-trimethyltridecanoate has eight possible stereoisomers; the absolute configuration of methyl 2,6,10 trimethyltridecanoate was determined by olfactometer bioassays using the eight stereoisomers separately, which were synthesized by Mori and Murata (1994). Costa et al. (2000) reported that females responded better to isomer 2*S*,6*R*,10*S* than the other isomers, when compared to the use of solvent as control. In addition, they showed that one of the isomers, 2*R*,6*S*,10*S*, did not attract the insects, and that most of the insects responded to the solvent when this isomer was used as a stimulus, suggesting a possible repellent action. The presence of this component in a racemic mixture, however, did not have an antagonistic effect in the field traps (Borges et al. 1998, 2011, Silva et al. 2014). *Euschistus heros* females are attracted to methyl 2,6,10-trimethyldodecanoate and to methyl (2*E*,4*Z*)-2,4-decadienoate, but the attractiveness of these compounds did not increase when combined with methyl 2,6,10-trimethytridecanoate (Moraes et al. 2008c). In field experiments, using 1 mg of a racemic mixture of methyl 2,6,10-trimethyltridecanoate impregnated in a rubber septum, caught a higher number of *E. heros* individuals, along with other Pentatomidae like, *Piezodorus guildinii* and *E. meditabunda*, as compared to a trap containing only the septum with a solvent (Borges et al. 1998, 2011). Two recent experiments in the field that also used a racemic mixture showed that the sex pheromones of *E. heros,* have great potential for the monitoring of this species on soybean crops (Borges et al. 2011, Silva et al. 2014). This is the first pheromone of a Neotropical stink bug that has been commercialized. Laboratory and field tests showed that this pheromone is a sex pheromone that attracts females (see Chapter 10).

The capture of *E. heros* adult females with pheromone-baited traps was significantly correlated to the mean population density during the initial to medium reproductive stages of the soybean (R1-R5). Additionally, the pheromone-baited trap contributed to a 50% reduction in pesticide application for stink bugs (Borges et al. 2011, Silva et al. 2014).

The compound methyl (2*E*,4*Z*)-2,4-decadienoate is the major component of the *E. heros* pheromone and it is present as a pheromone in all other species of *Euschistus*, but its role for *E. heros* is not clear. Recently, Moraes et al. (2008c) showed that this compound has its production enhanced in biotic and abiotic stress situations, such as humidity, temperature, and food supply stresses. When *E. heros* males were aerated with food (*Phaseolus vulgaris* pods), all three male-specific pheromone components were released in the ratio reported by Zhang et al. (2003), across seven consecutive days. When the insects were kept in aeration without food, after 48 hr they stopped producing the main male-specific pheromone component, methyl 2*S*,6*R*,10*S*-trimethyltridecanoate, and started to produce, in increasing amounts, methyl (2*E*,4*Z*)-2,4-decadienoate; this phenomenon is still not fully understood.

Halyomorpha halys *(Stål) [Pentatomidae: Pentatominae: Cappaeini]*

The Brown Marmorated Stink Bug (BMSB) *Halyomorpha halys* is native to North and South Korea, Japan, and China. It is considered to be an invasive species in the U.S. and Europe; the first record in America was reported in Allentown, Pennsylvania in 1996, and in Europe, near Zürich, Switzerland in 2008 (Leskey et al. 2012a, Zhu et al. 2012). Currently, *H. halys* is well-established in the mid-Atlantic region and has spread across most of the continental United States. This insect has caused huge economic losses in fruit trees, vegetable field crops, and ornamentals (Leskey et al. 2012a). The aggregation pheromone of *H. halys* is composed of two sesquiterpenes, *cis*-10,11-epoxy-1-bisabolen-3-ol and its *trans* isomer, in a ratio of 35:1 (Khrimian et al. 2014a) (Table 5.1). The absolute configuration was determined using stereoisomer libraries of 1-bisabolen-3-ol with chiral columns in GC analysis as (3*S*,6*S*,7*R*,10*S*)-10,11-epoxy-1-bisabolen-3-ol and

(3*R*,6*S*,7*R*,10*S*)-10,11-epoxy-1-bisabolen-3-ol. To evaluate the role of these male-specific compounds on male and female attraction, a field experiment was conducted. Lures were impregnated with individual compounds or with a mixture in the natural ratio of 35:1 and evaluated. The lure containing the two components in the natural ratio attracted a higher number of insects than those containing individual compounds; also, the natural ratio lure caught both adults and nymphs. Further field-testing, using a 3:1 racemic mixture of *cis* and *trans*-10,11-epoxy-bisabolen-3-ols, showed that the presence of both components is important, but that the presence of additional stereoisomers did not hinder *H. halys* attraction (Khrimian et al. 2014a, Weber et al. 2014a, Leskey et al. 2012b, 2015a,b). The field results suggest that the *cis* and *trans*-10,11-epoxy-bisabolen-3-ols work as an aggregation pheromone that attracts males, females, and nymphs.

Similar to *E. heros*, *H. halys* was not caught in pheromone-baited traps during late season (Khrimian et al. 2014a, Leskey et al. 2015a). Leskey et al. (2015a) suggested that diapause and individual reproductive status alter *H. halys* responses to the aggregation pheromone; *H. halys* responds, however, to another pheromone, methyl (2*E*,4*E*,6*Z*)-2,4,6-decadienoate, the pheromone of the oriental stink bug *P. stali*, during late season (Leskey et al. 2012a, 2015a). This ester was never identified in the chemical analysis of *H. halys*, and the reason why this insect responds to this pheromone during late season remains to be elucidated. Pheromone cross-attraction is common in Pentatomidae and it has been demonstrated that other species of Pentatomidae, such as *Piezodorus guildinii*, *E. meditabunda*, and *C. hilaris*, are attracted to *E. heros* pheromone-baited traps (Borges et al. 1998, 2011, Leskey et al. 2012a, 2015a, Aldrich et al. 2007, 2009).

Murgantia histrionica *(Hahn) [Pentatomidae: Pentatominae: Strachiini]*

Males of *Murgantia histrionica* produce, as an aggregation pheromone, two isomers of 10,11-epoxy-1-bisabolen-3-ol (Zhan et al. 2008, Khrimian et al. 2014a). Laboratory bioassays conducted with live males and females as odour sources showed that the males and females responded to male odour but not to female odour, indicating that the males produced pheromone and that this pheromone was probably an aggregation pheromone. Like other stink bugs, the highest peak production of the pheromone was during the middle of the day (Zahn et al. 2008, 2012). The absolute configuration of the two isomers was (3*S*,6*S*,7*R*,10*S*)-10,11-epoxy-1-bisaolen-3-ol and (3*S*,6*S*,7*R*,10*R*)-10,11-epoxy-1-bisaolen-3-ol (Table 5.1), produced in a 1.4:1 ratio (Khrimian et al. 2014b). Field experiments conducted with lures containing the synthetic pheromone showed that males, females, and nymphs were attracted to the isomers *SSRS* and *SSRR*, and that the other isomers did not affect negatively insect attraction (Weber et al. 2014b).

Nezara viridula *(L.) and* Nezara antennata *Scott [Pentatomidae: Pentatominae: Nezarini]*

Nezara viridula is a cosmopolitan, polyphagous pest of a wide variety of crops, including soybeans, cotton, potatoes, and others (Panizzi 1997). The *Nezara* group includes more than 100 species with eight genera and they are cosmopolitan, with higher diversity in the Afrotropical and Neotropical regions. For the species within the genus *Nezara,* studied in different geographic areas, the major pheromone components were identified as: *trans*-(*Z*)-bisabolene epoxide (*trans*-*Z*-EBA) ((*Z*)-(1′*S*,3′*R*,4′*S*)(–)-2-(3′,4′-epoxy-4′-methylcyclohexyl)-6-methyhepta-2,5-diene) and the corresponding *cis*-isomer (*cis*-*Z*-EBA) (Table 5.1) (Baker et al. 1987, Aldrich et al. 1987, 1993, Miklas et al. 2000, McBrien et al. 2001). The *Nezara* and *Chinavia* genera have the same components in their pheromone blends, including the same stereochemistry. Pheromone specificity is guaranteed by the different ratios of the two components produced by the different species (Aldrich et al. 1989, 1993, McBrien et al. 2001, Moraes et al. 2008a, Blassioli-Moraes et al. 2012). The laboratory bioassays showed that females are attracted to both the odour of live males and to the synthetic solutions containing the *cis* and *trans*-(*Z*)-bisabolene epoxide (Borges et al. 1987, Aldrich et al. 1987, Brezot et al. 1994, Borges 1995). Whereas field-testing with *Nezara* from a Nearctic population showed that the pheromone works as an aggregation pheromone that attracts males, females, and nymphs. Only two species of the genus *Nezara, N. antennata* and *N. viridula,* have had their sex pheromone identified, but nine *N. viridula* populations were studied from different geographical areas. All of these populations released the *trans* isomer in higher quantities than the *cis* isomer. Laboratory bioassays carried out with the *N. viridula* population from Londrina, Brazil, showed that only *N. viridula* females were attracted to synthetic EBA when both isomers were present in the correct ratio (Borges 1995). Later, Borges (1995) reported that females did not show the same level of response (attraction) to the racemic mixture as to the (1′*S*,3′*R*,4′*S*) enantiomer.

Oebalus poecilus *(Dallas) [Pentatomidae: Pentatominae: Carpocorini]*

Oebalus poecilus, Oebalus ypsilongriseus (Dallas), and *Tibraca limbativentris* (Stål) are the major rice pests in Brazil. The sex pheromone of *O. poecilus* was identified as one of the isomers of 1,10-bisaboladien-3-ol (zingiberenol) (Table 5.1); the absolute configuration was determined as (3*R*,6*R*,7*S*)-1,10-bisaboladien-3-ol (Oliveira et al. 2013). The authors did not assess male pheromone production throughout the day, but reported that females responded to males only during the morning (08:00–12:00). In laboratory bioassays, females responded to a synthetic solution containing the racemic mixture of all eight possible isomers of 1,10-bisaboladien-3-ol and to a mixture of diastereoisomers with 7*S* stereochemistry; females did not respond to diastereosisomers with 7*R* stereochemistry. Males released the sex pheromone component at a very low amount, 1.19 ± 1.12 ng/bug/day. The other rice-pest stink bug *T. limbativentris*, produced 1,10-bisaboladien-3-ol as a sex pheromone. These rice-pest stink bugs species share the same sex pheromone component but produce different isomers of 1,10-bisaboladien-3-ol. *Tibraca limbativentris* presents two isomers of 1,10-bisaboladien-3-ol.

Pallantia macunaima *Grazia [Pentatomidae: Pentatominae: Pentatomini]*

Pallantia macunaima is a stink bug with sporadic presence in soybean fields; there are no reported host plants for this species (Grazia and Frey-da-Silva 2001). The sex pheromone is produced by males and was identified as (6*R*,10*S*)-6,10,13-trimethyltetradecan-2-one (Table 5.1) (Fávaro et al. 2013). *Pallantia macunaima* females responded to male air-entrainment extracts and to a synthetic solution of 6,10,13-trimethyltetradecan-2-one (racemic mixture), whereas the males did not respond to male air-entrainment extracts nor to a synthetic solution with the racemic pheromone component, suggesting that this pheromone is a sexual pheromone. The authors did not conduct bioassays with the natural isomer produced by the males, and the results of field trials of 6,10,13-trimethyltetradecan-2-one, as a pure isomer or as a mixture of isomers, have not yet been reported.

Pellaea stictica *(Dallas) [Pentatomidae: Pentatominae: Pentatomini]*

Pellaea stictica is not currently considered to be an economically important stink bug in crops around the world, but has been recorded in South America, from Argentina to Guyana, and in North America (Henry 1984). This stink bug can be collected from *Ligustrum lucidum* Ait. (Oleaceae) from April to July, which corresponds to the autumn and winter seasons in Brazil (Panizzi and Grazia 2001). The sex pheromone of this species was identified as 2,4,8-trimethylpentadecan-1-ol (Table 5.1), and a synthetic solution containing all the isomers of the compound was attractive to females in Y-tube olfactometer bioassays; the male response to a synthetic compound was not evaluated (Fávaro et al. 2015). The results suggest that this compound works as a sex pheromone for this species. The absolute configuration of the natural compound was not determined, and field tests have not been conducted.

Piezodorus guildinii *(Westwood) [Pentatomidae: Pentatominae: Piezodorini]*

The small, green stink bug, *Piezodorus guildinii*, is a Neotropical pentatomid found in Central and South America, and is considered to be one of the most important soybean pests in Brazil (Schaefer and Panizzi et al. 2000), and it is already established in the U.S. *P. guildinii* was first reported in the U.S. in 1964 by Genung et al. (1964) and more recently, it has been described as an important soybean pest in Louisiana and other Southern states (Panizzi 2015, and reference therein). Borges et al. (1999a) identified two components of the *E. heros* sex pheromone in the sex pheromone blend of *P. guildinii,* methyl-2,6,10-trimethyltridecanoate and methyl 2,6,10- trimethyldodecanoate (Table 5.1). The amount of these two compounds present in the air-entrainment extracts was very low (Borges M., unpublished data). The function of these compounds and whether or not they are pheromones remain to be determined; however, *P. guildinii* females were caught in traps in the field tests using lures containing the sex pheromone of *E. heros*, the compound methyl 2,6,10-trimethyltridecanoate, suggesting that this compound has a role in the cross-attraction of stink bugs (Borges et al. 1998, 2011). In a more recent study, Borges et al. (2007) identified a new compound specific to males; this compound was not detected in the aerations of sexually immature females or males, and its retention time and mass spectra matched the sesquiterpene β-sesquiphellandrene (Table 5.1). This identification was confirmed using an authentic standard, and the absolute configuration was determined as (7*R*)-β-sesquiphellandrene (Borges et al. 2007). Males of *P. guildinii* produce approximately 40 ng/day of the sexual pheromone. The bioassays showed that *P. guildinii* females responded preferentially to (7*R*)-β-sesquiphellandrene, but also responded to the isomer (7*S*)-β-sesquiphellandrene. A racemic mixture of β-sesquiphellandrene was tested in a small field experiment in Brasilia, Brazil, and females were caught in the pheromone-baited traps; no stink bugs were caught in the control traps (Laumann et al., unpublished data).

Piezodorus hybneri *(Westwood) [Pentatomidae: Pentatominae: Piezodorini]*

The first pheromone study on the genus *Piezodorus* was carried out by Leal et al. (1998). These authors worked with *P. hybneri,* a species distributed in Thailand, Korea, Taiwan, Japan, and some regions of India and Australia (Panizzi et al. 2000). The authors reported three compounds, β-sesquiphellandrene, (*R*)-15-hexadecanolide, and methyl (8*Z*)-8-hexadecenoate in a 10:4:1 ratio (Table 5.1), as a sex pheromone blend for this species; laboratory bioassays revealed that both sexes were attracted to this blend, suggesting that it was an aggregation pheromone. Further studies conducted by Endo et al. (2007, 2010, 2012) showed that chemical communication within this species is more complex and dependent on the physiological state of the insects, with the insects changing the blend of the pheromone as they age. The production of β-sesquiphellandrene reaches peak-levels around 12 to 16 days after the final moult, whereas the other two components decrease with age. Field tests conducted in the summer in Japan (June-July) using traps baited with the *P. hybneri* aggregation pheromone and with a pheromone component of the *Riportus pedestris* (F.) (Hemiptera: Alydidae), the compound (*E*)-2-hexenyl-(*E*)-2-hexenoate, showed that the traps with the *R. pedestris* pheromone attracted more *P. hybneri* individuals than its own aggregation pheromone (Endo et al. 2010). In late autumn (October-November), when the insects were in reproductive diapause, both the aggregation pheromone and (*E*)-2-hexenyl-(*E*)-2-hexenoate (Table 5.1) attracted immature males and females; the aggregation pheromone also attracted nymphs. The fact that *P. hybneri* adults (male and female) and nymphs responded to the synthetic aggregation pheromone in the field only during autumn led the authors to conclude that this pheromone has another role in addition to its reproduction-related function.

Plautia stali *Scott [Pentatomidae: Pentatominae: Antestiini]*

Plautia stali males produce, as a pheromone, methyl (2*E*,4*E*,6*Z*)-2,4,6-decatrienoate (Table 5.1) (Sugie et al. 1996). The synthetic pheromone was tested in field conditions and attracted both genders, suggesting that this pheromone is an aggregation pheromone; this pheromone also attracted other stink bugs and natural enemies (see Chapter 10).

Thyanta custator *(F.),* Thyanta pallidovirens *(Stål), and* Thyanta perditor *(F.) [Pentatomidae: Pentatominae: Antestiini]*

Thyanta perditor males produce a different ester isomer of the *P. stali* pheromone, methyl-(2*E*,4*Z*,6*Z*)-2,4,6-decatrienoate (Table 5.1), as the main sex pheromone component (Moraes et al. 2005a). The same component was also identified in the pheromone blends of the two Nearctic species *Thyanta pallidovirens* and *Thyanta custator acerra* (Millar 1997, McBrien et al. 2002). Three species of *Thyanta* share this ester as the major compound in their pheromone blends, suggesting that close species of stink bugs share the same or similar blends of compounds as sex pheromones and that specificity is due to releasing different ratios of these compounds. Laboratory bioassays showed that 4 μg of methyl (2*E*,4*Z*,6*Z*)-2,4,6-decatrienoate was attractive to *T. perditor* females (Moraes et al. 2005a). Field tests using lures containing 1 mg methyl (2*E*,4*Z*,6*Z*)-2,4,6-decatrienoate, protected and unprotected from sunlight, were more efficient in capturing *T. perditor* than traps baited with the isomer methyl-(2*E*,4*E*,6*Z*)-2,4,6-decatrienoate, the pheromone of *P. stali*, and control traps. Additionally, traps baited with the sex pheromone captured a significantly higher number of insects than the sampling cloth technique, and also captured some Tachinid parasitoids of stink bugs

(Laumann et al. 2011). The attraction of Tachinid wasps to different isomers of methyl (2,4,6)-decatrienoate was reported by Aldrich et al. (2007).

Tibraca limbativentris *(Stål) [Pentatomidae: Pentatominae: Carpocorini]*

The Brazilian rice stalk stink bug, *Tibraca limbativentris*, has two sesquiterpenoid isomers of 1,10-bisaboladien-3-ol (Table 5.1) (zingiberenol) as the male-specific compounds (Borges et al. 2006). Bioassays with live insects and male extracts showed that females are attracted to males, but that males and females are not attracted to females, and that males are not attracted to males, indicating that 1,10-bisaboladien-3-ol acts as a sex pheromone. Non-selective synthesis produced two groups of isomers: (3*RS* 6*RS*, 7′*R*)-1,10-bisaboladien-3-ol and of (3*RS* 6*RS*, 7′*S*)-1,10-bisaboladien-3-ol. Y-tube olfactometer bioassays using the synthetic group of isomers showed that *T. limbativentris* females responded better to (3*RS* 6*RS*, 7′*S*) isomers than to (3*RS* 6*RS*, 7′*R*) (Borges et al. 2006). Female attraction to males, to male extract, and to synthetic sex pheromone was stronger during the night (18:00–24:00 hr), indicating that this species has a crepuscular/nocturnal activity cycle (Borges et al. 2006). These studies did not quantify the pheromone release rate by males. Field tests are being carried out to prove its effectiveness in trapping this species.

Defensive compounds

Defensive compounds are produced by stink bug nymphs and adults. These compounds have been described in several different species (see Millar 2005, Moraes et al. 2008a) and they are produced and stored in the DAGs in nymphs and in the MTGs of adults in a large orange reservoir (Aldrich et al. 1978, Aldrich 1988). In general, the defensive compounds have low chemical variability (Pavis 1987, Aldrich 1995, Moraes et al. 2008a, Fávaro et al. 2011) and are comprised of (*E*)-2-alkenals, 4-oxo-(*E*)-2-alkenals ($C_{6,}$ C_8 and C_{10}), and linear hydrocarbons (mostly C_{11} and C_{13}) (Table 5.2). Borges and Aldrich (1992) showed that the production of these compounds changes with the nymphal development. First instar nymphs produce a higher amount of 4-oxo-(*E*)-2-decenal than the other instar stages, and the latter nymphal stages produce higher quantities of 4-oxo-(*E*)-2-octenal and 4-oxo-(*E*)-2-hexenal than the first instar nymphs. Adults produce a blend of defensive compounds similar to those found in fifth instar nymphs, but the major compound, in general, is a linear hydrocarbon. In most of the species studied so far, *n*-tridecane is the main hydrocarbon, and only *E. meditabunda* and *E. rufomarginata* were reported to produce *n*-undecane as the major hydrocarbon (Pavis 1987, Pavis et al. 1994, Moraes et al. 2008a). In addition to these compounds, adults also produce some alcohols (C_6–C_8) and their esters, (*E*)-2-hexenyl, (*E*)-2-octenyl, and (*E*)-2-decenyl acetates (Table 5.2). Stink bugs also produce a series of minor compounds, such as alcohols and aldehydes with longer chains; for example, tetradecanal is produced by *T. limbativentris* and *E. heros,* and *D. melacanthus* and adults of *E. heros* produce linalool (Pareja et al. 2007, Blassioli-Moraes, unpublished data) (Table 5.2).

How stink bugs use their defensive compounds is not yet clear. There are studies stating the function of these compounds as a defensive strategy, and few studies report the conspecific roles of these chemicals as pheromones or their heterospecific roles as allomones. Recently, Eliyabu et al. (2012) reported the effect of several compounds from the defensive secretion of stink bugs against the jumping ant *Harpegnathos saltator* Jerdon (Hymenoptera: Formicidae). Noge et al. (2012) reported the effect of (*E*)-2-octenyl acetate, 4-oxo-(*E*)-2-hexenal, and (*E*)-2-hexenyl acetate against the praying mantis *Tenodera aridofolia sinensis* (Saussure) (Mantodea: Mantidae). MTG secretions also had a negative effect on entomopathogenic fungi. Borges et al. (1993) stated the action of two aldehydes common in stink bug defensive secretions, (*E*)-2-decenal and (*E*)-2-hexenal, against *Metarhizium anisopliae* (Metschnikoff) Sorokin (Hypocreales: Clavicipitacea). Sosa-Gomez and Moscardi (1998) showed that (*E*)-2-decenal had a negative effect on *M. anisopliae*. Recently, Lopes et al. (2015) reported the inhibitory effect of *N. viridula* defensive secretion on *Beauveria bassiana* (Bals.-Criv.) Vuill. sensu, whereas *D. melacanthus* was highly susceptible to fungal infections by *B. bassiana*. When synthetic solutions containing these compounds were evaluated against *B. bassiana*, both solutions inhibited (85% to 65%) conidial germination. The low effect of the volatiles from *D. melacanthus'* defensive secretion on conidial production might be explained by the great variability in the production of these compounds by the adults (Lopes et al. 2015).

The aldehydes and esters are responsible for the strong odour that stink bugs emit when stressed; these compounds are produced in relatively high quantities in such situations. Gilby and Waterhouse (1967) showed that three of the common allomones of stink bugs, (*E*)-2-octenal, (*E*)-2-decenal, and (*E*)-2-hexenal, act as defensive compounds against predators, but Pavis et al. (1994) and Lockwood and Story (1986, 1987) showed that these compounds can also function as alarm pheromones. Lockwood and Story (1985) found that tridecane could work as both an aggregation pheromone (at low concentrations) and an alarm pheromone (at high concentrations) for *N. viridula* in a dose-dependent manner. Although Fucarino et al. (2004) did not find any biological activity for tridecane in *N. viridula*, Aldrich (1988) proposed that one of the functions of the aliphatic hydrocarbons in the MTG is to serve as a solvent to modulate the evaporation of the other compounds, which can explain the high quantities of these compounds found in stink bug gland extracts. Most pentatomids aggregate upon hatching and disperse as later instars (Borges and Aldrich 1992). The compound 4-oxo-(*E*)-2-decenal was identified only from the DAGs of the first instar nymphs, and studies have proposed that this compound might be involved in their aggregation behaviour (Borges and Aldrich 1992, Fucarino et al. 2004). The 4-oxo-(*E*)-2-alkenals are compounds that have only been reported in Heteroptera; there is currently no information on the biological activity of the compounds in this order.

Semiochemicals from Stink Bugs as Kairomones for Natural Enemies

The chemical communication between stink bugs and parasitoids is well-documented with regard to the use of allomones as kairomones; there is no information so far about stink bugs using chemical information directly emitted from parasitoids and predators. On the other hand, several studies have shown that natural enemies use the chemical information of volatile and non-volatile compounds from their hosts (Borges et al. 1999b, Colazza et al. 1999, Conti et al. 2003). Laumann et al. (2009) showed that (*E*)-2-hexenal, a defensive compound from the MTG of *E. heros*, modifies the foraging behaviour of the egg parasitoid *T. podisi*, the main parasitoid of this stink bug. *Trissolcus basalis* (Wollaston) (Hymenoptera: Platygastridae) also has its foraging behaviour modified, but the chemical involved is (*E*)-2-decenal, another defensive compound from the MTGs of *N. viridula* and *Chinavia* sp. (Laumann et al. 2009). Bin et al. (1993) reported that *Tr. Basalis'* eggs have some chemical cues that are used by *N. viridula* to recognize the right host. *Telenomus podisi* also uses volatiles from the eggs of different stink bugs, such as *E. heros* and *T. limbativentris* (Tognon et al. 2014, Michereff et al. 2016). Cuticular lipids and footprints of stink bugs are used by natural enemies to locate their hosts at a short distance (Borges et al. 2003, Colazza et al. 2007). Colazza et al. (2007) reported that *n*-nonadecane is present in *N. viridula* cuticle and the egg parasitoid *Tr. basalis* uses this compound to distinguish residues from the male and female stink bugs. Egg parasitoids can also recognize the footprints of their hosts, for example, Borges et al. (2003) reported that *T. podisi* from a Brazilian population can recognize its preferred host's—the stink bug *E. heros* females' footprints, whereas *T. podisi* from an American population cannot. And in a more recent study, Salerno et al. (2009) showed that *Trissolcus brochymenae* (Ashmead) (Hymenoptera: Platygastridae) prefers the footprints left by *M. histrionica*'s mated females, as compared to the footprints from virgin females, males or parous host females. However, in both studies the chemicals involved in these interactions remain to be clarified. Natural enemies also use the sex and aggregation pheromones of stink bugs as cues to locate their hosts. In field experiments conducted in Brazil using traps baited with the *E. heros* pheromone (1 mg of methyl 2,6,10-trimethyltridecanoate per trap), two egg parasitoids from the Platygastridae family were captured, *T. podisi* and *Tr. basalis* (Borges et al. 1998). These results were supported by laboratory bioassays showing that *T. podisi* was attracted to synthetic racemic methyl 2,6,10-trimethytridecanoate in arena bioassays (Silva et al. 2006). Recently, Tognon and co-workers (2014) reported that *Tr. basalis,* in semi-field experiments, responded to a racemic mixture of 1,10 bisabolen-3-ol, the sex pheromone of *T. limbativentris*. In laboratory bioassays, the generalist egg parasitoid *Ooencyrtus telenomicida* (Vassiliev) (Hymenoptera: Encyrtidae) was attracted to the volatiles emitted by the virgin males of *N. viridula* (Peri et al. 2011). In addition to egg parasitoids, Tachinid flies also use the chemicals from adult stink bugs as cues to locate their hosts. Aldrich and co-workers (2007) showed that *Euclytia flava* (Townsend) (Diptera: Tachinidae) was caught in traps baited with the sex pheromones of *Thyanta* sp., methyl (2*E*,4*Z*,6*Z*)-2,4,6-decatrienoate, and *P. stali,* the isomer methyl (2*E*,4*E*,6*Z*)-2,4,6-decatrienoate; meanwhile, *Trichopoda pennines* (F.) (Diptera: Tachinidae) used the *N. viridula* sex pheromone to find its host (Harris and Todd 1980, Aldrich et al. 1987, Tillman et al. 2010).

Although, there is no report about the use of chemical signals directly from natural enemies by stink bugs. Several studies have shown that stink bugs are not attracted to the volatiles emitted by plants that were damaged by herbivory or by herbivory with oviposition by conspecifics (Colazza et al. 2004, Moraes et al. 2005b, 2008b, Conti et al. 2010, Michereff et al. 2011). These volatiles attract their main natural enemy, the egg parasitoids *T. podisi* and *Tr. basalis*; therefore, stink bugs avoid plants damaged by conspecific feeding in order to minimize competition or to avoid their natural enemies (Michereff et al. 2011, 2014). Recently, Melo-Machado et al. (2014) showed that *T. limbativentris* females prefer the volatiles of undamaged rice plants as compared to the volatiles emitted by rice plants damaged by the feeding activities of conspecifics.

Pheromone Cross-attraction in Stink Bugs

An interesting point, which is difficult to explain, is the cross-attraction of stink bugs by sex and aggregation pheromones. Some species are cross-attracted to blends with very similar chemical compositions; for example, *N. antennata* responds to the sex pheromone of *N. viridula*. In this case, the blend of the sex pheromones is the same, the difference is in the ratio of the components and therefore cross-attraction is plausible.

There are some species, however, that are cross-attracted to pheromone blends with a completely different chemistry, e.g., *N. viridula* is attracted to the sex pheromone of *T. perditor,* while *H. halys* and *C. hilaris* are attracted to the sex pheromone of *P. stali* (Aldrich et al. 2007, 2009) (Table 5.1). Recently, Tillman and co-workers (2010) showed that *E. servus* was attracted to traps that contained the *P. stali* pheromone combined with the *E. servus* pheromone. Borges et al. (1998, 2011) showed that *P. guildinii* and *E. meditabunda* were attracted to the *E. heros* pheromone (methyl 2,6,10-trimethyltridecanoate)-baited traps in field conditions. Although *E. meditabunda*, like *E. heros*, produces a methyl acetate as its sex pheromone, the molecules are quite different in size and methyl branch position (Table 5.1). Males of *E. meditabunda* produce the compound methyl 4,8,12-trimethypentadecanoate as a sex pheromone, and *P. guildinii* males produce β-sesquiphellandrene. Therefore, for these species, there is no overlapping pheromone chemistry to explain the cross-attraction.

This phenomenon was first observed in traps baited with the *E. heros* pheromone in 1998 in Brazil, attracting a similar cohort of stink bugs (Borges et al. 1998); this was also observed in several other stink bug species (Endo et al. 2006, 2010, Funayama 2008, Aldrich et al. 2007, 2009, Tillman et al. 2010) (Table 5.1). One possible explanation for the stink bug pheromone cross-attraction has already been proposed in several studies (Borges et al. 1998, 2011, Aldrich et al. 2007). These authors suggested that the pheromone cross-attraction is a means to find food and to aggregate as a passive defence against natural enemies. Borges et al. (2011) discussed that, in addition to the sex pheromones released from the lures inside the traps, other semiochemicals released by the first insects caught in the pheromone trap, such as defensive compounds from MTG, might lead the stink bugs to form a cluster of immature or adult insects. Future studies should consider these effects in order to evaluate if the other stink bugs are really attracted to the heterospecific sex or aggregation pheromones, or whether other chemical, vibrational, or visual cues are playing a role in this attraction (Čolk and Millar 2009).

Concluding Remarks

The sex and sex-aggregation pheromones and defensive chemicals of 45 species of stink bugs were described in this chapter, and 30 species had studies conducted proving the role of the semiochemicals that were identified as pheromones. Of these 30 species, 14 species produce pheromone compounds that are probably biosynthesized from acetate and proprionate units (Morgan 2010). The other 16 species described in this chapter produce pheromones that derive from the terpene pathway. All of the sex and sex-aggregation pheromones of Pentatomidae comprise molecules with 11 to 19 carbons, and for the most part (24 species), the molecule has 15 to 19 carbons. Only one species, *Bagrada hilaris*, produces a smaller molecule as a sex-aggregation pheromone, (*E*)-2-octenyl acetate; however, this compound is found in several other species of stink bugs in their MTG glands (Pavis

1987, Pareja et al. 2007, Moraes et al. 2008a). In general, the molecules have a very complex chemistry, presenting isomers with up to four chiral centres. The chemistry of the Pentatomidae pheromones' complexity and the lack of information about their behavioural ecology have delayed the extensive and intensive use of these pheromones for the monitoring and control of stink bugs and their natural enemies in crops. Field studies with Neotropical and Nearctic stink bugs have reported different behaviours in response to sex and sex-aggregation pheromones (see Chapter 10). For Nearctic species, the male-produced pheromone has been reported as a sex-aggregation pheromone, attracting males, females, and nymphs of the species. On the other hand, for the Neotropical species, the males produce sex pheromones, attracting only females in field and laboratory tests (see Chapter 10). This controversy may be explained either by the different system used to test the pheromone-baited trap, the chemical compound itself, the dispenser used to lure the insects, or by the methodology applied to count the sampled insects. For instance, the criteria used by Aldrich and co-workers (2007) included the trapped insects and insects within a 1-m radius of the traps.

Additionally, the behavioural differences can also be observed from the different types of traps used in field experiments for Neotropical and Nearctic stink bugs. For Neotropical stink bugs, in general, the traps included 2-litre transparent plastic bottles; these traps were placed at the level of the top of the plants in the field, and only the insects that were found inside the traps were considered (Borges et al. 1998). For Nearctic species, in general, a pyramidal trap was used (see Chapter 10), which is placed on the soil and the insects inside and in a 1-m radius from the trap were considered. The pyramidal trap was tested several times in Brazil in soybean crops to capture *E. heros* stink bugs, but without success (Pires C.S.S., personal communication). Therefore, to succeed in the use of Pentatomidae pheromones on crops in order to monitor and control stink bugs, it is necessary to conduct a detailed behavioural study that considers the mating behaviour, how stink bugs locate their host plants, and the pheromone production, including the biosynthesis of these semiochemicals. In brief, we suggest that the above-mentioned requirements may be achieved through renewed and more detailed studies of the stink bugs' behavioural ecology together with their chemical ecology.

Acknowledgments

We thank the National Council for Scientific and Technological Development (CNPq), the Brazilian Corporation of Agricultural Research (EMBRAPA) and the Research Support Foundation of the Federal District (FAP-DF).

References

Aldrich, J.R., M.S. Blum, H.A. Lloyd and H.M. Fales. 1978. Pentatomid natural products. Chemistry and morphology of the III-IV dorsal abdominal glands of adults. J. Chem. Ecol. 4: 161–172.

Aldrich, J.R., J.E. Oliver, W.R. Lusby, J.P. Kochansky and J.A. Lockwood. 1987. Pheromone strains of the cosmopolitan pest *Nezara viridula* (Heteroptera: Pentatomidae). J. Exp. Zool. 244: 171–175.

Aldrich, J.R. 1988. Chemical ecology of the Heteroptera. Annu. Rev. Entomol. 33: 211–238.

Aldrich, J.R., W.R. Lusby, B.E. Marron, K.C. Nicolaou, M.P. Hoffmann and L.T. Wilson. 1989. Pheromone blends of green stink bugs and possible parasitoid selection. Naturwissenschaften 76: 173–175.

Aldrich, J.R., M.P. Hoffmann, J.P. Kochansky, W.R. Lusby, J.E. Eger and J.A. Payne. 1991. Identification and attractiveness of a major pheromone component for Nearctic *Euschistus* spp. stink bugs (Heteroptera: Pentatomidae). Environ. Entomol. 20: 477–483.

Aldrich, J.R., H. Numata, M. Borges, F. Bin, G.K. Waite and W.R. Lusby. 1993. Artifacts and pheromone blends from *Nezara* spp. and other stink bugs (Heteroptera: Pentatomidae). Z. Naturfors. (Ser: C) 48: 73–79.

Aldrich, J.R., J.E. Oliver, W.R. Lusby, J.P. Kochansky and M. Borges. 1994. Identification of male-specific volatiles from Neartic and Neotropical stink bugs (Heteroptera: Pentatomidae). J. Chem. Ecol. 20: 1103–1111.

Aldrich, J.R. 1995. Chemical communication in the true bugs and parasitoid exploitation. pp. 318–363. *In*: Cardé, R.T. and W.J. Bell (eds.). Chemical Ecology of Insects II, Chapman & Hall, New York. 433 pp.

Aldrich, J.R., A. Khrimian and M.J. Camp. 2007. Methyl 2,4,6-decatrienoates attract stink bugs and tachinid parasitoids. J. Chem. Ecol. 33: 801–815.

Aldrich, J.R., A. Khrimian, M.J. Camp and X. Chen. 2009. Semiochemically based monitoring of the invasion of the brown marmorated stink bug and unexpected attraction of the native green stink bug (Heteroptera: Pentatomidae) in Maryland. Fla. Entomol. 92: 483–491.

Baker, R., M. Borges, N.G. Cooke and R.H. Herbert. 1987. Identification and synthesis of (Z)-(1′S,3′R,4′S)-(–)-2-(3′,4′-epoxy-4′-methylcyclohexyl)-6-methylhepta-2,5-diene, the sex pheromone of the southern green stink bug, *Nezara viridula* (L.). J. Chem. Soc. D 6: 414–416.

Bin, F., S.B. Vinson, M.R. Strand, S. Colazza and W.A. Jones Jr. 1993. Source of an egg kairomone for *Trissolcus basalis*, a parasitoid of *Nezara viridula*. Physiol. Entomol. 18: 7–15.

Blassioli-Moraes, M.C., R.A. Laumann, M.W.M. Oliveira, C.M. Woodcock, P. Mayon, A. Hooper et al. 2012. Sex pheromone communication in two sympatric Neotropical stink bug species *Chinavia ubica* and *Chinavia impicticornis*. J. Chem. Ecol. 38: 836–845.

Blassioli-Moraes, M.C., M. Borges and R.A. Laumann. 2013. Chemical ecology of insect parasitoids. pp. 225–244. *In*: Wajnberg, E. and S. Colazza (eds.). The Application of Chemical Cues in Arthropod Pest Management for Arable Crops. John Wiley & Sons, New York, NY, USA.

Borges, M., P.C. Jepson and P.E. Howse. 1987. Long-range mate location and close range courtship behavior of the green stink bug, *Nezara viridula* and its mediation by sex pheromones. Entomol. Exp. Appl. 44: 205–212.

Borges, M. and J.R. Aldrich. 1992. Instar-specific defensive secretions of stink bugs (Heteroptera: Pentatomidae). Experientia 48: 893–896.

Borges, M., S.C.M. Leal, M.S. Tigano-Milani and M.C.C. Valadares. 1993. Efeito do feromônio de alarme do percevejo verde, *Nezara viridula* (L.) (Hemiptera: Pentatomidae), sobre o fungo entomopatogênico *Metarhizium anisopliae* (Metsch.) Sorok. An. Soc. Entomol. Bras. 22: 505–512.

Borges, M. and J.R. Aldrich. 1994. Attractant pheromone for Nearctic stink bug, *Euschistus obscurus* (Heteroptera: Pentatomidae): insight into a Neotropical relative. J. Chem. Ecol. 20: 1095–1102.

Borges, M. 1995. Attractant compounds of the southern green stink bug, *Nezara viridula* (L.) (Heteroptera: Pentatomidae). An. Soc. Entomol. Bras. 24: 215–225.

Borges, M., F.G.V. Schmidt, E.R. Sujii, M.A. Medeiros, K. Mori, P.H.G. Zarbin et al. 1998. Field responses of stink bugs to the natural and synthetic pheromone of the neotropical brown stink bug, *Euschistus heros* (Heteroptera: Pentatomidae). Physiol. Entomol. 23: 202–207.

Borges, M., P.H.G. Zarbin, J.T.B. Ferreira and M.L.M. da Costa. 1999a. Pheromone sharing: Blends based on the same compounds for *Euschistus heros* and *Peizodorus guildinii*. J. Chem. Ecol. 25: 629–634.

Borges, M., M.L.M. Costa, E.R. Sujii, M. Das, G. Cavalcanti, G.F. Redígolo et al. 1999b. Semiochemical and physical stimuli involved in host recognition by *Telenomus podisi* (Hymenoptera: Scelionidae) toward *Euschistus heros* (Heteroptera: Pentatomidae). Physiol. Entomol. 24: 227–233.

Borges, M., S. Colazza, P. Ramirez-Lucas, K.R. Chauhan, J.R. Aldrich and M.C.B. Moraes. 2003. Kairomonal effect of walking traces from *Euschistus heros* (Heteroptera: Pentatomidae) on two strains of *Telenomus podisi* (Hymenoptera: Scelionidae). Physiol. Entomol. 28: 349–355.

Borges, M., M. Birkett, J.R. Aldrich, J.E. Oliver, M. Chiba, Y. Murata et al. 2006. Sex attractant pheromone from the rice stalk stink bug, *Tibraca limbativentris* Stal. J. Chem. Ecol. 32: 2749–2761.

Borges, M., J.G. Millar, R.A. Laumann and M.C.B. Moraes. 2007. A male-produced sex pheromone from the neotropical redbanded stink bug, *Piezodorus guildinii* (W.). J. Chem. Ecol. 33: 1235–1248.

Borges, M., M.C.B. Moraes, M.F. Peixoto, C.S.S. Pires, E.R. Sujii and R.A. Laumann. 2011. Monitoring the neotropical brown stink bug *Euschistus heros* (F.) (Hemiptera: Pentatomidae) with pheromone-baited traps in soybean fields. J. Appl. Entomol. 135: 68–80.

Brezot, P., C. Malosse, K. Mori and M. Renou. 1994. Bisabolene epoxides in sex pheromone in *Nezara viridula* (L.) (Heteroptera: Pentatomidae): role of *cis* isomer and relation to specificity of pheromone. J. Chem. Ecol. 20: 3133–3147.

Buxton, G.M., D.B. Thomas and R.C. Froeschner. 1983. Revision of the species of the sayi group of *Chlorochroa* Stal (Hemiptera: Pentatomidae). Occasional papers in Entomology. State of California, Dept. of Food and Agriculture, Division of Plant Industry, Laboratory Services (1983) (Sacramento, Calif) 29: 23.

Čokl, A. and J.G. Millar. 2009. Manipulation of insect signaling for monitoring and control of pest insects. pp. 279–316. *In*: Ishaaya, I. and A.R. Horowitz (eds.). Biorational Control of Arthropod Pests: Application and Resistance Management. Springer, New York, USA.

Colazza, S., G. Salerno and E. Wajnberg. 1999. Volatile and contact chemicals released by *Nezara viridula* (Heteroptera: Pentatomidae) have a kairomonal effect on the egg parasitoid *Trissolcus basalis* (Hymenoptera: Scelionidae). Biol. Control 16: 310–317.

Colazza, S., J.S. McElfresh and J.G. Millar. 2004. Identification of volatile synomones, induced by *Nezara viridula* feeding and oviposition on bean spp., that attracts the egg parasitoid *Trissolcus basalis*. J. Chem. Ecol. 30: 945–964.

Colazza, S., G. Aquila, C. De Pasquale, E. Peri and J.G. Millar. 2007. The egg parasitoid *Trissolcus basalis* uses *n*-nonadecane, a cuticular hydrocarbon from stink bug host *Nezara viridula*, to discriminate between female and male hosts. J. Chem. Ecol. 33: 1405–1420.

Conti, E., G. Salerno, F. Bin, H.J. Williams and S.B. Vinson. 2003. Chemical cues from *Murgantia histrionica* eliciting host location and recognition in the egg parasitoid *Trissolcus brochymenae*. J. Chem. Ecol. 29: 115–130.

Conti, E., G. Salerno, B. Leombruni, F. Frati and F. Bin. 2010. Short-range allelochemicals from a plant–herbivore association: a singular case of oviposition-induced synomone for an egg parasitoid. J. Exp. Biol. 213: 3911–3919.

Costa, M.L.M., M. Borges and E.F. Vilela. 2000. Effect of stereoisomers of the main component of the sex pheromone of *Euschistus heros* (F.) (Heteroptera: Pentatomidae) in the attactiveness of female. An. Soc. Entomol. Bras. 29: 413–422.

Dicke, M. and M.W. Sabelis. 1988. Infochemical terminology: based on cost-benefit analysis rather than origin of compounds? Funct. Ecol. 2: 131–139.

Eisner, T., M. Eisner and M. Deyrup. 1991. Chemical attraction of kleptoparasitic flies to heteropteran insects caught by orb-weaving spiders. Proc. Natl. Acad. Sci. USA 88: 8194–8197.

Endo, N., T. Wada, Y. Nishiba and R. Sasaki. 2006. Interspecific pheromone cross-attraction among soybean bugs (Heteroptera): does *Piezodorus hybneri* (Pentatomidae) utilize the pheromone of *Riptortus clavatus* (Alydidae) as a kairomone? J. Chem. Ecol. 32: 1605–1612.

Endo, N., T. Yasuda, K. Matsukur, T. Wada, S.E. Muto and R. Sasaki. 2007. Possible function of *Piezodorus hybneri* (Heteroptera: Pentatomidae) male pheromone: Effects of adult age and diapause on sexual maturity and pheromone production. Appl. Entomol. Zool. 42: 637–641.

Endo, N., R. Sasaki and S. Muto. 2010. Pheromonal cross-attraction in true bugs (Heteroptera): attraction of *Piezodorus hybneri* (Pentatomidae) to its pheromone versus the pheromone of *Riptortus pedestris* (Alydidae). Environ. Entomol. 39: 1973–1979.

Endo, N., T. Yasuda, T. Wada, S.E. Muto and R. Sasaki. 2012. Age-related and individual variation in male *Piezodorus hybneri* (Heteroptera: Pentatomidae) pheromones. Psyche 2012: ID 609572.

Fávaro, C.F., M.A.C. de, M. Rodrigues, J.R. Aldrich and P.H.G. Zarbin. 2011. Identification of semiochemicals in adults and nymphs of the stink bug *Pallantia macunaima* Grazia (Hemiptera: Pentatomidae). J. Braz. Chem. Soc. 22: 58–94.

Fávaro, C.F., T.B. Santos and P.H.G. Zarbin. 2012. Defensive compounds and male-produced sex pheromone of the stink bug, *Agroecus griseus*. J. Chem. Ecol. 38: 1124–1132.

Fávaro, C.F., R.A. Soldi, T. Ando, J.R. Aldrich and P.H.G. Zarbin. 2013. (6*R*,10*S*)-Pallantione: the first ketone identified as a sex pheromone in stink bugs. Org. Lett. 15: 1822–1825.

Fávaro, C.F., J.G. Millar and P.H.G. Zarbin. 2015. Identification and synthesis of the male-produced sex pheromone of the stink bug, *Pellaea stictica*. J. Chem. Ecol. 41: 859–868.

Fucarino, A., J.G. Millar, J.S. McElfresh and S. Colazza. 2004. Chemical and physical signals mediating conspecific and heterospecific aggregation behavior of first instar stink bugs. J. Chem. Ecol. 30: 1257–1269.

Funayama, K. 2008. Seasonal fluctuations and physiological status of *Halyomorpha halys* (Stål) (Heteroptera: Pentatomidae) adults captured in traps baited with synthetic aggregation pheromone of *Plautia crossota stali* Scott (Heteroptera: Pentatomidae). App. Entomol. Zool. 52: 69–75.

Gassen, D.N. 1996. Manejo de pragas associadas a cultura do milho. Aldeia Norte, Passo Fundo.

Genung, W.G., V.E. Green, Jr. and C. Wehlburg. 1964. Inter-relationship of stinkbugs and diseases to Everglades soybean production. Soil Crop Sci. Soc. Fla. Proc. 24: 131–137.

Gilby, A.R. and D.F. Waterhouse. 1967. Secretion from lateral scent glands of the green vegetable bug *Nezara viridula*. Nature 216: 90–91.

Guarino, S., C. De Pasquale, E. Peri, G. Alonzo and S. Colazza. 2008. Role of volatile and contact pheromones in the mating behavior of *Bagrada hilaris* (Heteroptera: Pentatomidae). Eur. J. Entomol. 105: 613–617.

Grazia, J. and A. Frey-da-Silva. 2001. Descrição dos imaturos de Loxa deducta Walker e *Pallantia macunaima* Grazia (Heteroptera: Pentatomidae) em ligustro, *ligustrum lucidum* Ait. Neo. Entomol. 30: 73–80.

Harris, V.E. and J.W. Todd. 1980. Male-mediated aggregation of male, female, and 5th instar southern green stink bugs and concomitant attraction of a tachinid parasite, *Trichopoda pennipes*. Entomol. Exp. Appl. 27: 117–126.

Henry, T.J. 1984. New United States records for two heteroptera: *Pellaea stictica* (Pentatomidae) and *Rhinacloa pallides* (Miridae). Proc. Entomol. Soc. Wash. 86: 519–520.

Ho, H.-Y. and J.G. Millar. 2001a. Identification and synthesis of a male-produced sex pheromone from the stink bug *Chlorochroa sayi*. J. Chem. Ecol. 27: 1177–1201.

Ho, H.-Y. and J.G. Millar. 2001b. Identification and synthesis of male-produced sex pheromone components of the stink bugs *Chlorochroa ligata* and *Chlorochroa uhleri*. J. Chem. Ecol. 27: 2067–2095.

Khrimian, A., A. Zhang, D.C. Weber, H.-Y. Ho, J.R. Aldrich, K.E. Vermillion et al. 2014a. Discovery of the aggregation pheromone of the brown marmorated stink bug (*Halyomorpha halys*) through the creation of stereoisomeric libraries of 1-bisabolen-3-ols. J. Nat. Prod. 77: 1708–1717.

Khrimian, A., S. Shirali, K.E. Vermillion, M.A. Siegler, F. Guzman, K. Chauhan et al. 2014b. Determination of the stereochemistry of the aggregation pheromone of harlequin bug, *Murgantia histrionica*. J. Chem. Ecol. 40: 1260–1268.

Laumann, R.A., M.F. Aquino, M.C.B. Moraes, M. Pareja and M. Borges. 2009. Response of egg parasitoids *Trissolcus basalis* and *Telenomus podisi* to compounds from defensive secretions of stink bugs. J. Chem. Ecol. 35: 8–19.

Laumann, R.A., M.C.B. Moraes, A. Khrimian and M. Borges. 2011. Field capture of *Thyanta perditor* with pheromone baited traps. Pesqui. Agropecu. Bras. 46: 113–119.

Leal, W.S., S. Kuwahara, X. Shi, H. Higuchi, C.E. Marino, M. Ono et al. 1998. Male-released sex pheromone of the stink bug *Piezodorus hybneri*. J. Chem. Ecol. 24: 1817–1829.

Leskey, T.C., G.C. Hamilton, A.L. Nielsen, D.F. Polk, C. Rodriguez-Saona, J.C. Berg et al. 2012a. Pest status of the brown marmorated stink bug, *Halyomorpha halys* (Stål), in the USA. Outlooks on Pest Manage. 23: 218–226.

Leskey, T.C., S.E. Wright, B.D. Short and A. Khrimian. 2012b. Development of behaviorally based monitoring tools for the brown marmorated stink bug, *Halyomorpha halys* (Stål) (Heteroptera: Pentatomidae) in commercial tree fruit orchards. J. Entomol. Sc. 47: 76–85.

Leskey, T.C., J.A. Agnello, C. Bergh, G.P. Dively, G.C. Hamilton, P. Jentsch et al. 2015a. Attraction of the invasive *Halyomorpha halys* (Hemiptera: Pentatomidae) to traps baited with semiochemical stimuli across the United States. Environ. Entomol. DOI 10.1093/ee/nvv049.

Leskey, T.C., A. Khrimian, D.C. Weber, J.R. Aldrich, B.D. Short, D.H. Lee et al. 2015b. Behavioral responses of the invasive *Halyomorpha halys* (Stal) (Hemiptera: Pentatomidae) to traps baited with stereoisomeric mixtures of 10, 11-epoxy-1-bisabolen-3-ol. J. Chem. Ecol. 41: 418–429.

Lockwood, J.A. and R.N. Story. 1985. Bifunctional pheromone in the first instar of the southern green stink bug, *Nezara viridula* (L.) (Hemiptera: Pentatomidae): its characterization and interaction with other stimuli. Ann. Entomol. Soc. Am. 78: 474–479.

Lockwood, J.A. and R.N. Story. 1986. Adaptive functions of nymphal aggregation in the southern green stink bug, *Nezara viridula* (L.) (Hemiptera: Pentatomidae). Environ. Entomol. 15: 739–749.

Lockwood, J.A. and R.N. Story. 1987. Defensive secretion of the southern green stink bug (Hemiptera: Pentatomidae) as an alarm pheromone. Ann. Entomol. Soc. Am. 80: 686–691.

Lopes, R.B., R.A. Laumann, M.C. Blassioli-Moraes, M. Borges and M. Farias. 2015. The fungistatic and fungicidal effects of volatiles from metathoracic glands of soybean-attacking stink bugs (Heteroptera: Pentatomidae) on the entomopathogen *Beauveria bassiana*. J. Invertebr. Pathol. 132: 77–85.

McBrien, H.L. and J.G. Millar. 1999. Phytophagous bugs. pp. 237–304. *In*: Hardie, J. and A. Minks (eds.). Pheromone of Non-Lepidoptera Insects Associated with Agriculture Plants. Cabi Publishing, New York, NY, USA.

McBrien, H.L., J.G. Millar, L. Gottlieb, X. Chen and R.E. Rice. 2001. Male-produced sex attractant pheromone of the green stink bug, *Acrosternum hilare* (Say). J. Chem. Ecol. 27: 1821–1839.

McBrien, H.L., J.G. Millar, R.E. Rice, J.S. McElfresh, E. Cullen and F.G. Zalom. 2002. Sex attractant pheromone of the red-shouldered stink bug *Thyanta pallidovirens*: a pheromone blend with multiple redundant components. J. Chem. Ecol. 28: 1797–1818.

Melo-Machado, R.C., J. Sant'Ana, M.C. Blassioli-Moraes, R.A. Laumann and M. Borges. 2014. Herbivory-induced plant volatiles from *Oryza sativa* and their influence on chemotaxis

behaviour of *Tibraca limbativentris* Stål (Hemiptera: Pentatomidae) and egg parasitoids. Bull. Entomol. Res. 104: 347–356.
Michereff, M.F.F., R.A. Laumann, M. Borges, M. Michereff Filho, I.R. Diniz, A. Faria Neto et al. 2011. Volatiles mediating plant-herbivory-natural enemy interaction in resistant and susceptible soybean cultivars. J. Chem. Ecol. 37: 273–285.
Michereff, M.F.F., M. Michereff Filho, M.C. Blassioli-Moraes, R.A. Laumann, I.R. Diniz and M. Borges. 2014. Effect of resistant and susceptible soybean cultivars on the attraction of egg parasitoids under field conditions. J. App. Entomol. 139: 207–216.
Michereff, M.F.F., M. Borges, M.A. Santos, R.A. Laumann, A.C.M.M. Gomes and M.C. Blassioli-Moraes. In press. The influence of volatile semiochemicals from stink bug eggs and oviposition-damaged plants on the foraging behaviour of the egg parasitoid *Telenomus podisi*. Bulletin of Entomological Research, 2016.
Miklas, N., M. Renou, I. Malosse and C. Malosse. 2000. Repeatability of pheromone blend composition in individual males of the southern green stink bug, *Nezara viridula*. J. Chem. Ecol. 26: 2473–2485.
Millar, J.G. 1997. Methyl (2*E*,4*Z*,6*Z*)-deca-2,4,6-trienoate, a thermally unstable, sex-specific compound from the stink bug *Thyanta pallidovirens*. Tetrahedron Lett. 38: 7971–7972.
Millar, J.G. 2005. Pheromones of true bugs. pp. 37–84. *In*: Schulz, S. (ed.). The Chemistry of Pheromones and other Semiochemicals II—Topics in Current Chemistry, Volume 240. Springer-Verlag Berlin, Heidelberg, Germany.
Moraes, M.C.B., J.G. Millar, R.A. Laumann, E.R. Sujii, C.S.S. Pires and M. Borges. 2005a. Sex attractant pheromone from the neotropical red-shouldered stink bug, *Thyanta perditor* (F.). J. Chem. Ecol. 31: 1415–1427.
Moraes, M.C.B., R.A. Laumann, C.S.S. Pires, E.R. Sujii and M. Borges. 2005b. Induced volatiles in soybean and pigeon pea plants artificially infested with the neotropical brown stink bug, *Euschistus heros*, and their effect on the egg parasitoid, *Telenomus podisi*. Entomol. Exp. App. 115: 227–237.
Moraes, M.C.B., M. Pareja, R.A. Laumann and M. Borges. 2008a. The chemical volatiles (semiochemicals) produced by neotropical stink bugs (Hemiptera: Pentatomidae). Neotrop. Entomol. 37: 489–505.
Moraes, M.C.B., M. Pareja, R.A. Laumann, C.B. Hoffmann-Campo and M. Borges. 2008b. Response of the parasitoid *Telenomus podisi* to induced volatiles from soybean damaged by stink bug herbivory and oviposition. J. Plant Interact. 3: 1742–1756.
Moraes, M.C.B., M. Borges, M. Pareja, H.G. Vieira, F.T. de Souza Sereno and R.A. Laumann. 2008c. Food and humidity affect sex pheromone ratios in the stink bug, *Euschistus heros*. Physiol. Entomol. 33: 43–50.
Morgan, E.D. 2010. Biosynthesis in Insects. RSC Publishing, Cambridge, UK.
Mori, K. and M. Murata. 1994. Synthesis of all eight stereoisomers of methyl 2,6,10 trimethyltridecanoate, the male-produced pheromone of the stink bugs, *Euschistus heros* and *E. obscurus*. Liebigs Ann. Chem. 1153–1160.
Noge, K., K.L. Prudic and J.X. Becerra. 2012. Defensive roles of (*E*)-2-alkenals and related compounds in Heteroptera. J. Chem. Ecol. 38: 1050–1056.
Oliveira, M.W.M., M. Borges, C.K.Z. Andrade, R.A. Laumann, J.A.F. Barrigossi and M.C. Blassioli-Moraes. 2013. Zingiberenol, (1*S*,4*R*,1′*S*)-4-(1′,5′-dimethylhex-4′-enyl)-1-methylcyclohex-2-en-1-ol, identified as the sex pheromone produced by males of the rice stink bug *Oebalus poecilus* (Heteroptera: Pentatomidae). J. Agric. Food Chem. 61: 7777–7785.
Panizzi, A.R. 1997. Wild hosts of pentatomids: ecological significance and role in their pest status on crops. Ann. Rev. Entomol. 42: 99–122.
Panizzi, A.R., J.E. McPherson, D.G. James, M. Javahery and R.M. McPherson. 2000. Stink bugs (Pentaomidae). pp. 421–474. *In*: Schaefer, C.W. and A.R. Panizzi (eds.). Heteroptera of Economic Importance. CRC Press, Boca Raton, Florida, USA.
Panizzi, A.R. and J. Grazia. 2001. Stink Bugs (Heteroptera, Pentatomidae) and an unique host plant in the Brazilian subtropics. Iheringia 90: 21–35.

Panizzi, A.R. 2015. Growing problems with stink bugs (Hemiptera: Heteroptera: Pentatomidae): Species invasive to US and potential Neotropical invaders. Am. Entomol. 61: 223–233.

Pareja, M., M. Borges, R.A. Laumann and M.C.B. Moraes. 2007. Inter- and intraspecific variation in defensive compounds produced by five neotropical stink bug species (Hemiptera: Pentatomidae). J. Insect Physiol. 53: 639–648.

Pavis, C. and P.H. Malosse. 1986. Mise en evidence d'un attractif sexuel produit par les males de *Nezara viridula* (L.) (Heteroptera: Pentatomidae). C.R. Acad. Sci. Series III 7: 272–276.

Pavis, C. 1987. Les secretions exocrines des hétéroptères (allomones et pheromones). Une mise au point bibliographique. Agronomie, EDP Sciences 7: 547–561.

Pavis, C., C. Malosse, P.H. Ducrot and C. Déscoins. 1994. Dorsal abdominal glands in nymphs of southern green stink bug, *Nezara viridula* (L.) (Heteroptera: Pentatomidae): chemistry of secretions of five instars and role of (*E*)-4-oxo-2-decenal, compound specific to first instars. J. Chem. Ecol. 20: 2213–2227.

Peri, E., A. Cusumano, A. Agrio and S. Colazza. 2011. Behavioral response of the egg parasitoid *Ooencyrtus telenomicida* to host-related chemical cues in a tritrophic perspective. BioControl 56: 163–171.

Salerno, G., F. Frati, E. Conti, C. De Pasquale, E. Peri et al. 2009. A finely tuned strategy adopted by an egg parasitoid to exploit chemical traces from host adults. J. Exp. Biol. 212: 1825–1831.

Schaefer, C.W. and A.R. Panizzi. 2000. Economic importance of Heteroptera: a general view. pp. 3–8. *In*: Schaefer, C.W. and A.R. Panizzi (eds.). Heteroptera of Economic Importance. CRC Press, Boca Raton, London, New York, Washington D.C.

Silva, C.C., M.C.B. Moraes, R.A. Laumann and M. Borges. 2006. Sensory response of the egg parasitoid *Telenomus podisi* to stimuli from the bug *Euschistus heros*. Pesqui. Agropecu. Bras. 41: 1093–1098.

Silva, W.P., M.J.B. Pereira, L.M. Vivan, M.C.B. Moraes, R.A. Laumann and M. Borges. 2014. Monitoramento do percevejo marrom *Euschistus heros* (Hemiptera: Pentatomidae) por feromônio sexual em lavoura de soja. Pesqui. Agropecu. Bras. 49: 844–852.

Silva, C.C.A., R.A. Laumann, M.C. Blassioli-Moraes, M.F.S. de Aquino and M. Borges. 2015. Biologia comparada de dois percevejos, *Chinavia impicticornis* e C. ubica (Hemptera: Pentatomidae) Pesqui. Agropecu. Bras. 50: 1–10.

Sosa-Gomez, D.R. and F. Moscardi. 1998. Laboratory and field studies on the infection of stink bugs, *Nezara viridula*, *Piezodorus guildinii*, and *Euschistus heros* (Hemiptera: Pentatomidae) with *Metarhizium anisopliae* and *Beauveria bassiana* in Brazil. J. Invert. Pathol. 71: 115–120.

Sugie, H., M. Yoshida, K. Kawasaki, H. Noguchi, S. Moriya, K. Takagi et al. 1996. Identification of the aggregation pheromone of the brown-winged green bug, *Plautia stali* Scott (Heteroptera: Pentatomidae). Appl. Entomol. Zool. 31: 427–431.

Tillman, P.G., J.R. Aldrich, A. Khrimian and T.E. Cottrell. 2010. Pheromone attraction and cross-attraction of *Nezara*, *Acrosternum*, and *Euschistus* spp. stink bugs (Heteroptera: Pentatomidae) in the field. Environ. Entomol. 39: 610–617.

Tognon, R., J. Sant'Ana and S.M. Jahnke. 2014. Influence of original host on chemotaxic behaviour and parasitism in *Telenomus podisi* Ashmead (Hymenoptera: Platygastridae). Bull. Entomol. Res. 104: 781–787.

Weber, D.C., T.C. Leskey, G. Cabrera Walsh and A. Khrimian. 2014a. Synergy of aggregation pheromone with methyl (*E,E,Z*)-2,4,6-decatrienoate in attraction of *Halyomorpha halys* (Hemiptera: Pentatomidae). J. Econ. Entomol. 107: 1061–1068.

Weber, D.C., G. Cabrera Walsh, A.S. DiMeglio, M.M. Athanas, T.C. Leskey and A. Khrimian. 2014b. Attractiveness of harlequin bug, *Murgantia histrionica* (Hemiptera: Pentatomidae), aggregation pheromone: field response to isomers, ratios and dose. J. Chem. Ecol. 40: 1251–1259.

Weber, Donald C., A. Khrimian, M.C. Blassioli-Moraes and J.G. Millar. Semiochemistry of Pentatomoidea. Chapter 15 *In*: Biology of Invasive Stink Bugs and Related Species (J. McPherson et al., eds.). CRC Press, in press.

Zahn, D.K., J.A. Moreira and J.G. Millar. 2008. Identification, synthesis, and bioassay of a male-specific aggregation pheromone from the harlequin bug, *Murgantia histrionica*. J. Chem. Ecol. 34: 238–251.

Zahn, D.K., J.A. Moreira and J.G. Millar. 2012. Erratum to: Identification, synthesis, and bioassay of a male-specific aggregation pheromone from the harlequin bug, *Murgantia histrionica*. J. Chem. Ecol. 38: 126.

Zarbin, P.H.G., C.F. Fávaro, D.M. Vidal and M.A.C.M. Rodrigues. 2012. Male-produced sex pheromone of the stink bug *Edessa meditabunda*. J. Chem. Ecol. 38: 825–835.

Zhang, A.J., M. Borges, J.R. Aldrich and M. Camp. 2003. Stimulatory male volatiles for the neotropical brown stink bug, *Euschistus heros* (F.) (Heteroptera: Pentatomidae). Neotr. Entomol. 32: 713–717.

Zhu, G., W. Bu, Y. Gao and G. Liu. 2012. Potential geographic distribution of brown marmorated stink bug invasion (*Halyomorpha halys*). Plos One. DOI: 10.1371/journal.pone.0031246.

CHAPTER 6

Substrate-borne Vibratory Communication

Andrej Čokl,[1,*] *Raul Alberto Laumann*[2] and *Nataša Stritih*[1]

Introduction

Plant-dwelling stink bugs from the subfamily Pentatominae (Heteroptera: Pentatomidae) are economically important pests that have attracted the attention of the scientific community, which wishes to investigate their biology along with the different aspects of their communication as a basis for developing environmentally friendly pest control management. Studies on communications by plant-dwelling Pentatominae, which started more than 40 years ago, were accelerated by the development of non-contact vibration recording techniques that enabled research on plants under field and semi-field conditions.

Signals that efficiently carry information under different environmental conditions are designed to fit the properties of the transmission media. In the field, a stink bug male-emitted pheromone attracts conspecific females to meet the male on a plant (Chapter 5). Dense vegetation is a specific medium that constrains communication through chemical, visual and airborne vibratory signals. Leaves prevent longer-range optical contact, wind and local air currents decrease the efficiency of communication with olfactory signals, and high-frequency airborne vibrations that can only be

[1] Department of Organisms and Ecosystems Research, National Institute of Biology, Večna pot 111, SI-1000 Ljubljana, Slovenia.
Email: natasa.stritih@nib.si

[2] Laboratório de Semioquímicos, Embrapa Recursos Genéticos e Biotecnologia, Avda W5 Norte (Final), 71070-917, Brasilia, DF, Brazil.
Email: raul.laumann@embrapa.br

* Corresponding author: andrej.cokl@nib.si

produced efficiently by small insects are strongly dampened (Markl 1983). Consequently, stink bugs, like the members of most of the other insect groups under study (Cocroft and Rodriguez 2005), communicate with the substrate-borne component of vibrations produced on plants using different mechanisms. Communication through plants via vibratory signals is limited by plant dimensions, architecture and mechanical properties (Bennet-Clark 1998, Čokl 2008, Michelsen 2014). Nevertheless, at shorter distances, stink bugs also communicate using airborne, contact mechanical, chemical and visual signals.

Long-lasting research on stink bug vibratory communication has responded to many open questions. Over the past decade, the number of species found to engage in vibrational communication has significantly increased; vibratory signals produced by mechanisms other than abdomen vibration have been described, and new knowledge has been gained with regard to insect/plant interactions during plant-borne vibratory communication. More than thirty years of data on the morphology and function of receptor and higher-order vibratory neurons have recently been upgraded with intracellular recording techniques, and interspecific communication contact has been documented between the members of different Pentatomidae subfamilies.

Despite the progress that has been made in researching communication by plant-dwelling stink bugs, many issues require further and more detailed studies in both the field and laboratory. Do regular variations in spectra and amplitudes that are determined at different distances from the source give reliable information on the distance to the calling mate? Is the neuronal system capable of processing a frequency modulation and the distance-dependent ratio of spectral peak amplitudes? What triggers males to start calling, and why is it that this emission has no effect on female signalling for several species? Which signal modalities support recognition in the absence of male pheromones? Is mate location in the field restricted to the plant in this case? What is the role of the species non-specific signals produced by the tremulation of the whole body and by the vibration of lifted wings? How do several females on a plant react to a single conspecific-calling and -smelling male? These and many other questions represent a highly interesting challenge for future research, and their answers will increase interest in the successful application of basic science research into praxis in the field.

Vibratory Communication and Reproductive Behaviour

Mating behaviour investigations up until the present in Pentatomine species have run through different phases, as exemplified by the different studies on the southern green stink bug *Nezara viridula* (L.) (Borges et al. 1987, Čokl and Virant-Doberlet 2003). At long distances, the male pheromone attracts scattered mates from the field to gather on the same plant (Borges et al. 1987). The male pheromone motivates the female to start calling (Zgonik and Čokl 2014), by emitting vibratory signals that trigger males to search for the calling female, and the males respond to her with different types of songs (Čokl 2008). Female calling signals modulate the amount of male-produced pheromone (Miklas et al. 2003a) and mediate the vibrational directionality on plant crossings (Ota and Čokl 1991, Čokl et al. 1999). When mates meet, the courtship phase of their mating behaviour follows their calling; the short-range exchange of information by vibratory, visual, contact chemical and tactile signals leads to copulation (Blassioli-Moraes et al. 2014). A rival male song has been regularly recorded in different species when several males court the same female (Čokl and Virant-Doberlet 2003). The female accepts the courting male by lifting her abdomen, or she rejects him by emitting a repellent song, which is accompanied by the vigorous shaking of her body. In several species, males in copula produce long-lasting copulatory vibratory signals (Bagwell et al. 2008).

Vibratory communication signals have been described in 36 stink bug species. The list of species and songs, together with the authors who have described them is shown in Table 6.1. Generally, we can divide Pentatominae stink bug songs according to the mating behaviour phase into calling, courtship, rival, repellent and copulatory songs. The other approach is to determine the songs according to the order of their appearance in a sequence of different emissions as the male or female first (MS-1, FS-1) or second (MS-2, FS-2) song, etc. Signals are further described by their temporal, frequency and amplitude characteristics. Their statistically determined values vary within the same species also because different authors have used different techniques to record the signals emitted by the insects on different substrates and under different experimental conditions. Furthermore, the transition from one to another song type runs over several non-regular intermediate forms, and different authors had different criteria for separating them from the songs that characterize specific phases of mating behaviour.

Shestakov (2015) recently published a comparative analysis of the male vibrational signals in 16 sympatric pentatomine species from European Russia. Vibratory emissions were recorded in the field by piezoelectric cartridges that were positioned on the species-characteristic host plants. The author described male vibrational signals with calling functions in *Eurydema oleracea* (L.), *Graphosoma lineatum* (L.), *Pentatoma rufipes* (L.) and *Palomena prasina* (L.), and male rival singing was attributed to *P. rufipes, Dolycoris baccarum* (L.) and *Aelia acuminata* (L.). Male courtship has been recorded in *Eurydema ornata* (L.), *Carpocoris pudicus* (Poda), *Chlorochroa juniperina* (L.), *Chlorochroa pinicola* (Mls.), *Holcostethus vernalis* (Wolff.), *P. prasina, A. acuminata* and *Codophila varia* (F.). The vibratory signals emitted

Table 6.1. Vibrational songs described in Pentatomine stink bugs in the calling, courtship, rivalry, copulatory and repellent behavioural contexts with references.

Species	References	Calling		Courtship		Riv	CPS	RP
		F	M	F	M	M	M	F
Aelia acuminata	Shestakov 2015		•		•			
Carpocoris fuscispinus	Shestakov 2015		•			•		•
Carpocoris purpuriepennis	Shestakov 2015		•					
Carpocoris pudicus	Shestakov 2015				•			
Chinavia hilare	Čokl et al. 2001	•	•	•	•	•		
Chinavia ubica	Laumann et al. 2016	•	•		•			
Chinavia impicticornis	Blassioli-Moraes et al. 2005	•	•	•	•	•		
Chlorochroa juniperina	Shestakov 2015		•		•			
Chlorochroa ligata	Bagwell et al. 2008	•	•		•	•	•	
Chlorochroa pinicola	Shestakov 2015		•		•			•
Chlorochroa uhleri	Bagwell et al. 2008	•	•		•		•	
Chlorochroa sayi	Bagwell et al. 2008	•	•		•	•	•	
Codophila varia	Shestakov 2015		•					
Dichelops melacanthus	Blassioli-Moraes et al. 2014	•	•	•	•	•		
Dolycoris baccarum	Shestakov 2015					•		•
Edessa meditabunda	Silva et al. 2012	•	•		•	•		
Eurydema ornata	Shestakov 2015		•		•			
Eurydema oleracea	Shestakov 2015		•					
Eushistus conspersus	McBrien and Millar 2003	•	•	•	•		•	
Eushistus heros	Blassioli-Moraes et al. 2005	•	•	•	•	•	•	
Graphosoma semipunctatum	Shestakov 2015		•		•			•
Graphosoma lineatum	Shestakov 2015		•					•
Holcostethus strictus	Pavlovčič and Čokl 2001	•	•	•	•		•	
Holcostethus vernalis	Shestakov 2015				•			•

Table 6.1 contd. ...

...Table 6.1 contd.

Species	References	Calling		Courtship		Riv	CPS	RP
		F	M	F	M	M	M	F
Murgantia histrionica	Čokl et al. 2004	•	•		•	•		
Nezara antennata	Kon et al. 1988	•	•	•	•			
Nezara viridula	Čokl et al. 2000	•	•	•	•	•		•
Palomena prasina	Čokl et al. 1978	•	•		•			
Palomena viridissima	Čokl et al. 1978	•	•		•			
Pentatoma rufipes	Shestakov 2015		•		•			•
Piezodorus guildinii	Blassioli-Moraes et al. 2005	•	•	•		•		
Piezodorus lituratus	Gogala and Razpotnik 1974		•		•	•		
Staria lunata	Shestakov 2015		•		•			
Thyanta custator accera	McBrien et al. 2002	•	•	•	•			
Thyanta pallidovirens	McBrien et al. 2002	•	•	•	•			
Thyanta perditor	Blassioli-Moraes et al. 2005	•	•	•	•	•	•	

CPS = copulatory songs, F = female songs, M = male songs, Riv = rival songs, RP = repellent songs.

by *Carpocoris fuscispinus* (Boh.), *Carpocoris purpuriepennis* (Des.) and *E. ornata* have calling and rivalry functions, and those of *Graphosoma semipunctatum* (F.) and *Staria lunata* (Hahn) were recorded during the calling and courtship phases of mating behaviour. These data significantly enlarged the number of known Pentatomine species with described vibratory emissions. These findings provide a solid basis for further investigations to identify the female emissions in these species and to describe the behavioural context of duets under more controlled conditions.

Calling

The female calling song

The female calling song (FCS), which is also known as the female first song (FS-1 or FS1), is characteristic of the calling phase of mating behaviour. It triggers the first song of a duet in most of the Pentatomine stink bugs that have been studied to date, and it elicits various male responses.

The female calling song is produced (a) as the first vibratory emission in a duet by *Chinavia impicticornis* (Stål), *Chinavia ubica* (Rolston), *Dichelops melacanthus* (Dallas), *Euschistus heros* (Fabr.), *P. prasina*, *Palomena viridissima* (Poda), *Piezodorus guildinii* (Westwood), *Thyanta custator accera* (McAtee), *Thyanta pallidovirens* (Stål) and *Thyanta perditor* (Fabr.), (b) as a response to male calling in *Chlorochroa sayi* (Stål), *Chlorochroa uhleri* (Stål), *Edessa meditabunda* (Fabr.), *Holcostethus strictus* (Fabr.) and *Murgantia histrionica* (Hahn) and (c) as the first song in a calling duet or as a response to male calling in *Chinavia hilaris* (= *Acrosternum hilare*) (Say), *Chlorochroa ligata* (Say), and *Euschistus conspersus* (Uhler).

The time and frequency characteristics of the female calling songs from different species are shown in Table 6.2. This song is generally characterized by long-lasting sequences of low-frequency, narrowband units with a stable duration and a highly independent repetition rate (Fig. 6.1). Species-specificity is primarily expressed in the time pattern of song pulses or pulse trains. The female calling song is composed of readily repeated pulse trains in *C. impicticornis*, *C. ubica*, *C. ligata*, *C. uhleri*, *C. sayi*, *N. viridula* and *T. perditor*. The duration of the pulse trains depends on the number of pulses per group. The longest pulse trains were recorded in *T. perditor*; pulse trains

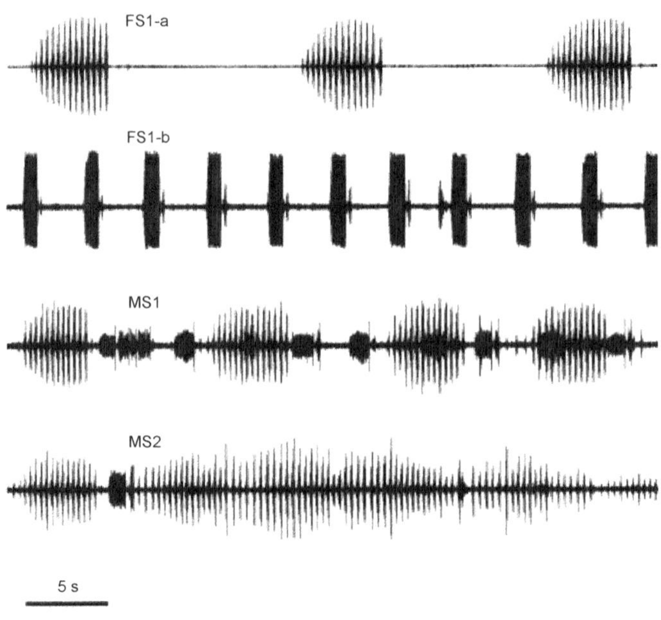

Figure 6.1. Vibratory songs of *C. impicticornis* that were recorded on the plant. The female calling song of the first (FS1-a) and second (FS1-b) types, the male calling song pulse trains (MS-1) emitted in a duet with FS-1b signals, and the male courtship song pulse sequence triggered by FS1-b.

Table 6.2. Mean values (± SD, N/n) of Pentatominae stink bug female calling song signal pulse or pulse train duration (ms), repetition time (ms) and dominant frequency (Hz).

Species	R	Duration		Repetition time		Dominant frequency	
		Pulse train	Pulse	Pulse train	Pulse	Pulse train	Pulse
Chinavia impicticornis	LO	811 ± 34 (150/6)	560 ± 50 (159/6)–112 ± 11 (159/6)	1731 ± 54 (112/3)–2024 ± 43 (48/2)	960 ± 79 (89/4)	133 ± 4 (180/4)–105 ± 8 (110/2)	125 ± 2 (90/4)–100 ± 5 (78/4)
Chinavia impicticornis FS-1a	LA	3886 ± 250 (47/3)	195 ± 32 (49/3)–234 ± 33 (62/3)	11008 ± 856 (15/3) –11676 ± 485 (12/3)	233 ± 29 (151/3)	92.3 ± 1.2 (12/3)–94.2 ± 1.1 (20/3)	-
Chinavia ubica FS-1a	LA	1878 ± 153 (15/3)–3004 ± 721 (10/3)	142 ± 22.7 (30/3)–175 ± 175 (53/3)	4191 ± 1130 (14/3) –6848 ± 1125 (15/3)	196 ± 54 (148/4)	106.5 ± 2.5 (12/3) –109.3 ± 1.9 (15/3)	-
Chinavia hilare	LO	–	638 ± 167 (104/7)	–	3756 ± 900 (13/3)	–	82 ± 2 (66/6)
*Chlorochroa ligata-cal-SF**	LO	4917 ± 3657 (380/19)	59 ± 12 (380/19)	19609 ± 13793 (380/19)	–	98 ± 14 (380/19)	87 ± 13 (380/19)
Chlorochroa ligata-reply	LO	1710 ± 261 (20/1)	–	27670 ± 11394 (20/1)	–	89 ± 5 (20/1)	-
Chlorochroa sayi	LO	5217 ± 2395 (37/7)	84 ± 20 (37/7)	Transient reply to male	–	93 ± 9 (37/7)	87 ± 6 (37/7)
Chlorochroa uhleri	LO	1433 ± 1017 (55/3)	392 ± 108 (100/5)	3588 ± 3047 (55/3)	766 ± 107 (100/5)	82 ± 7 (100/5)	91 ± 8 (57/3)
Dichelops melacanthus	LO	–	347 ± 90 (181/22)	–	879 ± 439 (229/22)		145 ± 5 (278/22)
	LA	–	349 ± 95 (415/26)	–	417 ± 398 (393/26)	–	134 ± 15 (414/20)

Table 6.2 contd. ...

...Table 6.2 contd.

Species	R	Duration		Repetition time		Dominant frequency	
		Pulse train	Pulse	Pulse train	Pulse	Pulse train	Pulse
Edessa meditabunda	LO	–	433 ± 78.8 (80/11)	–	677.2 ± 104.5 (80/11)	–	77.6 ± 6 (80/11)
Eushistus conspersus	LO	–	352 ± 105 (186/16)	–	591 ± 172 (168/16)	–	131 ± 14 (186/16)
Eushistus heros	LO	–	948 ± 183 (79/5)	–	3500 ± 1043 (74/5)	–	145 ± 7 (77/5)
E. heros-emitted without male response	LA	–	817 ± 68 (20/5) – 1382 ± 150 (20/5)	–	2748 ± 718 (20/5)– 3098 ± 535 (20-5)	–	111 ± 2 (20/5)– 130 ± 4 (20/5)
E. heros-emitted in duet with MS-1	LA	–	925 ± 186 (20–5) – 1605 ± 295 (20/5)	–	2475 ± 329 (20/5) –3098 ± 535 (20/5)	–	107 ± 4 (20/5)– 134 ± 2 (20/5)
Holcostethus strictus	LO	–	93 ± 26.6 (291)	–	340 ± 124.6 (199)	–	111 ± 8 (54)
Murgantia histrionica	LO	–	628 ± 128 (248/8)	–	1196 ± 205 (165/4)	–	95 ± 1 (30/2)
Nezara antennata	MI	–	145 ± 19 (27/1)	–	1488 ± 487 (27/1)	±	50 ± 14 (27/1)
N. viridula-NB prepulse	LO	–	196 ± 37 (100/5)	–	Fused	–	111 Hz

N. viridula-NB long pulse	LO	–	1667 ± 365 (342/8)	–	5060 ± 787 (342/8)	–	94
N. viridula-BB pulse	LO	-	127 ± 45 (284/8)	–	172 ± 54 (149/8)	–	109
Palomena prasina	MI	–	398 (217–630) (321)	–	1010 (268 – 2476) (556)	–	90.4 ± 3.5 (16/1)
Palomena viridissima	MI	–	351 ± 115	–	–	–	< 200
Piezodorus guildinii DF 1st part-DF 2nd part	LO	–	4908 ± 641 (70/3)	–	Not applicable	–	168 ± 10 (35/3) –120 ± 6 (35/3)
Thyanta custator acc.	LO	–	239 ± 65 (135/5)	–	337 ± 65 (184/9)	–	102 ± 4 (319/12)
Thyanta pallidovirens	LO	–	209 ± 64 (159/11)	–	386 ± 103 (128/11)	–	100 ± 7 (159/11)
Thyanta perditor	LO	13410 ± 194 (52/7)	210 ± 40 (117/7)	18810 ± 2343 (52/7)	370 ± 50 (100/6)	Not determined	79 ± 6 (117/7)

**Chlorochroa ligata* FS1 calling song type PT/SF; R = type of registration; LO = recorded on the loudspeaker, LA = recorded with the laser vibrometer, MI = recorded with the microphone; *Nezara viridula*-NB = narrowbanded FS-1, *Nezara viridula*-BB = broadbanded FS-1.

of approximately thirteen s are composed of approximately sixty readily repeated pulses which are approximately two hundred ms long. *N. viridula* females produce two types of female calling song signals (Čokl et al. 2000). The non-pulsed type (FCS-np) is composed of a long pulse preceded by a short pre-pulse, and the pulsed type (FCS-p) is constituted of readily repeated pulses that are approximately 130 ms long. The female calling song of the other Pentatomine species is characterized by readily repeated pulses with durations ranging from a few hundred ms in *T. pallidovirens* to several s as measured in *P. guildinii*. The dominant frequency of all the female calling songs as described until the present ranges between seventy and 180 Hz.

The male calling song

The male calling song (MCS), which is also known as the male first song (MS-1 or MS1), is produced in different species either as the first emission in a duet or as the response to a calling female (Fig. 6.1, Table 6.3).

In *C. uhleri, C. sayi, H. strictus* and *M. histrionica*, the male calling song triggers female responses. In *C. impicticorne, E. heros, P. guildinii, T. perditor, C. ligata* and *C. hilaris* (= *A. hilare*), the song is produced either spontaneously or in response to female calling. In most of the investigated species (*N. viridula, Nezara antennata* Scott, *P. prasina, P. viridissima, D. melacanthus, E. meditabunda, E. conspersus, T. pallidovirens* and *T. custator accera*), the male calling song is emitted spontaneously in the presence or absence of a female, as a response to female calling or as the first emission in a calling duet. *P. guildinii* males emit fast repeated pulses of their first male song after approaching and antennating females. In *E. conspersus*, McBrien and Millar (2003) differentiated among three different male song types (MS-1, MS-2 and MS-3) that were produced during the calling phase of mating behaviour. MS-1 and MS-2 song pulses of similar time and significantly different spectral properties change into a sequence of shorter pulses of the MS-3 song that gives rise to the fourth male song (MS-4) preceding copulation during the courtship phase. Differences between species are primarily expressed in terms of the time characteristics of pulses or pulse trains as exemplified in *C. impicticornis, C. ubica, C. hilaris, C. ligata, C. uhleri* and *T. perditor*. The dominant frequency of male calling songs ranges between means of approximately 74 Hz in *E. meditabunda* to 145 Hz in *E. heros*. *D. melacanthus* females and males emit two types of calling song pulses that differ in time characteristics (Blassioli-Moraes et al. 2014). The first type was recorded in male-female couples and the second type was recorded when they were placed individually or as a pair. MS-2 signals were produced randomly as a sequence of two to six pulses intercalated with MS-1, and FS-2 appeared as few pulses emitted within a sequence of 10–30 FS-1 pulses.

Male responses to female calling

In most (but not all) Pentatomine stink bugs, the calling female emits vibratory signals continuously from the same location on a plant, and the male responds by emitting the male calling song, engaging in a directional movement towards the female and increasing pheromone production.

Recently, Zgonik and Čokl (2014) investigated the early phase of calling behaviour in *N. viridula* with a special interest in signals that trigger the emission of the female or male calling song. Each test was conducted on a fresh plant in order to avoid pheromone and/or contact chemical signal residues. The results of experiments with a single male or female on a plant have shown that in the light or the dark, males spontaneously emit the calling song more often than females. Experiments using insect couples on a plant showed that optical contact with the female triggers the emission of a male calling song that is soon replaced by courtship song pulse trains, which are readily exchanged in a duet with female calling song pulse trains. Male calling was induced more often by presenting live females than males because the female-specific rotation of antennae was induced in the vicinity of a male or by the presence of the male pheromone. The female responded with the calling song exclusively to the presence of the male pheromone.

Miklas and co-authors (2003a) first showed the direct influence of vibratory communication signals on pheromone production. The stimulation of *N. viridula* males with the conspecific female calling song increased the amount of collected pheromone. We may hypothesize that increased amounts of pheromone give females information on the presence of more males on the plant, attract more mates to land on it, prolong female calling and increase their motivation for copulation. The question of potential rivalries between several females on a given plant in the presence of the male pheromone is still open.

In most species, males respond to calling females by emitting the first male song with a calling function. *N. viridula* males are exceptions in that they often respond to the calling female with the courtship song at distances that prevent visual or any other contact. In playback experiments, males responded to artificially synthesized female calling songs more often and with higher selectivity in the courtship than the calling song (Žunič et al. 2011). The authors observed that the male calling song was usually emitted in response to synthesized female calling songs with an inter-pulse interval and repetition time outside the range of natural female calling song characteristics. Males also responded exclusively with the calling song when stimulated with the female calling song of the sympatric species *P. prasina* and *P. viridissima*. The role of the *N. viridula* male calling song requires further investigation because signals of similar time and frequency patterns have also been recorded during antennation prior to copulation. In addition, it is not clear what the significance of different narrow and

Table 6.3. Mean values (± SD, N/n) of Pentatominae stink bug male calling song signal pulse or pulse train duration (ms), repetition time (ms) and dominant frequency (Hz).

Species	R	Duration		Repetition time		Dominant frequency	
		Pulse train	Pulse	Pulse train	Pulse	Pulse train	Pulse
Chinavia impicticornis MS-2 (LO) = MS-1 (LA)	LO	938 ± 54 (34/3) 2459 ± 331 (114/5)	130 ± 5 (336/2) 82 ± 2 (271/4)	3598 ± 1703 (87/7)	197 ± 23 (632/9)	130 ± 2 (65/5)–92 ± 2 (43/5)	Not applicable
Chinavia impicticornis	LA	4971 ± 535 (30/3)	214 ± 34 (185/3)	Not determined	231 ± 23 (65)–272 ± 72 (53)	91 ± 2 (10)–98 ± 1 (10)	Not determined
Chinavia ubica	LA	1740 ± 193 (10)–3332 ± 110 (10)	163 ± 27 (54)–187 ± 37 (19)	Not determined	190 ± 22 (48)–225 ± 31 (36)	105 ± 2 (10) – 120 ± 10 (10)	Not determined
Chinavia hilare	LO	2309 ± 464 (88/11)	–	Not determined	–	–	77 ± 2 (8)^{S} 94 ± 2 (10)^{L}
Chlorochroa ligata	LO	1710 ± 261 (201/1)	–	27670 ± 11394 (20/1)	–	89 ± 5 (20/1)	–
Chlorochroa sayi	LO	1090 ± 204 (300/15)	–	11754 ± 611 (300/15)	–	108 ± 16 (300/15)	–
Chlorochroa uhleri	LO	–	1716 ± 295 (351/19)	–	22117 ± 10612 (336/19)	–	93 ± 9 (353/19)
Dichelops melacanthus	LO	–	581 ± 164 (91/26)	–	1065 ± 268 (195/26)	–	110 ± 10 (201/26)
	LA	–	591 ± 238 (410/28)	–	874 ± 516 (389/28)	–	121 ± 19 (410/28)
Edessa meditabunda	LO	–	1667 ± 147 (76/11)	–	4590 ± 1834 (65/11)	–	74 ± 3 (76/11)

Eushistus conspersus	LO	–	476 ± 119 (239/24) [a]322 ± 23 (39/6)[d]	–	693 ± 156 (278/25)	–	130 ± 14 (278/25)
Eushistus heros	LO	–	948 ± 183 (79/5)	–	3566 ± 1043 (74/5)	–	145 ± 7 (77/5)
	LA	–	6839 ± 1873 (153/8)	–	16105 ± 9103 (20/8) –34070 ± 8859 (20/8)	–	107 ± 4 (20/8)– 159 ± 4 (20/8)
H. strictus-prepulse	LO	–	197.8–110.2 (42)	–	fused	–	104 ± 7 (38)
H. strictus-pulse	LO	–	153.1 ± 21 (56)	–	31200 ± 17200 (43)	–	155 ± 6 (39)
Murgantia histrionica	LO	–	9235 ± 3341 (10)	–	22756 ± 6474 (10)	–	96 ± 7 (10)
Nezara antennata	MI	–	59 ± 14 (92/6)	–	991 ± 270 (92/6)	–	50–240
N. viridula-NB	LO	–	263 ± 96 (191/5)	–	1202 ± 467 (191/5)	–	96
N. viridula-BB	LO	–	121 ± 34 (209/5)	–	455 ± 222 (207/5)	–	109
Palomena prasina	LA	–	196 (101–369) (683)	–	311 (152–680) (626)	–	95 ± 8.1 (194)
Piezodorus guildinii	LO	–	24 ± 5 (43/3)	–	NA	–	117 ± 10 (45/3)
Thyanta custator acc.	LO	–	465 ± 171 (139/11)	–	1072 ± 276 (91/9)	–	97 ± 7 (139/11)
Thyanta pallidovirens	LO	–	339 ± 82 (63/7)	–	715 ± 93 (46/7)	–	101 ± 3 (63/7)
Thyanta perditor	LO	7100 ± 696 (27/7)	522 ± 176 (127/7)	ND	959 ± 411 (90/6)[alone] 903 ± 89 (21/1)[duet]	ND	122 ± 15 (21/1)[a]10 ± 11 (106/6)[d]

S—indicates short; L—indicates long; a—indicates alone; d—indicates duet broadband types of the male calling song pulses is in temporal and spectral terms (Čokl et al. 2000).

The most obvious reaction of a male is his oriented movement towards a female that is calling from one place on the plant. Vibrational directionality mediated by the female calling song was first described in *N. viridula* (Ota and Čokl 1991, Čokl et al. 1999). Searching males stop moving on the stem crossings with side branches and stalks, spread their legs in different "ways" and walk towards the source of the vibrations after receiving several female calling song signals. The amplitude difference between the *N. viridula* female calling song signals recorded on different branches at distances of 1 cm from the crossing with the stem is high enough to orient the male in the direction of the leg that was vibrated at the highest amplitude (Stritih et al. 2000). Nevertheless, the amplitude difference does not always provide reliable directionality information because of the non-linear and frequency-dependent decrease in the vibratory signal amplitude. The position of the peak maxima and minima (Chapter 7) varies according to the travelling distance, and consequently, the searching male may be faced with higher signal amplitudes on a branch leading in the wrong direction and away from the source. More relevant information represents the time difference between the signal arrival to different legs spread over a crossing. The male that has his legs spread over different stalks or branches will move in the direction of the first stimulated leg. A difference between the vibratory signal times of arrival to spatially separated legs as short as 0.2 ms triggers a directional response in scorpions (Brownell and Farley 1979). The low frequency vibratory signals of Pentatomine stink bugs propagate through green host plants with velocities of up to 50 m/s (Chapter 7), creating arrival time differences of at least 0.5 ms at distances of 2.5 cm between the legs. The neuronal processing of relevant amplitude and time differences has not yet been investigated in Pentatominae or Heteroptera in general.

Courting

The female courtship song

The female courtship song (FCrS), which is also known as the female second song (FS2, FS-2), was recorded in 15 species (Table 6.4) during the courtship phase of mating behaviour. In comparison to the calling song, this female vibratory emission type produced a less regular and shorter response to male courting. The emission of the female courtship song is usually accompanied by visual and contact chemical or tactile signals that are exchanged between mates at close and contact distances. The function of the *T. perditor* female second song that is emitted in the presence or absence of a male is not clear yet. The primary spectral energy of the female second (courtship) song signals is produced within the characteristic Pentatominae range, with the lowest dominant frequencies being approximately 78 Hz in *C. impicticornis* and the highest being approximately 152 Hz in *E. heros*. The female courtship song develops from the calling song and is characterized in *C. impicticornis* and *C. ubica* by pulses or pulse trains of different durations or of similar duration as measured in *N. viridula*. Sub-groups of greater or fewer fused pulses form the complex pulse train pattern of the *C. uhleri* female courtship song. *P. guildinii, T. custator accera* and *T. pallidovirens* emit several s-long amplitude modulated pulses during courtship, and female courtship song pulses less than 0.5 s long have been described in *D. melacanthus, E. conspersus, H. strictus* and *N. antennata* (Kon et al. 1988).

The male courtship song

The male courtship song (MCrS) has also been described by different authors as the male second song (MS-2, MS2), and it is the typical male emission in most of the species that have been recorded during courtship (Fig. 6.1). All male courtship songs described to date (Table 6.5) have typical Pentatomine frequency characteristics, with the dominant frequency ranging from approximately 62 Hz in *C. ligata* to 175 Hz in *E. heros*. Species-specific differences in the male courtship song with respect to temporal characteristics and amplitude modulation patterns are more frequently expressed in comparison with the calling song.

The male courtship song develops from the calling song over various transitional forms. We can differentiate between three basic types of the male courtship song.

(a) The *N. viridula* type is characterized by the fusion of calling song pulses into a several s-long pulse train. The latter are triggered and exchanged by female calling song pulse trains. Similar courtship duetting was described in *P. prasina, P. viridissima, T. custator accera* and *T. pallidovirens*. *P. prasina* males produce courtship song pulse trains that are composed of pulses with different amplitude and temporal characteristics. Within the song characteristic form, there are longer pulses of higher amplitude preceded by lower amplitude short and fused pulses (Polajnar et al. 2013). The authors classified the latter as a special courtship song type (MS-3) because they may also appear as separate emissions. The role of MS-3 pulse trains is not yet clear. The species-specificity of the male courtship song is expressed primarily in the time pattern of constitutive units.

(b) The *Chlorochroa* male courtship song type is characterized by a transition from the male calling song pulse train to a sequence of more or less readily repeated single pulses. This type of courtship signalling has been described in *C. hilaris, C. ligata, C. sayi, C. uhleri, C. impicticornis, M. histrionica, D. melacanthus, E. meditabunda* and *H. strictus*. Species-specificity is determined by the different pulse durations and repetition

Table 6.4. Mean values (± SD, N/n) of Pentatominae stink bug female courtship song signal pulse or pulse train duration (ms), repetition time (ms) and dominant frequency (Hz).

Species	R	Duration		Repetition time		Dominant frequency	
		Pulse train	Pulse	Pulse train	Pulse	Pulse train	Pulse
Chinavia impicticornis	LO	1129 ± 190 (22/4) 1802 ± 454 (69/4)	113 ± 29 (334/5)	5 ± 2 (22/4) 12 + 3 (69/4)	Not applicable	109 ± 6 (178/3) 123 ± 7 (58/2)	106 ± 6 (70/3)
Chinavia impicticornis FS-1b	LA	1048 ± 112 (13/6) –2709 ± 253 (20/6)	235 ± 59 (58/3)– 955 ± 99 (10/4)	3078 ± 204 (14/6) –9340 ± 635 (20/6)	–	77.9 ± 0.6 (20/6) – 88.8 ± 0.7 (20/6)	–
Chinavia ubica FS-1b	LA	1350 ± 204 (30/3)	138.8 ± 28.2 (18/3)–841 ± 132.6 (33/3)	3020 ± 888 (32/33)	169 ± 21 (10/3)– 898 ± 137 (33/3)	109.2 ± 0.7 (18/3)– 115 ± 4.9 (8/3)	–
Chlorochroa uhleri	LO	1433 ± 1017 (55/3)	-	3588 ± 3047 (35/3)	–	91 ± 8 (57/3)	–
Dichelops melacanthus variation of FS-1	LA	–	199 ± 46 (156/14)	–	256 ± 165 (144/14)	–	132 ± 18 (157/14)
Dichelops melacanthus-FS-3/MS-3 in courtship	LA	–	221 ± 63 (157/7)	–	109 ± 84 (152/7)	–	FM 179 ± 12– 146 ± 5 (57/7)
Eushistus conspersus	LO	–	163 ± 36 (68/12)	–	765 ± 422 (45/11)	–	111 ± 13 (68/12)
Eushistus heros	LO	5663 ± 1198 (44/6)	–	Not applicable	–	152 ± 7 (44/6)	–
	LA	Laser recorded signals show that previous FS-2 is overlapped FS-1 with MS1					

Holcostethus strictus	LO	–	208 ± 61.2 (705)	–	429.5 ± 151 (697)	–	133 ± 15 (54)
Nezara antennata	MI	–	85.6 ± 17 (57/3)	–	488.4 ± 178.8 (57/3)	–	110–190 (57/3)
N. viridula	LO	3904 ± 1179 (135/9)	–	irregular	–	90	–
Piezodorus guildinii	LO	–	4630 ± 200 (41/3)	–	18630 ± 1780 (41/3)	–	131 ± 2 (41/3)
Thyanta custator acc.	LO	–	3078 ± 2152 (38/8)	–	Not applicable	–	97 ± 12 (38/8)
Thyanta pallidovirens	LO	–	3401 ± 1339 (73/13)	–	6635 ± 5365 (23/5)	–	105 ± 5 (73/13)
Thyanta perditor	LO	–	710 ± 290^{S} (73/4) 1462 ± 630^{L} (14/4)	–	1897 ± 1102^{S} (44/4) 3166 ± 1762^{L} (14/4)	–	83 ± 4^{S} (81/4) 84 ± 4^{L} (14/4)

Chinavia impicticornis FS-1b = FS-2; *Chinavia ubica* FS-1b = FS-2; ^{S} = short pulses; ^{L} = long pulses.

rates. The *C. hilaris* pulses and the *M. histrionica* pulse trains of male courtship songs are emitted as a response to female calling, and the pulse trains of *C. impicticornis* are emitted as a response to the female second song. The connection between male courtship song emissions and female signals is not clear in *C. hilare, C. impicticornis* and *C. ubica*. The male courtship song of *H. strictus* was recorded together with the female courtship song without the regular pattern of signal exchanges, and the pulses of the second male song of *D. melacanthus* were intercalated within those of the calling song and developed into a male third song in the vicinity of a female as a response to female third song signals. Pulses of the *E. meditabunda* male courtship song alternate and later superimpose those of the female first song.

(c) The *Euschistus* type of male courtship song is characterized by pulse trains originating from the longer pulses of the male calling song. Male courtship vibratory emissions of this type have been described in *E. heros, E. conspersus, C. ubica* and *T. perditor.* The species-specificity is expressed in the form of temporal song unit characteristics. The *E. conspersus* male emitted its fourth song (MS-4) during the courtship phase that developed from the second one (MS-2); the latter was produced in duets with the female first and second songs. The *C. ubica* second male song is produced as a response to the female song signals, and *E. heros* male courtship song signals are triggered by those of the female first song. The latter overlap the MS-2 responses and cause significant changes in the amplitude pattern by interference (Chapter 7) (Čokl et al. 2015). The second male song signals of *T. perditor* were recorded as a response to female calling signals, and they develop into the third male song that is emitted during butting and antennation prior to copulation.

Males and females significantly change their mating behaviour and signalling modalities when they are close together. Recognition and motivation for copulation are enhanced by vision and by the detection of tactile signals during mutual antennation. The role of these signal modalities in courtship has not yet been investigated in detail. Close-range courtship behaviour was first described in *C. ligata* and *Cosmopepla bimaculata* (Thomas) by Fish and Alcock (1973) as the male antennation of the female's body, the male stroking with antennae on the underside of the female's abdomen and the head butting of the female abdomen to lift and turn it 180° in order to adopt a copulatory position. A similar sequence of courtship events has been described in many other Pentatomine stink bugs such as *N. antennata, N. viridula* (Mitchel and Mau 1969, Harris and Todd 1980, Borges et al. 1987), *C. hilaris, T. custator accera* and *T. pallidovirens, E. conspersus, E. meditabunda, D. melacanthus, C. impicticornis* and *C. ubica*.

Table 6.5. Mean values (± SD, N/n) of Pentatominae stink bug male courtship song signal pulse or pulse train duration (ms), repetition time (ms) and dominant frequency (Hz).

Species	R	Duration		Rep. time	Trigger/origin	Dominant frequency	
		Pulse train	Pulse	PT or pls		Pulse train	Pulse
Chinavia impicticornis MS-1 (LO) = MS-2 (LA)	LO	943 ± 5 /79/4)[S]– 2534 ± 389 (45/5)[L]	280 ± 12 (50/3)	PT 8419 ± 717 (47/3)* –12102 ± 1287 (38/3)** Pulse 410 ± 23 (50/3)	Close contact with the female	120 ± 8 (45/5)	94 ± 6 (57/3)
	LA	–	196 ± 16 (30)–227 ± 28 (30)	244 ± 18 (30)–286 ± 43 (30)	Close contact with the female	–	97–104
Chinavia ubica	LA	–	387 ± 339 (22)– 1057 ± 611 (18	485 ± 253 (34)–1212 ± 648 (17)	Close contact with the female	–	104–135
Chinavia hilare	LO	–	105 ± 21 (133/7)	367 ± 74 (114/6)	FS-1/MS1 pulse train to single pulses	–	123 ± 9 (40/4)
Chlorochroa ligata long pulse	LO	–	625 ± 223 (394/20)	1875 ± 884 (377/20)	MS-1 pulse train to a sequence of MS-2 long and short pulses	–	62 ± 10 (304/20)
Chlorochroa ligata short pulse	LO	–	114 ± 29 (400/20)	302 ± 86 (400/20)			82 ± 16 (400/20)
Chlorochroa sayi	LO	–	116 ± 35 (275/13)	fused	MS-1 pls train to short almost fused pulses	–	82 ± 13 (275/13)
Chlorochroa uhleri long pulse	LO	–	567 ± 241 (240/12)	1010 ± 593 (239/12)	MS-1 pulse train to a sequence of MS-2 long and short pulses	–	70 ± 13 (240/12)
Chlorochroa uhleri short pulse	LO	–	129 ± 40 (317/16)	285 ± 77 (317/16)		–	91 ± 19 (317/16)

Table 6.5 contd. ...

...Table 6.5 contd.

Species	R	Duration		Rep. time	Trigger/ origin	Dominant frequency	
		Pulse train	Pulse	PT or pls		Pulse train	Pulse
Dichelops melacanthus variation of MS-1	LA	–	195 ± 6 (100/11)	297 ± 259 (100/11)	FS-1/MS1 to MS-1 variations	–	118 ± 17 (100/11)
Dichelops melacanthus MS-3/ MCrS function	LO	Not recorded on the loudspeaker membrane					
	LA	–	929 ± 781 (160/7)	NA	FS-3/MS-3 has the MCrS function	–	114 ± 16 (160/7)
Edessa meditabunda	LO	5–11 pls/pls train	174 ± 84 (75/12)	306 ± 211 (64/12)	FS-1/long pls to pls trains	–	80 ± 6 (75/12)
Eushistus conspersus MS-4 is the true male courtship song	LO	312 ± 102 (67/7)	108 ± 20 (174/7)	Pls train 623 ± 142 (65/7) pls 119 ± 20 (66/7)	MS-3 pls trains prior to copulation; MS-2, MS-3 variations of MS-1	106 ± 10 (67/7)	111 ± 7 (51/9)
Eushistus heros MS-3 (LO) = MS-2 (LA)	LO	–	54 ± 9 (58/2)	81 ± 12 (53/2)	FS-1/LO MS-3 from MS-1	–	175 ± 5 (25/2)
	LA	–	58 ± 9 (50/6)– 131 ± 19 (50/6)	141 ± 40 (50/6)–289 ± 51 (50/6)	FS-1/splitting of MS-1 pulse into sequence of short pulses	–	105 ± 3 (50/6)– 153 ± 10 (50/6)
Holcostethus strictus pulse type 1	LO	–	45 ± 9 (678)	1743 ± 1279 (690)	FCrS/MCS to MCrS pulse train (phrase) that consisted of 3 types of pulses (echemes 1, 2, 3)	–	90 ± 10 (54)
Holcostethus strictus pulse type 2	LO	–	47 ± 8 (700)	131 ± 30 (700)		–	104 ± 6 (51)
Holcostethus strictus pulse type 3	LO	–	372 ± 91 (293)	NA		–	FM107 ± 7 (54) –212 ± 21 (54)

Murgantia histrionica MS-2 pulse train composed of the 1st, other and trill pulses parts	LO	10056 ± 3368 (81/5)	1st pulse 6366 ± 2087 (104/8) other pulses 2026 ± 15326 (145/5) trill 88 ± 12 (172/3)	NA 1st pulse NA other pulses 60 ± 19 (102/6) trill NA	FS-1/MS1 pulse to MS-2 pulse train	trill 87 ± 3 (12)	ND
Murgantia histrionica MS-3 pulses	LO	–	199 ± 35 (80/4)	551 ± 74 (60/3)	MS-3 pulses prior to copulation	–	94 ± 5 (12)
Nezara antennata	MI	ND					
Nezara viridula	LO	3110 ± 749 (50/5)	–	–	FS-1/Fusion of MS-1 pulses	90	–
Palomena prasina	MI	±	±			±	±
P. prasina pre-pulse	LA	–	413 (109–1173)	ND	FS-1/pulses to pulse trains	–	ND
P. prasina other pulses		–	268 (87–825)	369 (159–999)	FS-1/pulses to pulse trains	–	ND
P. prasina pulse train		1689 (557–23445) (N = 419)	–	Highly variable	FS-1/pulses to pulse trains	95.5 ± 7.6 (N = 109)	–
Palomena viridissima	MI	ND			FS-1/single MS-1 pulses to 4 pulses pulse train	ND	
Piezodorus guildinii	LO	MS-1 recorded in the calling and courtship phase of mating behaviour					

Table 6.5 contd. ...

...Table 6.5 contd.

Species	R	Duration		Rep. time	Trigger/ origin	Dominant frequency	
		Pulse train	Pulse	PT or pls		Pulse train	Pulse
Thyanta custator acc. Alone	LO	3353 ± 1024 (80/11)	±	24760 ± 2513 (52/8)	No FS/short pulses to long pulse	95 ± 6 (195/16)	–
Thyanta custator acc. with FS-2		2241 ± 788 (115/7)	±	2876 ± 657 (103/7)	FS-2/pulse or pulse train (2–11 fused pls)	75 ± 4 (20/7)–98 ± 6 (20/7)	–
Thyanta pallidovirens	LO	3359 ± 681 (66/14)	42 ± 8 (66/14)	4802 ± 1028 (56/13) pulses fused	FS-2/MS-1 pulses into pulse trains	96 ± 5 (66/14)	NA
Thyanta perditor	LO	–	3390 ± 990 (40/7)* 4490 ± 1290 (8/1)**	NA	FS-1/sequence of MS-2 pulses	–	109 ± 12 (40/8)
Thyanta perditor MS-3 is probably MS-2	LO	2110 ± 460 (138/12)	–	5290 ± 1380 (138/12)	Emitted when butting the female, MS-2 to MS-3	70 ± 7 (138/12)	–

NA = Not applicable, ND = Not determined the female courtship song signals prior to copulation.

*singing alone and **singing in a duet.

Rivalry

Rivalry accompanied by the emission of the male rival song (MRS) (Table 6.6) has been described when several males were calling and courting the same female in *C. impicticornis, C. ligata* and *C. sayi, D. melacanthus, E. heros, M. histrionica, N. viridula, P. guildinii* and *T. perditor*. The song develops either from the male calling or courtship song and is terminated when one male stops alternating. The emission of the male rival song silences the female until the winning male changes the rival song into a calling or courtship song. The rival song is characterized by pulses of different durations ranging from 360 ms in *N. viridula* to above 1,330 ms in *C. impicticornis*. Male *C. sayi* emits rival song pulses with mean durations of approximately 516 ms or 173 ms. The mean dominant frequency ranges between 96 Hz in *D. melacanthus* to 149 Hz in *E. heros*. The rival song pulses in most of the investigated stink bug species are characterized by frequency modulation. The regular alternation of the pulses produced by two rival males is characteristic for all of the above mentioned species except for *E. heros* and *C. ligata*. The emission of the rival song by one *E. heros* male triggered the other to respond with the first male song pulses, forming an alternating duet with a regular a-a-a-b-a-a-a-b- pattern. The male rival song of *C. ligata* consists of individual pulses that are exchanged between rival males in a non-alternating fashion for up to twelve hours with only short, half-to-two minute-long breaks.

Copulation

The characteristic pattern of courtship songs is disrupted prior to copulation into randomly produced pulses that are accompanied by high-amplitude substrate vibrations from the vigorous shaking of the whole body (tremulation) (Table 6.7). In *N. viridula*, the female courtship song changes the stable male courtship song pulse train pattern into a sequence of irregularly repeated pulses that silence the female and trigger her to accept or reject the male's copulatory attempts (Čokl 2008). The female's decision is potentially influenced by visual cues because females prefer males with longer antennae (McLain 1998). The *N. viridula* female rejects the male by tremulating her whole body and emitting the female repellent song (FRS) that is characterized by several s-long sequences of short pulses with broadband spectral characteristics. Courtship alternation in *M. histrionica* is terminated by the transition of the male courtship song pulse trains with regular time and amplitude modulation patterns into approximately 200 ms-long sequences of readily repeated pulses that either precede copulation or trigger rivalry singing. Immediately prior to copulation, *T. perditor* males change the courtship song pulse trains into a sequence of pulse trains; the sequence is characterized by fused pulses with irregular amplitude

Table 6.6. Mean values (± SD, N/n) of Pentatominae stink bug male rival song signal pulse or pulse train duration (ms), repetition time (ms) and dominant frequency (Hz).

Species	R	Duration		Repetition time		Dominant frequency	
		Pulse train	Pulse	Pulse train	Pulse	Pulse train	Pulse
Chinavia impicticornis	LO	–	1330 ± 400 (101/3)	–	1493 ± 490 (101/3)	–	107 ± 2 (101/3)
Chlorochroa ligata	LO	–	1284 ± 332 (400/20)	–	NA	–	126 ± 18 (400/20)
Chlorochroa sayi	LO	–	516 ± 172 (240/12)[L] 173 ± 48 (240/12)[S]	–	NA	–	127 ± 9 (240/12)[L] 97 ± 13 (240/12)[S]
Dichelops melacanthus	LO	–	1102 ± 675 (65/5)	–	ND	–	96 ± 5 (65/5)
Eushistus heros	LO	–	984 ± 275 (40/1)	–	1151 ± 298 (40/1)	–	149 ± 9 (40/1)
Murgantia histrionica	LO	–	366 ± 55 (90/3)	–	586 ± 118 (90/3)	–	99 ± 2 (30/1)
Nezara viridula	LO	–	353 ± 105 (63/5)	–	950 ± 362 (61/5)	–	118
Piezodorus guildinii	LO	–	980 ± 50 (33/3)	–	1640 ± 80 (33/3)	–	135 ± 8 (33/3)
Thyanta perditor	LO	–	680 ± 50 (53/4)	–	1250 ± 360 (53/4)	–	102 ± 6 (53/4)

NA = not applicable, ND = not determined.

modulation that silence the female and prepare her for copulation. McBrien and co-authors (2002) described three different male songs in *T. custator accera* that are produced prior to copulation. Pulses that form the second male song pulse trains fuse into approximately 4 s-long pulses of the male third song, which changes into the fourth male song prior to copulation and is characterized by a sequence of approximately 130 ms-long pulses. The male copulatory song (MCpS) of *H. strictus* has two pulse train types (E-1 and E-2) and has a more complex time pattern. The low amplitude E-1 pulse trains typically last for around 30 ms to more than one minute and are composed of fused pulses with a few single pulses at the end. The sequence terminates with 1 to 3 higher-amplitude E-2 pulse trains, each of which consists of 5–7 pulses. Prior to copulation, *D. melacanthus* females emit the female third song that triggers male responses. The song differs from

Table 6.7. Mean values (± SD, N/n) of Pentatominae stink bug copulatory signal pulse or pulse train duration (ms), repetition time (ms) and dominant frequency (Hz).

Species	R	Duration		Repetition time		Dominant frequency	
		Pulse train	Pulse	Pulse train	Pulse	Pulse train	Pulse
Chinavia hilare	LO	–	40 ± 6 (57/3)	–	209 ± 59 (75/4)	-	92–129
Chlorochroa uhleri	LO	–	127 ± 24 (120/6)	–	246 ± 47 (120/6)	–	94 ± 23 (120/6)
Chlorochroa sayi	LO	–	78 ± 36 (20/1)	–	134 ± 34 (20/1)	–	106 ± 16 (20/1)
Chlorochroa ligata	LO	–	106 ± 24 (60/3)	–	103 ± 8 (12)	–	92 ± 3 (12)
Murgantia histrionica	LO	–	83 ± 7 (12)	–	103±8 (12)	–	92 ± 3 (12)
Thanta perditor	LO	2110 ± 460 (138/12)	–	5290 ± 1380 (138/12)	–	70 ± 7 (138/12)	±
T. custator accera	LO	–	126 ± 22 (176/14)	–	808 ± 321 (142/14)	–	99 ± 9 (176/14)
H. strictus E-1 PT	LO	0.03–6000	–	–	–	117 ± 7 (54)	–
H. strictus E-2 PT	LO	786 ± 101 (16/5)– 1154 ± 129 (13/7)	–	2400 ± 300 (28)	–	FM sweep 165 ± 13 –113 ± 11 (54)	–

other female vibratory emissions by an extensive frequency modulation of approximately 200 ms-long pulses. Male *C. hilaris, C. uhleri, C. sayi* and *C. ligata, H. strictus, M. histrionica, T. custator accera* and *T. perditor* individuals emit several minute-long sequences of regularly repeated pulses that are less than 100 ms long in copula, with the dominant frequency ranging between 92 and 129 Hz. The role of the post-copulation male vibratory signal emission is not yet clear.

Multifunctional Vibratory Signals

The above-described calling, courtship and rival songs are produced by vibrating the abdomen. These signals were recorded in many species together with low amplitude percussion signals, high amplitude vibrations produced by the vigorous tremulation of the whole body (tremulatory

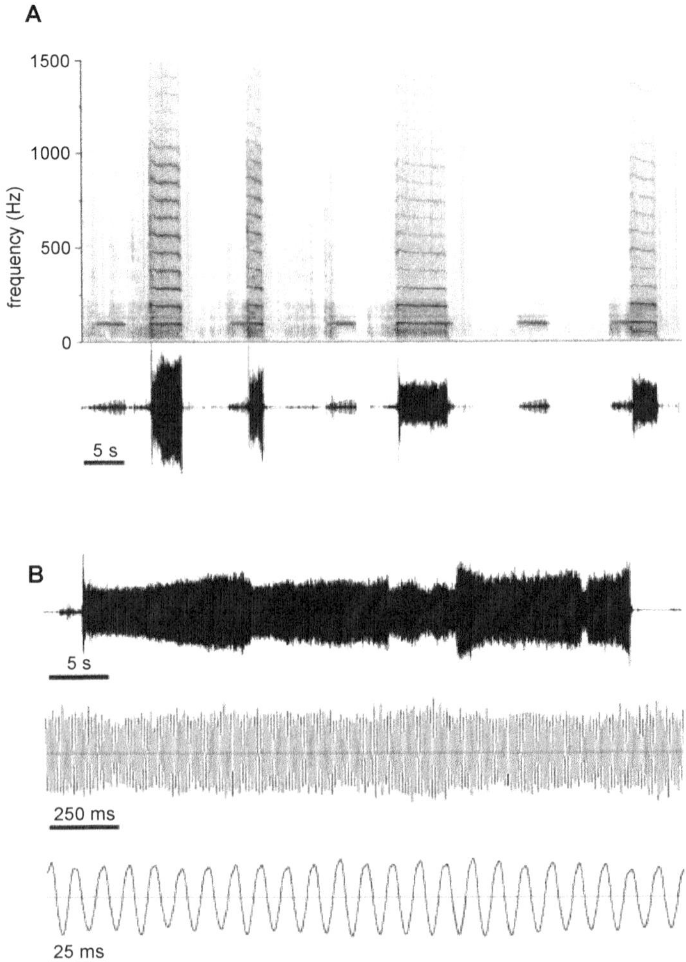

Figure 6.2. *C. impicticornis* buzzing signals recorded on plants. (A) Sonograms with oscillograms for a sequence with four high amplitude buzzing signals were recorded together with FS-1 pulse trains. (B) Long buzzing signals are shown at different time scales.

signals) or by the vibration of lifted wings (buzzing signals) (Kavčič et al. 2013). Low species-specific percussion, tremulatory and buzzing signals were emitted in less-defined behavioural contexts.

Tremulatory signals were first described in Pentatominae for *H. strictus* as E-1 and E-2 components within the male courtship pulse trains. Their time, amplitude and spectral characteristics are comparable to those described in the Asopinae predatory stink bug species *Podisus maculiventris* (Say) (Heteroptera: Pentatomidae: Asopinae) (Žunič et al. 2008). Recently, Kavčič and co-authors (2013) described tremulatory signals in *E. heros* of up to 5 mm/s in velocity that exceeded the values of simultaneously recorded abdomen-produced signals by 15 to 20 dB. The sonograms of plant-recorded tremulatory signals are characterized by a high frequency onset that induces a low-frequency plant resonance tail. On plants, male and female *E. heros* tremulated spontaneously during calling and prior to copulation. The role of tremulatory signals during the pair formation and copulation process requires detailed studies on different stink bug species.

Buzzing signals have been described in *E. heros* (Kavčič et al. 2013) and recently in *C. impicticornis* and *C. ubica* (Fig. 6.2). The primary characteristic of randomly repeated and several s-long buzzes is their high velocity, which reaches up to 40 mm/s, exceeding the values of the signals produced by abdomen vibration by approximately 20 dB. The spectra of buzzing signals are characterized by an extremely narrow dominant frequency peak at approximately 110 Hz and distinct higher harmonic peaks at frequencies of up to 1500 Hz.

Percussion signals produced by tapping on the substrate (plant) with front legs were described by Kavčič and co-authors (2013) in *E. heros* males and females. Their primary spectral energy lies below 200 Hz, with a velocity below 0.5 mm/s. The role of these percussion signals is not yet clear. Insect tapping on the substrate was recorded either in the absence of other vibratory emissions during searching, or it was recorded simultaneously with or as a response to vibratory signals produced by abdomen vibration. Recent investigations by Shestakov (2015) significantly increased the number of known Pentatomine species that produce percussion signals. Female *C. fuscispinus*, *C. pinicola*, *G. lineatum*, *P. rufipes*, *H. vernalis* and *D. baccarum* produce "protest" calls by striking (tapping) against the substrate with their abdomen when repelling the silent insects that are trying to copulate. The time and frequency characteristics of these protest signals were evaluated in *P. rufipes*. Male abdomen tapping on the substrate was observed only during *A. acuminata* courtship, when composite signals from abdomen vibration and percussion elements are produced.

Intra- and Interspecific Diversity of Songs and their Recognition

The population diversity of the song repertoire has been investigated in *N. viridula* (Čokl et al. 2000). Genetic studies of geographically separated populations confirmed the species origin as being in Africa followed by its dispersion via different lineages to all the continents except the Arctic and Antarctica (Kavar et al. 2006). Despite their genetically confirmed differences, males and females from different continents communicate during mating by using the same vibratory song repertoire (Čokl et al. 2000). The dominant frequency and spectra of other spectral characteristics show no population specificity. However, the authors observed significant differences in the duration of female and male calling song signals. The mean duration values for the pulse trains of female calling songs were approximately 1.7 s for a population from Slovenia and approximately 0.8 s for an Italian population. The female calling song pulse train duration also varies within populations; in the population from Slovenia, the mean values were approximately 1.7 s during the 1999/2000 season and approximately 0.8 s in the population that was sampled from the same place in 2011 (Žunič et al. 2011). The mean durations of MS-1 pulses were approximately 0.26 s in the males from Slovenia and approximately 0.08 s for the males from Italy.

Despite the above shown differences, normal calling behaviour and communication could be recorded between the members of different geographically isolated populations. Furthermore, hybrids of the Brazilian and Slovenian populations produced the same song types as the parental populations but with distinctly different song parameters that can be attributed to genetic factors (Virant-Doberlet et al. 2000). Several parameters are sex-linked, and the values of certain hybrid song parameters are intermediate between those of the parental types.

Significant differences have been recorded in the song repertoire of the Australian and Slovenian populations of *N. viridula* (Ryan et al. 1996). The non-pulsed type of the Australian population's female calling song was emitted as a response to male calling, which carries characteristics of the female courtship song signals that were described in the females from Slovenia. Jeraj and Walter (1998) showed that the females from Australia did not respond to the males from Slovenia, and the researchers observed very low frequency of copulation. However, Slovenian females readily exchanged calling song signals with the males from Australia and regularly copulated with them.

Miklas and co-authors (2003b) compared the variability of vibratory signals and mate choice selectivity in *N. viridula* individuals from France and Guadeloupe. These investigators found significant differences between both populations in terms of the time parameters for the female calling song and in the spectra of the male courtship song signals. In mixed pairs, the female calling songs of both populations regularly induced searching by the males of their own or of the alien population. However, the males showed a preference for responding by courtship song to the calling song of its own population.

Communication via species-specific signals represents one of the mechanisms that prevent hybridization. The efficiency of species isolation depends on the level of synergy between the signals of different modalities that mediate processes from calling, searching and courting to copulation. Vibrational communication is limited by the mechanical properties and dimensions of the substrate, which decrease the informational value of the vibratory signal with increasing distance between the sender and receiver. Miklas and co-authors (2001) used playback experiments to show that *N. viridula* males differentiate between the pulsed and non-pulsed types of the female song when it is reproduced on a non-resonant substrate. On a plant, males responded equally to signals of both types because fast repeated pulses fused through the resonant substrate during transmission to such an extent that they were perceived as a longer non-pulsed unit.

Interspecific vibrational communication shows that this process does not perfectly isolate species, and that other mechanisms must be employed to prevent hybridization. Kiritani and co-authors (1963) described interspecific copulation between sympatrically distributed *N. antennata* and *N. viridula* mates that on one hand follow the same basic phase sequence of mating behaviour and on the other produce significantly different songs (Kon et al. 1988). The opposite case is true for sympatric *C. ubica* and *C. impicticornis* stink bugs that halt communication with species-specific vibratory signals during the very early stage of the calling phase (Laumann et al., unpublished data). In playback experiments, male *N. viridula* did not differentiate between the conspecific female calling song and the *T. custator accera* male second song (Hrabar et al. 2004).

The role of different parameters in song recognition has been investigated through the studies of stereotyped behavioural patterns. Žunič and co-authors (2011) analysed *N. viridula* male vibratory responses to played-back synthesized conspecific female calling songs of different time (duration of pulse trains, inter-pulse train intervals, repetition time, and duty cycle) and frequency (dominant frequency) characteristics. The pulse train duration and inter-pulse interval were the decisive and important parameters for female calling song recognition (Fig. 6.3). The best responses to different inter-pulse intervals were obtained for values of approximately two s, and the second responsiveness peak was recorded at intervals of approximately seven s. Intervals shorter than one s sharply decreased male responsiveness. Males showed a lower tolerance for duty cycle changes that were conducted by modified pulse duration than by modified inter-pulse intervals. In the broad effective range between 90 and 180 Hz, males responded best at the characteristic dominant frequency of female calls, which was 105 Hz. Males responded best to synthesized signals of constant, species-characteristic repetition rates and dominant frequency at signal durations of approximately 0.7 s, with a sharp decrease at values below 0.6 and above 1.5 s.

A study by Žunič and co-authors (2011) confirmed the importance of relative temporal parameters for song recognition. The authors tested the native Slovenian population that was characterized during a particular season as having pulse trains that were approximately 0.7 s long for female calling songs. For example, the mean duration for the female calling song pulse in the year 2000 was approximately 1600 ms (Čokl et al. 2000). The test males

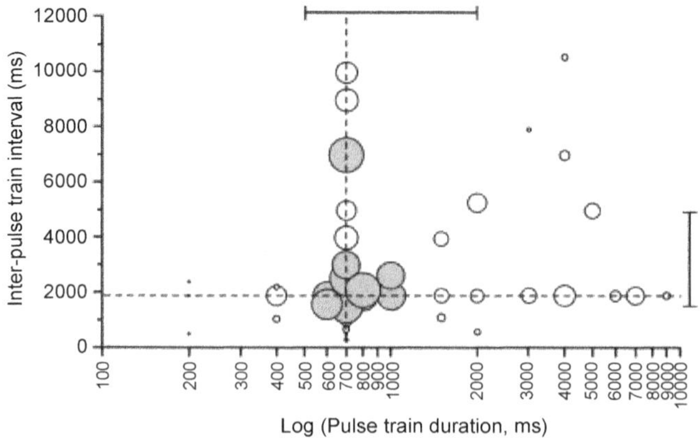

Figure 6.3. Male *N. viridula* vibratory responses (the circles show the percentage of male responses to the specific parameter combination) to female calling song pulse trains of different pulse train durations and inter-pulse train intervals. The reference value (dashed line) shows the means of the parameters that were calculated from the values measured in five females from the Slovenian population. The responses that were significantly lower according to the reference values are indicated by white circles. Black lines indicate the ranges of each parameter distribution for geographically isolated Australian, French, Japanese and Slovenian populations (With permission from Žunič et al. 2011).

tuned their responses to the signal duration range that was characteristic of the female population for the same season.

Mate recognition under natural conditions is largely influenced by environmental noise, which is induced by wind, raindrops and vibratory signals, among others, that are produced by other conspecific or alien species signallers that stay within the sensitivity range of the communicating insect (Chapter 7).

Vibration-producing Mechanisms

Stink bugs from the subfamily Pentatominae produce vibratory signals by vibrating their abdomens, tremulating their whole bodies, buzzing their lifted wings and by percussion with their front legs or abdomen on the substrate.

Males and females produce species- and gender-specific calling, courtship, repellent and rival song signals through the vertical vibration of their abdomens. Using their anatomical knowledge on the muscles that constitute the skeletal motor mechanism of the thorax in *N. viridula* (Malouf 1932), Kuštor (1989) and Amon (1990) described their function in the production of vibratory signals by implanting electrodes in different muscles to record electromyograms in freely moving bugs simultaneously with the vibratory signals produced by abdomen vibration.

The thorax and abdomen are dorsally connected by the first and second abdominal tergits, which are fused into a chitinous plate. The anterior chitinous membrane connects the plate to the thorax and the posterior membrane to the third abdominal tergit. The broad lateral parts of the plate are fused with pleurits at both sites. The metaphragma of the thorax is connected to the plate's antecostal ridge by a pair of strong tergal longitudinal muscles (TLI). Another set of tergal muscles (TLII) is fixed anteriorly on the plate's ridge and posteriorly on the anterior ridge of the third abdominal tergit. The anterior and posterior movements of the plate are caused by the contraction of the TLI and TLII muscles, respectively. Two pairs of lateral compressor muscles (LCrI and LCrII) and the depressor tymbali (DrTy) are laterally fixed to the plate. The TLI, TLII and LCrI muscles contract synchronously and then vertically vibrate the abdomen against the thorax, which is fixed by the legs on to the substrate. The amplitude of the substrate vibration cycles increases with the increasing amplitude of the total EMG signals. The electrical activity of LCrII and other abdominal muscles was not connected to signal production. The role of the small DrTy muscle in vibration production is not yet clear.

The synchronized muscle contractions phase coupled with cycles of vibrations was recorded simultaneously from the abdomen and the plant below to confirm the above-described mechanism of signal production in *N. viridula* and other pentatomine stink bug species. The mean velocity of *N. viridula* FS-1 signals that were produced by abdominal vibration and recorded on the thorax of females standing on a *Cyperus alternifolius* (L.) leaf varied individually between 0.3 and 0.8 mm/s (Čokl et al. 2007). During transmission over the thorax and legs to the substrate, the velocity of the vibrations produced by the abdomen does not change significantly. The velocity measured on the leaf immediately below the singing bug varied between 1.6 dB below and 5.6 dB above the recorded thorax value.

Vibratory signals produced by other mechanisms have been described in Asopine (Žunič et al. 2008) and Pentatomine stink bugs (Kavčič et al. 2013). In *E. heros*, Kavčič and co-authors (2013) described high amplitude tremulatory signals produced by the vigorous shaking of the whole body in males and females, with their legs standing firmly on the substrate, together with high amplitude vibratory signals produced by the tremulation of lifted wings, and low amplitude percussion vibratory signals produced by tapping their front legs on the substrate (the plant). Detailed studies of these vibration-producing mechanisms are lacking.

The use of different vibration-producing mechanisms significantly enlarges the range of communication signals used for communication through plants in different behavioural contexts. The recent demonstration of high amplitude tremulation and buzzing signals in stink bugs poses a question about their possible transmission over air between adjacent plants in the field. Behavioural studies are needed in order to explain why percussion signals are emitted either alone or in concert with those produced by abdominal vibration. Signals produced by different mechanisms open new questions about their role in communication, their transmission through plants and their detection by different sensory organs.

Sensing and Processing of Vibratory Signals

Bimodal communication via chemical and substrate-borne signals also demands elaborate sensory and neuronal adaptations on the receiver side. At the receptor level, we can expect the precise coding of temporal characteristics for vibratory signals with high sensitivity at a frequency range below 200 Hz and at the higher neuronal level, information received over different inputs requiring multimodal processing. Comparative data on the neuronal basis of behaviour are lacking, and studies in *N. viridula* still represent the model not only for Pentatominae but also for the whole family.

Receptor organs

Vibrations are detected by different mechanoreceptors that are situated in and on the insect body. Leg scolopidial organs and campaniform sensilla are in direct contact with the substrate and thus represent the most sensitive sensory input for vibratory communication signals. Their sensitivity is almost 40 dB above the velocity values of the vibratory signals produced by different mechanisms in *N. viridula*. High amplitude tremulatory and buzzing signals indicate that less sensitive mechanoreceptors such as those situated in and on antennae may represent additional sensory input. Furthermore, hair sensilla on the body may detect the air particle movements produced by mates at short distances.

The morphology of the leg scolopidial mechanoreceptors in *N. viridula* was described by Michel and co-authors (1983). The subgenual organ is situated in the blood channel and is composed of two scolopidia, each of which has one sensory cell. The body of the organ is on one side fixed on the blood channel wall and on the other by a thin flag-like ligament formed by two cap cells distally situated at different points of the blood channel wall. The sensory cells of the organ are excited by vibrations that are transmitted through a solid (cuticle), a liquid (haemolymph within the blood channel) and air (the trachea running close to the blood channel wall). Detailed studies of the synergy between the vibrations that are transmitted through different media to the receptor cells are lacking.

The femoral, tibial distal and tarso-praetarsal chordotonal organs control the position and detect the angular velocity of the insect's leg joints. The most complex one is the femoral chordotonal organ, which is situated distally in the femur. This organ is made up of twelve scolopidia (each of which has two sensory cells); eight scolopidia are distributed in the condensed scoloparium and four are in the dispersed scoloparium. The body of the condensed scoloparium is fixed proximally to the inner cuticle wall. Three cap cells are distally fixed to the *M. levator* tibiae and the other five form the ligament that is fixed to the tibial apodeme. The four scolopidia of the dispersed scoloparium are distributed within the ligament. The tibial distal chordotonal organ is situated in the blood channel and controls the tibio-tarsal joint by two scolopidia (one with two and one with one sensory cell). The tarso-praetarsal chordotonal organ consists of two scoloparia; the proximal scoloparium contains two scolopidia (one with one and the other with two sensory cells) and is distally fixed to the unguitractor tendon that moves the praetarsus, and the distal scoloparium is distally fixed to the inner wall of the anterior and posterior claws.

Other mechanoreceptors that are involved in detecting substrate vibrations are situated on the legs and in the antennae. The surface of the *N. viridula* legs is equipped with non-grouped campaniform sensilla, and for the antennae, Jeram and Pabst (1996) described the Johnston's and the central organ. The Johnston's organ is situated in the third antennal segment (distal pedicellite) and is composed of forty-five scolopidia distributed around the periphery and anchored in the invagination of the joint cuticle between the pedicel and flagellum. Each amphinematic scolopidium contains three sensory and three enveloping cells. The central organ is also anchored within the same joint, and it is composed of seven mononematic scolopidia (each with one or two sensory cells). The axons of seventeen scolopidia of Johnston's organ join one antennal nerve, and those of the others run within the antennal nerve on the other side.

We can expect that the high amplitude wing buzzing and body tremulation vibratory signals induce airflow of characteristics as were described in tethered flying flies (Barth 2002). The air currents generated during flying are detected by *Cupiennius salei* (Keyserlink) spiders, by the trichobotria, and they elicit prey-catching behaviour. The hair sensilla of *C. salei* spiders are highly sensitive and are involved in the detection of intraspecific, low-frequency body tremulation signals and the wing beat of flies. Trichobotria are common in Heteroptera, and their physiology was studied more than 40 years ago in *Pyrrhocoris apterus* (L.) (Drašlar 1973). Čokl (1984) showed the effect of abdominal trichobotria movement on the responsiveness of ventral cord vibratory interneurons in *N. viridula*. We could hypothesize that the high amplitude tremulatory and buzzing signals are detected by hair sensilla situated on the body, and as such, these sensilla represent an additional information input for stink bugs that are communicating at close range during courtship.

The receptor cells of leg mechanoreceptors finally terminate on the ipsilateral side of the corresponding segment of the ventral cord ganglion (Čokl and Amon 1980, Zorović 2005). Both subgenual receptor cells form dense terminal arborisations in the central part of the ganglion. The primary branch of the joint chordotonal organ and campaniform sensilla receptor cell axons arches anteriorly until the ganglion midline, with some side branches diverging from the main axon soon after entering the ganglion. Axons of antennal mechanoreceptors finally terminate in the lateral deutocerebrum of the brain, in the subaesophageal ganglion and in the central ganglion, with a few axons projecting into its abdominal neuromeres (Jeram and Čokl 1996).

The threshold curves of *N. viridula* leg vibratory receptor cells reveal a group of sensory cells tuned to frequencies below 100 Hz and two additional sensory cells tuned to frequencies of approximately 200 Hz and between 500 and 1000 Hz (Čokl 1983, Zorović 2005). The responses of the low-frequency receptor neurons originate in the activation of the campaniform sensilla and/or joint chordotonal organs. Their highest velocity sensitivity varies between 0.1 and 0.01 mm/s in the frequency range between 50 and 70 Hz. Their threshold curves follow the line of equal displacement values, with the highest amplitude sensitivity at approximately 0.01 μm as expressed in the front legs. The low frequency group is made up of receptor cells with different threshold sensitivities and different response patterns elicited by the different phases of leg movement during vibration. Phase-locked responses enable the precise coding of frequency modulation patterns below 100 Hz, at velocities between 0.1 and 1 mm/s.

The threshold curves of two leg receptor neuron types show the best sensitivity at frequencies above 100 Hz. The middle frequency receptor neuron responds best at 200 Hz with a threshold sensitivity of approximately 0.01 mm/s. The high frequency receptor neuron responds with peak sensitivity at approximately 0.001 mm/s at frequencies between 500 and 1000 Hz. The threshold curves of the middle and high frequency receptor neurons follow the line of equal acceleration value (approximately 10 mm/s^2) in the frequency range below the best frequency. In the above frequency range, the threshold lines of the middle and high frequency receptor neurons follow the lines of equal displacement values at approximately 10^{-7} for the middle and 10^{-9} m for the high frequency receptor neuron types. Tonic responses and hyperbolic intensity curves are characteristic of both neuron types. At 200 Hz, the middle and the high frequency receptor neurons respond with prolonged responses. Although direct evidence for receptor neuron origins is lacking, we assume that their responses originate in the activity of both subgenual organ sensory cells.

The sensitivity of the *N. viridula* Johnston's and central organ scolopidia fit well with the spectra of stink bug vibratory communication signals at and below 100 Hz (Jeram and Čokl 1996). The vibration of the flagellum perpendicular to its long axis evokes responses in the frequency range between 60 and 200 Hz at velocities of approximately 1 mm/s; a lower sensitivity was measured when the flagellum was vibrated in parallel with its long axis. The highest sensitivity (ca. 0.1 mm/s) was measured for 100 Hz signals; responses at this frequency are phase-locked.

Morphology and Function of Higher Order Vibratory Neurons

In an early investigation, Čokl and Amon (1980) described the functional properties of three types of vibratory interneurons in the ventral nerve cord of *N. viridula* that exhibited their best sensitivity between 100 and 600 Hz, and one type was tuned to a broad frequency range between 600 and 1500 Hz. In their detailed study using intracellular staining and recording techniques, Zorović and co-authors (2008) described the anatomy and function of 10 different types of thoracic ventral nerve cord vibratory interneurons. Based on their gross morphology, the authors grouped the interneurons into four categories. Five types of ascending neurons have their cell body located in the metathoracic neuromere of the central ganglion and the ascending axon running upwards to the brain on the contralateral side (Fig. 6.4). The cell body of the unpaired ascending type with two bilaterally running ascending axons is located centrally in the mesothoracic neuromere. The authors also described three different types of local interneurons with branches on both sides of the central ganglion. One type of descending interneurons was described as having its cell body in the prothoracic ganglion and a descending axon running contralaterally to the soma, forming terminal arborisations in the meso- and metathoracic neuromere. One group of the above-mentioned neurons is tuned to frequencies below 100 Hz and the other is tuned to a middle frequency range of approximately 200 Hz. In comparison with low and middle frequency receptor neurons, the interneurons show lower sensitivity; the acceleration threshold of low frequency-tuned interneurons lies at approximately 100 mm/s^2, and that of neurons tuned to frequencies around 200 Hz at approximately 1 mm/s^2. Compared to receptor neurons, the identified higher order neurons have more complex response patterns, confirming complex processing of excitatory and inhibitory inputs.

In *N. viridula*, Zorović (2011) correlated the responses of four types of previously described ascending interneurons (Zorović et al. 2008) with the time structure of the species vibratory communication signals (Fig. 6.4). The author showed that the temporal filtering of conspecific vibratory signals already takes place at the level of ventral nerve cord interneurons. The mean spike rate of all four types showed a preference for pulse durations below 600 ms as a characteristic for male calling song signals. Conversely, the mean spike rate showed no selectivity for the duration of intervals between pulses. The selectivity for either short pulse duration, long pulse interval duration or no selectivity at all was exhibited in the peak neuron

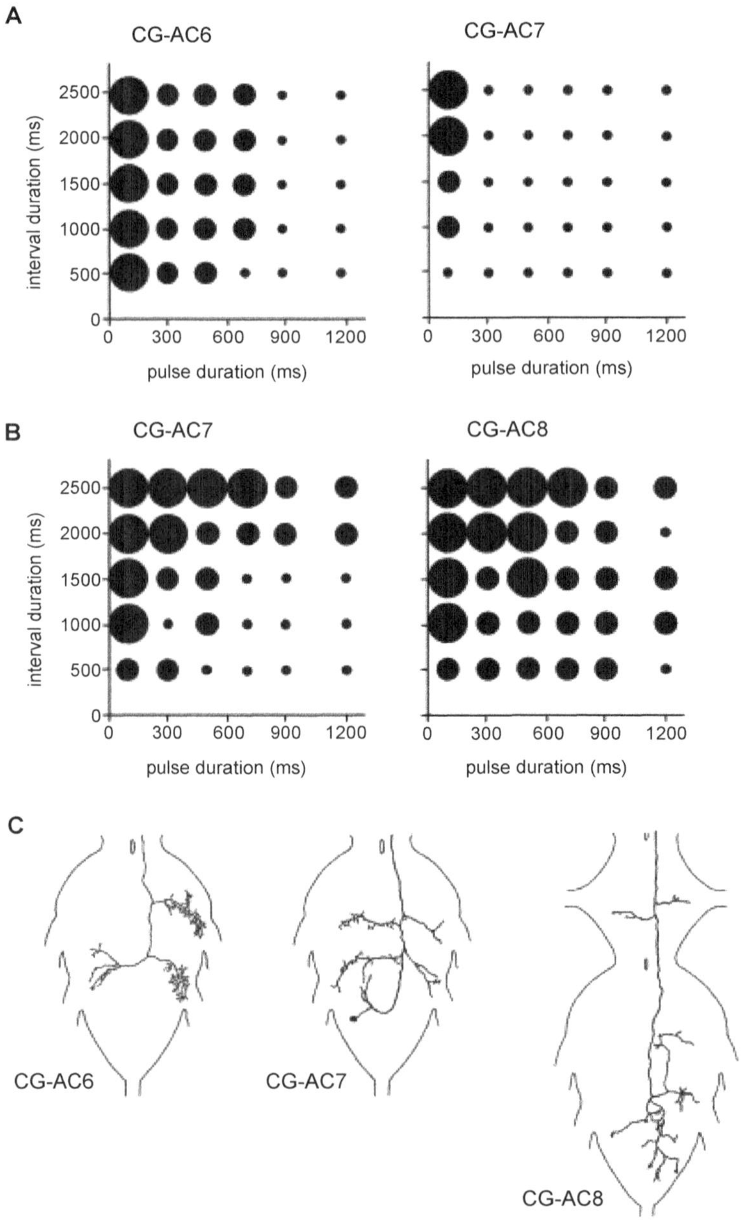

Figure 6.4. Responses of the ascending ventral cord vibratory interneurons to vibratory stimuli (100 Hz, 20 cm/s^2) of different interval and duration value combinations shown as the mean spike frequency response arrays. (A) or as peak instantaneous spike frequency response arrays (B). Large, medium or small circles indicate strong, intermediate or weak (or no) responses, respectively. (C) Whole-mount drawings of tested interneurons from types CG-AC6, CG-AC7 and CG-AC8 (Redrawn with permission from Zorović 2011).

spike rate. The weakest responses were obtained by combining long pulse and short pulse interval durations. No preference was detected for the constant period or duty cycle, as is characteristic for this species and most other stink bug female calling songs.

Comparative investigations of neuronal processing at different levels are lacking in stink bugs and Heteroptera. The neuronal basis of behaviour is a promising field of research but demands time-consuming experimental studies at the single cell level. These data are needed in order to understand the function of neuronal networks that process multimodal information to trigger proper behavioural responses. These complex studies are lacking not only in stink bugs but also in other insect groups.

Acknowledgments

We thank the Slovene Research Agency, the National Council for Scientific and Technological Development (CNPq), the Brazilian Corporation of Agricultural Research (EMBRAPA) and the Research Support Foundation of the Federal District (FAP-DF).

References

Amon, T. 1990. Electrical brain stimulation elicits singing in the bug *Nezara viridula*. Naturwissenschaften 77: 291–292.

Bagwell, G.J., A. Čokl and J.G. Millar. 2008. Characterization and comparison of the substrate-borne vibrational signals of *Chlorochroa uhleri*, *C. ligata* and *C. sayi*. Ann. Entomol. Soc. Am. 101(1): 235–246.

Barth, F.G. 2002. A spider's world: senses and behaviour. Springer, Berlin.

Bennet-Clark, H.C. 1998. Size and scale effects as constraints in insect sound communication. Phil. Trans. Roy. Soc. London B 353: 194–201.

Blassioli-Moraes, M.C., R.A. Laumann, A. Čokl and M. Borges. 2005. Vibratory signals of four Neotropical stink bug species. Physiol. Entomol. 30: 175–188.

Blassioli-Moraes, M.C., D.M. Magalhaes, A. Čokl, R.A. Laumann, J.P. Da Silva, C.C.A. Silva et al. 2014. Vibrational communication and mating behaviour of *Dichelops melacanthus* (Hemiptera: Pentatomidae) recorded from the loudspeaker membrane and plants. Physiol. Entomol. 39: 1–11.

Borges, M., P.C. Jepson and P.E. Howse. 1987. Long-range mate location and close-range courtship behaviour in the green stink bug, *Nezara viridula* and its mediation by sex pheromone. Entomol. Exper. Et Appl. 44: 205–212.

Brownell, P. and R.D. Farley. 1979. Orientations to vibrations in sand by the nocturnal scorpions *Paruroctonus mesaensis*: mechanisms of target localisation. J. Comp. Physiol. A 131: 31–38.

Cocroft, R.B. and R.L. Rodriguez. 2005. The behavioral ecology of insect vibrational communication. Bioscience 55(4): 323–334.

Čokl, A., M. Gogala and A. Blaževič. 1978. Principles of sound recognition in three pentatomide bug species (Heteroptera). Biološki Vestnik (Ljubljana) 26: 81–94.

Čokl, A. and T. Amon. 1980. Vibratory interneurons in the central nervous system of *Nezara viridula* L. (Pentatomidae, Heteroptera). J. Comp. Physiol. 139: 87–95.

Čokl, A. 1983. Functional properties of vibroreceptors in the legs of *Nezara viridula* (L.) (Heteroptera, Pentatomidae). J. Comp. Physiol. 150: 261–269.

Čokl, A. 1984. The effect of abdominal trichobotria movement on the responsiveness of the ventral cord vibratory interneurons in *Nezara viridula* L. Proc. Int. Congr. Entomol. 17: 186.
Čokl, A., M. Virant-Doberlet and A. McDowell. 1999. Vibrational directionality in the southern green stink bug, *Nezara viridula* (L.), is mediated by female song. Animal Beh. 58: 1277–1283.
Čokl, A., M. Virant-Doberlet and N. Stritih. 2000. The structure and function of songs emitted by southern green stink bugs from Brazil, Florida, Italy and Slovenia. Physiol. Entomol. 25: 196–205.
Čokl, A., H.L. McBrien and J.G. Millar. 2001. Comparison of substrate-borne vibrational signals of two stink bug species, *Acrosternum hilare* and *Nezara viridula* (Heteroptera: Pentatomidae). Ann. Entomol. Soc. Am. 94: 471–479.
Čokl, A. and M. Virant-Doberlet. 2003. Communication with substrate-borne signals in plant-dwelling insects. Ann. Rev. Entomol. 48: 29–50.
Čokl, A., J. Prešern, M. Virant-Doberlet, G.J. Bagwell and J.G. Millar. 2004. Vibratory signals of the harlequin bug and their transmission through plants. Physiol. Entomol. 29: 372–380.
Čokl, A., M. Zorović and J.G. Millar. 2007. Vibrational communication along plants by the stink bugs *Nezara viridula* and *Murgantia histrionica*. Behav. Processes 75: 40–54.
Čokl, A. 2008. Stink bug interaction with host plants during communication. J. Insect Physiol. 54: 1113–1124.
Čokl, A., R.A. Laumann, A. Žunič-Kosi, M.C. Blassioli-Moraes, M. Virant-Doberlet and M. Borges. 2015. Interference of overlapping insect vibratory communication signals: an *Euschistus heros* model. PloS One 10(6): 1–16.
Drašlar, K. 1973. Functional properties of trichobotria in the bug *Pyrrhocoris apterus* (L.). J. Comp. Physiol. 84: 175–184.
Fish, J. and J. Alcock. 1973. The behavior of *Chlorochroa ligata* (Say) and *Cosmopepla bimaculata* (Thomas) (Hemiptera: Pentatomidae). Entomol. News 84: 260–268.
Gogala, M. and R. Razpotnik. 1974. Metoda oscilografske sonagrafije za bioakustične raziskave (Method of oscillographic sonagraphy for bioacoustic research). Biološki Vestnik (Ljubljana) 22: 209–216.
Harris, V.E. and J.W. Todd. 1980. Temporal and numerical pattern of reproductive behavior in the southern green stink bug, *Nezara viridula* (Hemiptera: Pentatomidae). Entomologia experimentalis et Applicata 27: 105–116.
Hrabar, N., M. Virant-Doberlet and A. Čokl. 2004. Species specificity of male southern green stink bug *Nezara viridula* (L.) reactions to female calling song. Acta Zool. Sin. 50: 566–575.
Jeraj, M. and G.H. Walter. 1998. Vibrational communication in *Nezara viridula*: response of Slovenian and Australian bugs to one another. Behav. Processes 44: 51–58.
Jeram, S. and A. Čokl. 1996. Mechanoreceptors in insects: Johnston's organ in *Nezara viridula* (L.) (Pentatomidae, Heteroptera). Pfluegers Archiv-Eur. J. Physiol. 439 (Suppl.): R281.
Jeram, S. and A.M. Pabst. 1996. Johnston's organ and central organ in *Nezara viridula* (L.) (Heteroptera, Pentatomidae). Tissue Cell 28: 227–235.
Kavčič, A., A. Čokl, R.A. Laumann, M.C. Blassioli-Moraes and M. Borges. 2013. Tremulatory and abdomen vibration signals enable communication through air in the stink bug *Euschistus heros*. PloS One 8(2): 1–10.
Kavar, T., P. Pavlovčič, S. Sušnik, V. Meglič and M. Virant-Doberlet. 2006. Genetic differentiation of geographically separated populations of the southern green stink bug *Nezara viridula* (Hemiptera: Pentatomidae). Bull. Entomol. Res. 96: 117–128.
Kiritani, K., H. Hokyo and J. Yukawa. 1963. Coexistence of the two related stink bugs *Nezara viridula* and *Nezara antennata* under natural conditions. Res. Pop. Ecol. 5: 11–22.
Kon, M., A. Oe, H. Numata and T. Hidaka. 1988. Comparison of the mating behaviour between the sympatric species *Nezara antennata* and *N. viridula* (Heteroptera: Pentatomidae) with special reference to sound emission. J. Ethol. 6: 91–98.
Kuštor, V. 1989. Activity of muscles of the vibration producing organ in the bug *Nezara viridula*. MSc Thesis, University of Ljubljana, Ljubljana, Slovenia.

Laumann, R.A., A. Čokl, M.C. Blassioli-Moraes and Mi. Borges. 2016. Vibratory communication and its relevance to reproductive isolation in two sympatric stink bug species (Heteroptera: Pentatomidae: Pentatominae). J. Insect Behav. (submitted).
Maluf, N.S.R. 1932. The skeletal motor mechanism of the thorax of the "stink bug" *Nezara viridula* L. Bull. Soc. Roy. Entomol. Egypt 16: 161–203.
Markl, H. 1983. Vibrational communication. pp. 332–353. *In*: Huber, F. and H. Markl (eds.). Neuroethology and Behavioural Physiology. Springer, Berlin Heidelberg New York.
McBrien, H.L., A. Čokl and J.G. Millar. 2002. Comparison of substrate-borne vibrational signals of two consperse stink bug species *Thyanta pallidovirens* and *T. custator accera* (Heteroptera: Pentatomidae). J. Insect Behav. 15(6): 715–738.
McBrien, H.L. and J.G. Millar. 2003. Substrate-borne vibrational signals of the consperse stink bug (Heteroptera: Pentatomidae). The Canadian Entomologist 135: 555–567.
McLain, D.K. 1998. Non-genetic benefits of mate choice: fecundity enhancement and sexy sons. An. Behav. 55: 1191–1201.
Michel, K., T. Amon and A. Čokl. 1983. The morphology of the leg scolopidial organs in *Nezara viridula* (L.) (Heteroptera: Pentatomidae). Rev. Can. Biol. Exp. 42: 130–150.
Michelsen, A. 2014. Physical aspects of vibrational communication. pp. 199–213. *In*: Cocroft, R.B., M. Gogala, P.S.M. Hill and A. Wessel (eds.). Studying Vibrational Communication. Springer, Heidelberg New York Dordrecht London.
Miklas, N., N. Stritih, A. Čokl, M. Virant-Doberlet and M. Renou. 2001. The influence of substrate on male responsiveness to the female calling song in *Nezara viridula*. J. Insect Behav. 14(3): 313–332.
Miklas, N., T. Lasnier and M. Renou. 2003a. Male bugs modulate pheromone emission in response to vibratory signals of the conspecifics. J. Chem. Ecol. 29(3): 561–574.
Miklas, N., A. Čokl, M. Renou and M. Virant-Doberlet. 2003b. Variability of vibratory signals and mate choice selectivity in the southern green stink bug. Behav. Processes 61: 131–142.
Mitchel, W.C. and R.F.L. Mau. 1969. Sexual activity and longevity of the southern green stink bug, *Nezara viridula*. Ann. Rev. Entomol. Soc. Am. 62: 1246–1247.
Ota, D. and A. Čokl. 1991. Male location in the southern green stink bug *Nezara viridula* (Heteroptera: Pentatomidae) mediated through substrate-borne signals on ivy. J. Insect Behav. 4: 441–447.
Pavlovčič, P. and A. Čokl. 2001. Songs of *Holcostethus strictus* (Fabricius): a different repertoire among land bugs (Heteroptera: Pentatomidae). Behav. Proc. 53: 65–73.
Polajnar, J., A. Kavčič, A. Žunič and A. Čokl. 2013. *Palomena prasina* (Hemiptera: Pentatomidae) vibratory signals and their tuning with plant substrates. Central Eur. J. Biol. 8(7): 670–680.
Ryan, M.A., A. Čokl and G.H. Walter. 1996. Differences in vibratory sound communication between a Slovenian and Australian population of *Nezara viridula*. Behav. Processes 36: 183–193.
Shestakov, L.S. 2015. A comparative analysis of vibrational signals in 16 sympatric species (Pentatomidae, Heteroptera). Entomol. Rev. 95(3): 310–325.
Silva, C.C.A., R.A. Laumann, J.B.C. Ferreira, M.C. Blassioli-Moraes, M. Borges and A. Čokl. 2012. Reproductive biology, mating behavior, and vibratory communication of the brown-winged stink bug, *Edessa meditabunda* (Fabr.) (Heteroptera: Pentatomidae). Psyche 2012: 1–9.
Stritih, N., M. Virant-Doberlet and A. Čokl. 2000. Green stink bug *Nezara viridula* detects differences in amplitude between courtship song vibrations at stem and petiolus. Pfluegers Archiv-Eur. J. Physiol. 439(Suppl.): R190–R192.
Virant-Doberlet, M., A. Čokl and N. Stritih. 2000. Vibratory songs of hybrids from Brazilian and Slovenian populations of the green stink bug *Nezara viridula*. Pfluegers Archiv-Eur. J. Physiol. 439(Suppl.): R196–R198.
Zgonik, V. and A. Čokl. 2014. The role of signals of different modalities in initiating vibratory communication in *Nezara viridula*. Central Eur. J. of Biol. 9(2): 200–2011.

Zorović, M. 2005. Morphological and physiological properties of vibrational neurons in thoracic ganglia of the stink bug *Nezara viridula* (L.) (Heteroptera: Pentatomidae). Dissertation Thesis, University Ljubljana.

Zorović, M., J. Prešern and A. Čokl. 2008. Morphology and physiology of vibratory interneurons in the thoracic ganglia of the southern green stink bug *Nezara viridula* (L.). J. Comp. Neurol. 508: 365–381.

Zorović, M. 2011. Temporal processing of vibratory communication signals at the level of ascending interneurons in *Nezara viridula* (Hemiptera: Pentatomidae). PloS ONE 6: 1–8.

Žunič, A., M. Virant-Doberlet and A. Čokl. 2008. Communication with signals produced by abdominal vibration, tremulation and percussion in *Podisus maculiventris* (Heteroptera: Pentatomidae). Ann. Rev. Soc. Amer. 101(6): 1169–1178.

Žunič, A., M. Virant-Doberlet and A. Čokl. 2011. Species recognition during substrate-borne communication in *Nezara viridula* (L.) (Pentatomidae: Heteroptera). J. Insect Behav. 24: 468–487.

CHAPTER 7

Stink Bug Communication Network and Environment

Andrej Čokl, Alenka Žunič Kosi* and *Meta-Virant-Doberlet*

Introduction

Insects exchange information by signals transmitted between sender and receiver through air, solids or water. The privacy of multimodal communication is enabled by the various combinations of signal characteristics that are used in different behavioural contexts, at different times and in different ecological conditions. The environment significantly influences the communication processes: signals of different modalities are specifically modified by the transmission properties of the medium, temperature, humidity and by the noise produced by abiotic and biotic factors. Plants represent the natural environment of most insects giving them among other things food, shelter and substrate for reproduction. Despite many advantages, dense vegetation limits the communication distance: high frequency airborne sound is strongly attenuated, local air currents and complex plant architecture decrease the reliability of chemical signals for providing information on the location of the calling mate, visual and contact information exchange is efficient only at close distances. Polyphagous Pentatominae (Heteroptera: Pentatomidae) feed exclusively on herbaceous plants and like most other plant-dwelling insects communicate along the plants using the substrate-borne component of their vibratory emissions. In order to increase the communication distance and optimize signal-to-noise ratio, stink bugs emit vibratory signals of low frequency and narrow-band

Department of Organisms and Ecosystems Research, National Institute of Biology, Večna pot 111, SI-1000 Ljubljana, Slovenia.
Emails: Alenka.Zunic-Kosi@nib.si; Meta.Virant@nib.si
* Corresponding author: Andrej.cokl@nib.si

characteristics that are best tuned with the transmission properties of plants, and have adapted their sensory system to operate within the group-specific narrow frequency communication window. The aim of the present chapter is to describe the impact of plants and abiotic and biotic factors on the different parameters of vibratory communication signals together with the various adaptations that have evolved in stink bugs in order to optimize information exchange in such conditions.

Plants as Transmission Medium for Stink Bug Vibratory Communication Signals

Almost 70 years ago Ossiannilsson (1949) hypothesized that small insects communicate with signals transmitted through plants. Since the pioneering work of Ichikawa and Ishii (1974) in planthoppers and Gogala et al. (1974) in cydnid bugs, substrate-borne vibrational communication has been demonstrated in most insect groups (Cocroft and Rodriguez 2005). Because of their small body size, insects can produce only high frequency compressional sound waves of intensity relevant for airborne communication. Since the diameter of the body has to be in the range of or above one-third of the radiated sound wavelength (Markl 1983), stink bugs with a body size of about 1 cm or less can efficiently produce only air-borne sounds of frequencies above 10 kHz. Dense vegetation strongly attenuates high frequency sound and plant-dwelling insects can choose between the possibility of flying out of the relatively safe plant environment or of exchanging information via the substrate-borne component of their vibratory emissions. Insects singing in the open air are exposed to predators that are attracted by their calls and the communication distance of insects talking through plants is limited by the plant dimensions and architecture.

Although stink bugs and most other insects have chosen the latter option (Cocroft and Rodriguez 2005) they can, under specific conditions, also communicate with signals transmitted through the air. Casas et al. (1998) measured simultaneously leaf vibrations with the laser Doppler vibrometer and air velocities in the vicinity of the vibrated leaf with the laser Doppler anemometer. They recorded vibrations of leaves 1 cm away from the vibrated one at intensities well above the threshold of highly sensitive arthropod hairs (Shimozava et al. 2003). Eriksson et al. (2011) recorded the responses of the leafhopper *Scaphoideus titanus* (Ball) females to the male advertisement calls that were emitted on another plant separated by up to a 6 cm air gap between the partly overlapped leaves. The authors measured 20–40 dB attenuation of about 200 Hz vibrations transmitted between large grapevine leaves at distances of up to 11 cm. In the stink bug species *Euschistus heros* (Fabr.) Kavčič et al. (2013) demonstrated airborne sound communication with signals produced by abdomen vibration and tremulation of the body. Inter-plant communication with vibratory signals transmitted through air and even soil (Michelsen 2014) opens up an interesting field for future research with a high potential for field application.

Michelsen et al. (1982) first demonstrated that insects communicate through plants by bending waves (Cremer et al. 1973, 2005). Bending waves propagate with velocity, growing with increasing frequency: the group velocity is proportional to the square root of angular frequency. Casas et al. (2007) have described the non-dispersive characteristics of 3–4 mm diameter plant stems when vibrated in a broad frequency range above 5 kHz. High frequency signals are, on the one hand less distorted because they are carried by frequency independent propagation velocity, but on the other hand, they are subjected to high attenuation due to the low-pass filtering properties of the medium. High frequency substrate-borne components with prominent peaks of up to 10 kHz have been recorded simultaneously with Ensifera airborne sound communication signals. The plant recorded components of bushcrickets of the genera *Tettigonia*, *Decticus* and *Ephippiger*, stridulations on green plants were recorded at distances of up to 80 cm with damping values of 20–50 dB/m (Keuper and Kühne 1983, Keuper et al. 1985). Vibrations induced by stridulatory airborne sound signals improve mate localization (Latimer and Schatral 1983, Weidemann and Keuper 1987) and may influence male-male spacing behaviour (Schatral and Kalmring 1985). More details on the auditory-vibratory sensory system in bushcrickets have been reviewed by Rössler et al. (2006) and Stritih and Čokl (2014).

The impact of plants on vibratory signal frequency characteristics

Stink bugs like many other plant-dwelling insects communicate with vibratory signals of the main energy produced at frequencies below 500 Hz and with the dominant frequency ranging around 100 Hz (Chapter 6). Such frequency characteristics are tuned with the low-pass filtering properties of herbaceous plants. Michelsen et al. (1982) demonstrated in *Thesium bavarum* (Schrank), frequency filtering of sinusoidal vibration with its amplification being around 100 Hz at distances of 3 and 7 cm from the point of vibration. The authors confirmed efficient plant-borne communication in the low frequency range comparing the intensity of the airborne and substrate-borne components of insect vibratory emissions. The majority of the spectral components of the airborne component of *Euides speciosa* (Boheman) signals lie around 550 Hz (Traue 1978) and that of the substrate-borne component lie somewhere between 150 and 250 Hz. Spectra of the airborne sound component of the cydnid bug stridulatory signals extend up to 12 kHz with the peak between 3 and 4 kHz (Gogala et al. 1974). On the plant spectra narrow up to 3 kHz with the main energy measured in most cases below 500 Hz. Transmission curves of the vibrations transmitted through banana pseudostem show low attenuation at frequencies below 100 Hz and stronger damping at higher frequencies of up to 1000 Hz (Barth et al. 1988).

Similar frequency filtering of green plants has also been confirmed in natural conditions. Burrower bugs *Scaptocoris carvalhoi* (Becker) and *Scaptocoris castanea* (Party) (Cydnidae) (Čokl et al. 2006) feed and mate under soil on soybean roots. Stridulatory signals recorded from soil close to these bugs have broad-band characteristics with spectral elements extending up to 5 kHz and the dominant frequency being around 500 Hz. The spectra of signals recorded on the soybean stem show attenuation of the higher frequency components and shift of the dominant frequency below 500 Hz. At the stem height of 28 cm above the soil the dominant frequency ranges around 300 Hz and the subdominant peaks do not exceed 1000 Hz. Low-pass filtering properties have also been shown for plant-transmitted percussion signals. Tapping of the predatory stink bug *Podisus maculiventris* (Say) on the non-resonant loudspeaker membrane induces vibrations with the narrow dominant frequency peak being around 97 Hz and an extensive higher frequency component of around 2000 Hz (Žunič et al. 2008). Frequency characteristics of the plant recorded percussion signals differ significantly: narrow spectra below 1000 Hz are characterized by the 97 Hz dominant frequency peak and its first harmonic.

Bending wave group velocity in *Vicia faba* (L.) was 36 m/s for 200 Hz and 129 m/s for 2000 Hz vibration (Michelsen et al. 1982). Different plant vibrations are frequency modulated. Percussion signals (Chapter 6), leaves fluttering in the wind and rain drops (see below) induce in leaves vibratory signals with an irregular short high frequency onset followed by a low frequency and longer lasting regular part. Many stink bug species like *Chlorochroa uhleri* (Stål), *Chlorochroa ligata* (Say), *Chlorochroa sayi* (Stål) (Bagwell et al. 2008) and *Piezodorus lituratus* (Fabr.) (Gogala and Razpotnik 1974) produce frequency modulated vibratory signals with decreasing or increasing frequency sweeps. Gogala (2006) has shown that different *P. lituratus* male songs were emitted as a response to a conspecific male rival song played back with reversed frequency modulation or change of other spectral characteristics. Because of the frequency dependent propagation velocity, we may expect, at different distances from the source, different delays between faster and slower propagated subunits, which consequently create spectrally different signals. The distance-dependent pattern of frequency shifts accompanied by the low-pass filtering properties of plants may be used for evaluation of the distance to the source. This hypothesis needs to be experimentally confirmed.

Until now, all the described vibratory signals recorded on the non-resonant or resonant substrates, are characterized by frequencies tuned with the low-pass filtering properties of their host green plants (Čokl et al. 2005, Čokl 2014). Their dominant frequency ranges below 200 Hz and higher harmonics extend above 1000 Hz only in wing buzzing signals. We can conclude that stink bugs have adapted the frequency of their signals to be produced in a narrow frequency window that ensures transmission through plants with lowest attenuation.

Attenuation of vibratory signals during transmission through plants

Michelsen et al. (1982) measured the velocity of vibratory signals emitted naturally on plants by cydnid bugs and "small cicadas" (Michelsen et al. 1982). The values of the signals recorded on the plant's stem or leaves ranged between 0.1 and 1 mm/s and those on dry leaves between 0.3 and 2 mm/s. Comparable velocity values were also obtained in stink bugs. The mean velocity of *Nezara viridula* (L.) calling song signals, recorded from the body of females, standing on the *Cyperus alternifolius* (L.) leaf, varied individually between 0.3 and 0.8 mm/s (Čokl et al. 2007). The mean velocity of body recorded calling song signals produced by *Murgantia histrionica* (Hahn) males, standing on the loudspeaker membrane, ranged around 0.9 mm/s (Čokl et al. 2007).

Transmission over the legs to the substrate does not significantly change the velocity of the signals produced by abdomen vibration (Čokl et al. 2007). The value of *N. viridula* female calling song signals recorded from the leaf immediately below the singing bug varied between 5.6 above and 1.6 dB below the velocity measured simultaneously from the bug's body. Similar differences have been obtained for signals recorded from insects singing on a non-resonant (loudspeaker) or resonant (plant) substrate. Velocities of *M. histrionica* male calling song signals, measured below the calling bug differed in males standing on different substrates (Čokl et al. 2007). At the distance of one cm from the calling bug, the velocity values ranged around 0.2 mm/s on the loudspeaker membrane, around 0.5 mm/s on the bladderpot leaf, around 1.3 mm/s on ivy and around 2.3 mm/s on the London rocket leaf.

Comparison of the spectra of *N. viridula* female calling song recorded simultaneously from the pronotum and from the non-resonant substrate immediately below, has shown the same position of the dominant, the first and the second harmonic peaks. The spectra of plant-recorded signals differ by additional spectral components below 80 and above 350 Hz (Čokl et al. 2005).

Michelsen et al. (1982) were the first to measure the low internal damping of plant transmitted vibratory signals. The authors predicted reflections at the top and at the root that consequently created standing wave conditions with nodes and antinodes in the plant's rod-like structures like stems, stalks, side branches and leaf veins. The low damping of vibrations travelling through green plants was confirmed later in banana plants: values between 0.3 and 0.4 dB/cm were measured for individual frequencies of band-limited noise (Barth 2002).

Regular vibratory signal velocity variation with distance was first shown for 124 Hz pure tones transmitted through *C. alternifolius* stem (Čokl 1988). Induced plant vibrations contained four distinct frequency peaks: the 124 Hz dominant peak with its first harmonic at 250 Hz and the new peak at 84 Hz with its harmonic at 170 Hz. Amplitudes of each spectral peak varied along the stem with regularly repeated minima and maxima separated by the distance decreasing with increasing spectral peak frequency. Regular variation of the vibratory signal's velocity with distance was later confirmed under natural conditions in the model of *N. viridula* female calling song transmitted through sedge stem (Čokl et al. 2007). Low velocity changes of the signals transmitted from the body to the leaf surface (see above) were followed by 1.5–2.5 dB/cm damping during the transmission to the neighbouring leaves and by 15–19 dB/cm during the transmission from the leaves over the rosette to the stem. Much lower attenuation has been measured for signals transmitted through the sedge stem: damping values calculated along the 123 cm stem were 0.06 dB/cm (re. 0.07 mm/s) for the first and 0.1 dB/cm (re. 0.11 mm/s) for the second female. The velocity of naturally emitted female calling song signals decreased non-linearly by transmission through the sedge stem (Čokl et al. 2007) with minima (nodes) and maxima (antinodes) regularly repeated up to a distance of 70 cm from the reference point at the stem top. The distance between peaks was approximately 20 cm and maximal velocity difference between the neighbouring velocity peak maxima and minima reached 19 dB at a distance of 10 cm (1.9 dB/cm). Peak velocity values within the first 20 cm exceeded the reference value in both tests.

The relation between the spectral and velocity characteristics of signals recorded at different distances from the point of vibration was tested on a bean plant vibrated artificially with both types of *N. viridula* female calling song signals (Čokl et al. 2007). Signals of mean velocities between 2 and 3.6 mm/s were attenuated at a distance of 27 cm by 0.6 dB/cm. Attenuation values at a distance of 3 cm between the neighbouring peaks of minimal and maximal velocity values ranged between 5.6–5.8 dB/cm for signals of both types. Experiments with artificially induced pre-recorded female calling song signals confirmed frequency dependent distance between velocity peaks. The curve of velocity variation with distance fits well with the curve of relative dominant frequency peak amplitude variation. The intermodal distance of the first harmonic frequency peak is half of the distance between peaks of the dominant frequency. The relation between spectral peak amplitudes varies with distance and consequently at certain distances from the source, the amplitude of the first harmonic exceeds the value of the dominant frequency peak. As a result, the amplitude modulation pattern differs at different distances from the source with the most pronounced effect for frequency modulated signals.

Michelsen et al. (1982) predicted standing wave conditions with nodes and antinodes in plants activated by sine waves of long duration. Polajnar et al. (2012) tested the hypothesis that resonance causes regularly repeated peaks of velocity minima and maxima in low frequency vibratory signals transmitted along the herbaceous plant stem. The authors compared the patterns of different vibrational pulses (pure tones and pre-recorded vibratory signals of *N. viridula* and *P. maculiventris*) transmitted along the sedge stem with calculated spatial profiles of corresponding eigen frequencies. The measured distance between nodes matched the calculated values and confirmed that resonance causes signal velocity variation in the studied system.

Velocity variation with distance differs for signals transmitted through mechanically different substrates. The frequency characteristics of the *M. histrionica* male calling song differ from those of *N. viridula* in terms of broader spectra with distinct higher harmonics extending up to 1000 Hz (Čokl et al. 2004, 2007). The relation between spectral peak amplitudes is comparable for signals recorded on the dorsum of the calling male, ivy, London rocket and bladderpot. Signals recorded on the non-resonant loudspeaker membrane and ivy are characterized by spectra with lower relative amplitudes of harmonics below 700 Hz. Velocity variation with distance shows for the signals transmitted through London rocket the same characteristics as those measured for *N. viridula* female calling song signals transmitted through sedge or bean plants (see above). Transmission through bladderpot woody stem differs. The velocity of naturally emitted male calling song signals decreased by 10–20 dB at a distance of 10 cm away from the source and remained at this level without regularly repeated velocity minima and maxima at longer distances.

Magal et al. (2000) first demonstrated different vibration transmission properties of leaf veins and lamina. Experiments with *M. histrionica* signals transmitted through cabbage leaves (Čokl et al. 2004) confirmed their results. Lamina transmitted female calling song signals are attenuated by 4.6 dB/cm and those of males by 3.6 dB/cm. Damping values of around 0.8 and 0.4 dB/cm were measured for female and male signals transmitted through the main vein. Vein transmitted signals show non-linear variation of velocity with distance and lower damping of higher frequency spectral components as compared with those transmitted through the lamina.

The impact of plants on time characteristics of vibratory signals

Michelsen et al. (1982) recorded more than 20 ms long long vibrations induced in the plant by a few ms sine wave pulses as a consequence of the vibratory signal travelling with low attenuation up and down the plant several times. Such an impact may decrease the efficiency of song recognition based on the signal species- and gender-specific temporal characteristics.

The duration and repetition time of *N. viridula* female calling song pulse trains determine the level of male responses (Žunič et al. 2011). Males respond within the range of natural signal duration values that are relatively not significantly increased when recorded on different plants and at different distances from the source. Comparative analyses of female calling songs in most Pentatominae species have shown that they are generally characterized by pulses or pulse trains separated by intervals long enough to prevent their fusion caused by increased duration as a consequence of reflections at the phase borders. Courtship signals are mainly emitted at short distances where significant changes in their species-specific delicate and complex time structure cannot be expected.

An exception to this general conclusion represents the recognition of *N. viridula* pulsed type of female calling song (Miklas et al. 2001), the temporal characteristics of which differ significantly from those of the non-pulsed type (Čokl et al. 2000). Pulse trains of both types have similar duration and repetition time but differ in the number of pulses per pulse train. The non-pulsed type is composed of a short pre-pulse followed by a long one and the pulsed type is composed of several pulses of equal duration and short inter-pulse interval (Chapter 6). The behavioural context of the female pulsed calling song is not clear. Miklas et al. (2001) have demonstrated that on the non-resonant loudspeaker membrane, the males responded significantly more often to the non-pulsed than to the pulsed type but that on a plant, the males did not differentiate between them. Recording of the pulse type calling song transmitted through a plant revealed that the inter-pulse duration increased and that the pulses fused to such an extent that males did not differentiate anymore between pulse trains of both type.

Until now, all the investigated stink bug species of the subfamily Pentatominae have been known to communicate through a narrow frequency window predominantly with signals produced by abdomen vibration. Such spectral characteristics increase the signal-to-noise ratio and restrict song recognition to the discrimination of a variety of time and amplitude modulation patterns. Communication of stink bugs during calling and courtship is characterized by male-female song duets with precise timing of signals in order to prevent their overlapping.

The Impact of Abiotic Factors on Stink Bug Vibratory Communication

Wind represents the main natural source of high amplitude vibratory abiotic noise. The vibrations of banana plants and bromeliads' leaves induced by wind are characterized by peak frequencies below 10 Hz and higher frequency components at 40 to 60 dB lower level (Barth et al. 1988). The spectra of low frequency vibrations caused by leaves fluttering in high speed wind are extended up to 50 Hz at –20 dB level. Casas et al. (1998) showed that basic oscillations of apple leaves induced by wind were independent of wind speed: their frequencies ranged between 7 and 10 Hz for the parallel and between 7 and 14 Hz for the perpendicular flow. The velocity of leaf vibrations ranged between 30 and 60 mm/s at low and between 70 and 130 mm/s at high speed wind.

Raindrops falling on the leaf induce its vibration with the irregular and regular phases (Casas et al. 1998). The 9 to 29 ms long irregular phase is characterized by frequencies spanning up to 25 kHz and maximal velocity between 76 and 137 mm/s. The regular phase of leaf vibration follows the irregular one with a basic frequency between 6 and 11 Hz and exponentially decreasing amplitude with a half-life of around 163 ms. Characteristics of the vibrations induced by raindrops are similar to those recorded by insects landing on the leaf (Casas et al. 1998).

Although experimental data on the impact of wind, rain or other abiotic noise on stink bug mating behaviour and copulation success in the field are lacking, we cannot expect these factors to have significant influence. Stink bugs communicate with longer vibratory signals through a narrow frequency window that ranges outside the characteristics of the vibrations induced in plants by wind, falling raindrops or insects landing on them.

A stronger impact on stink bug communication may be expected in an environment with pure tone or white noise vibratory sources. The impact of pure tone vibratory background in the frequency range close to the values characteristic for stink bug communication signals, has been investigated in *N. viridula* (Polajnar and Čokl 2008). Male reaction to 100 Hz disturbance vibration was evaluated by measurement of the percentage of males moving towards the source of artificially induced female calling song and/or by monitoring the males responding to it by emission of the calling or courtship song. The disturbance vibration did not significantly change the proportion of males moving in the direction of the female calling song source but a higher proportion of them responded with the courtship song in the absence of 100 Hz background vibration. Females reacted differently to disturbance with the 100 Hz pure tone. A high proportion of them stopped calling before the end of the trial. Some females also changed the repetition rate of the calling song signals or they introduced some courtship song signals in the calling song sequence. In control tests the lowest coefficient of variation was calculated for the dominant frequency. In the presence of 100 Hz disturbance, females changed the dominant frequency of their calls in order to increase the frequency difference. At 5 Hz difference 77% (n = 7) changed the dominant frequency, at 5 to 10 Hz difference the figure was 29% (n = 4) and just 10% (n = 3) changed the dominant frequency when the difference exceeded 10 Hz.

The impact of white noise has been tested in *N. viridula* (Spezia et al. 2008). The authors correlated the directional response of males on a Y-shaped dummy plant to the source of played-back female calling song signals. With different intensities of the calling signals they determined the threshold level for male directionality response. In further tests they used combined stimulation with subthreshold play-back of the female calling song signals and acoustic Gaussian noise at different intensity levels. The proportion of insects that reacted to the subthreshold signals shows a non-monotonic behaviour with characteristic presence of a maximum for increasing levels of noise. The authors confirmed the non-dynamic stochastic resonance phenomenon underlying constructive interplay between the noise and attracting female calls. In this case, the environmental noise can have a constructive impact on insect behaviour, improving communication by amplifying the subthreshold deterministic signal to the level that is perceived by the searching mate.

The Impact of Biotic Factors on Stink Bug Vibratory Communication

In the field, mates are surrounded by signals emitted by conspecific and alien insect species (Cocroft 2003, Cocroft and Rodriguez 2005). All investigated stink bug species of the subfamily Pentatominae communicate through a narrow frequency window with signals produced by abdomen vibration, body tremulation, percussion and wing buzzing (Chapter 6). Communication with signals of narrow-band spectral characteristics increases the signal-to-noise ratio due to low sensitivity to vibrations outside the characteristic frequency band. On the other hand, such signals limit their species- and gender-specific character to a variation of different combinations of temporal and amplitude pattern parameters. Frequency modulation is less reliable information for calling song recognition because of the frequency dependent bending wave propagation velocity, which in standing wave conditions creates signals of very different amplitude modulation patterns. The main biotic noise is thus represented by the vibrations produced by conspecific and heterospecific individuals that emit signals at the same time, on the same plant and within the same narrow frequency range.

Mating in most Pentatomine species is characterized by male-female duets with precise timing of signals to prevent their overlapping. Nevertheless, overlapping of signals has been regularly recorded in male-female duets of *Dichelops melacanthus* (Dallas) (Blassioli-Moraes et al. 2014) and *E. heros* (Čokl et al. 2015). In both species the authors observed interference with the typical pulsed amplitude modulation pattern of masked vibrations. Discrimination of *D. melacanthus* overlapped MS-3 and FS-3 signals could be based on neuronal differentiation (recognition) of significantly different frequency characteristics of both signals. FS-3 spectra show well-expressed frequency modulation (dominant frequency with upper and lower limits of 179 and 146 Hz respectively) with three clear higher harmonic peaks, and MS-3 frequency non-modulated signals are characterized by broad-band spectra with the dominant frequency being around 111 Hz and just one higher harmonic peak (Blassioli-Moraes et al. 2014).

Reinvestigation of *E. heros* vibratory communication signals (Čokl et al. 2015), first described as loudspeaker recorded vibrations by Blassioli-Moraes et al. (2005), revealed the interference of overlapped signals, that triggers active reaction of males and females to avoid or to decrease its effect on the amplitude modulation pattern. Interference significantly modifies the compound vibration into a sequence of fused pulses of duration that increases with the decreasing frequency difference of overlapped vibrations. Males and females avoided or minimized the consequences of overlapping and interference by changing the temporal and frequency characteristics of duetting signals. Males adapted signal duration according to the interval between the consecutive female calls and, when overlapped, they changed the frequency of their emissions in order to increase the frequency difference; increased frequency difference decreases the duration of fused pulses and restores the amplitude pattern of both signals. Females increased the duration of their signal at the changed frequency level. Similar reaction has been described in *N. viridula* female song overlapped with 100 Hz background disturbance vibration (see above) (Polajnar and Čokl 2008).

Biotic noise produced by several males and females on the same plant needs further investigation. Borges et al. (1987) demonstrated that male pheromone attracts adult females to the male's vicinity. Females stimulated by the presence of male pheromone start calling via emission of the calling song (Zgonik and Čokl 2014) that increases pheromone production (Miklas et al. 2003). The female calling song in *N. viridula* and many other pentatomine stink bugs is characterized by the emission of signals with highly autonomous and constant duration and repetition time of song units. Rivalry between several conspecific calling females has not been investigated yet either in laboratory or in natural conditions. The typical pentatominae reaction of several males responding to a calling female is to change the calling and courting behaviour to rivalry characterized by duets alternating with the rival song pulses (Čokl et al. 2000). Rivalry between more than two males has not been investigated yet and synchronization of signals remains an interesting question.

Different stink bug species have been observed in the field on the same plant but to our knowledge, there are no experimental data on the communication contact between them obtained in the natural conditions. Additional sources of biotic noise relevant in terms of disturbing intraspecific communication in Pentatominae are airborne and substrate-borne signals and cues, whose frequency characteristics induce in herbaceous plants vibrations dominated by frequency components below 400 Hz. Such an instructive example is the transmission of stridulatory signals produced by soybean root-dwelling burrower bugs *Scaptocoris carvalhoi* (Becker) and *Scaptocoris castanea* (Perty), as their dominant higher frequency components disappear in signals recorded from upper soil plant parts (Čokl et al. 2006). Stridulatory signals recorded from the soybean stem 28 cm above the soil carry spectral characteristics similar to those of the plant recorded pentatomine stink bug emissions: strong frequency components below 50 Hz and the typical around 100 Hz dominant frequency peak.

The effect of biotic noise produced by conspecific and/or alien species' vibratory signals has been studied only under laboratory conditions (De Groot et al. 2010). The authors investigated the responsiveness of *N. viridula* males to conspecific female calling songs in the presence of either played-back conspecific or heterospecific female calling songs or synthesized vibratory signals of different temporal characteristics. Heterospecific and conspecific signals that obscured temporal characteristics of the conspecific female song decreased male responsiveness. Simultaneously presented two conspecific female calling song signals increased male responsiveness when the repetition rate of the perceived compound signals ranged within the species-specific values. Studies of the impact of background biotic noise at three different intensities demonstrated that increased signal to noise ratio restores the male responsiveness at the level recorded in control conditions in the absence of noise. Background biotic noise had a stronger effect on male vibratory responsiveness than on its searching behaviour.

In the following study De Groot et al. (2011) investigated the responsiveness of *N. viridula* males to a conspecific female calling song induced on a plant from two sources. Male responses were low because they perceived both songs as a unit with a temporal structure significantly different from the one characteristic for the species-specific female calling song. On the other hand, simultaneous stimulation with the conspecific and heterospecific (*Chinavia hilaris* Say) female calling songs did not significantly affect the searching behaviour of the male. When conspecific and heterospecific signals overlapped, males made orientation errors and most of them located the source of the heterospecific signals.

The impact of biotic noise needs further investigations based on relevant results obtained by field observations. Such studies will reveal the mechanisms used to avoid decreased signal-to-noise ratio that limits the distance and efficiency of vibratory communication through plants.

Acknowledgements

We thank the Slovenian Research Agency (Slovenia) for financial support of the insect communication research.

References

Bagwell, G.J., A. Čokl and J.G. Millar. 2008. Characterization and comparison of the substrate-borne vibrational signals of *Chlorochroa uhleri*, *C. ligata* and *C. sayi*. Ann. Entomol. Soc. Am. 101(1): 235–246.

Barth, F.G., H. Bleckmann, J. Bohnenberger and E.-A. Seyfarth. 1988. Spiders of genus *Cupiennius* SIMON 1891 (Aranea, Ctenidae). II. On the vibratory environment of a wandering spider. Oecologia 77: 194–201.

Barth, F.G. 2002. A Spider's World: Senses and Behavior. Springer, Berlin Heidelberg New York.

Blassioli-Moraes, M.C., R.A. Laumann, A. Čokl and M. Borges. 2005. Vibratory signals of four Neotropical stink bug species. Physiol. Entomol. 30: 175–188.

Blassioli-Moraes, M.C., D.M. Magalhaes, A. Čokl, R.A. Laumann, J.P. Da Silva, C.C.A. Silva et al. 2014. Vibrational communication and mating behaviour of *Dichelops melacanthus* (Hemiptera: Pentatomidae) recorded from loudspeaker membranes and plants. Physiol. Entomol. 39: 1–11.

Borges, M., P.C. Jepson and P.E. Howse. 1987. Long-range mate location and close-range courtship behaviour in the green stink bug, *Nezara viridula* and its mediation by sex pheromone. Entomol. Exper. Et Appl. 44: 205–212.

Casas, J., S. Bacher, J. Tautz, R. Meyhöfer and D. Pierre. 1998. Leaf vibrations and air movement in a leafminer-parasitoid system. Biocontrol 11: 147–153.

Casas, J., C. Magal and J. Sueur. 2007. Dispersive and non-dispersive waves through plants: implications for arthropod vibratory communication. Proc. R. Soc. B 274: 1087–1092.

Cocroft, R.B. 2003. The social environonment of an aggregating, ant-attended tree hopper. J. Insect Behav. 16: 79–95.

Cocroft, R.B. and R.L. Rodriguez. 2005. The behavioral ecology of insect vibrational communication. Bioscience 55(4): 323–334.

Cremer, L., M. Heckl and E.E. Ungar. 1973. Structure-borne Sound, Structural Vibrations and Sound Radiation at Audio Frequencies. Springer, Berlin Heidelberg New York.

Cremer, L., M. Heckl and B.A.T. Petersson. 2005. Structure-borne Sound: Structural Vibrations and Sound Radiation at Audio Frequencies. Springer, Berlin.

Čokl, A. 1988. Vibratory signal transmission in plants as measured by laser vibrometer. Periodicum Biologorum 90(2): 193–196.

Čokl, A., M. Virant-Doberlet and N. Stritih. 2000. The structure and function of songs emitted by southern green stink bugs from Brazil, Florida, Italy and Slovenia. Physiol. Entomol. 25: 196–205.

Čokl, A., J. Prešern, M. Virant-Doberlet, G.J. Bagwell and J.G. Millar. 2004. Vibratory signals of the harlequin bug and their transmission through plants. Physiol. Entomol. 29: 372–380.

Čokl, A., M. Zorović, A. Žunič and M. Virant-Doberlet. 2005. Tuning of host plants with vibratory songs of *Nezara viridula* L. (Heteroptera: Pentatomidae). J. Exp. Biol. 208: 1481–1488.

Čokl, A., C. Nardi, J.M.S. Bento, E. Hirose and A.R. Panizzi. 2006. Transmission of stridulatory signals of the burrower bugs, *Scaptocoris castanea* and *Scaptocoris carvalhoi* (Heteroptera: Cydnidae) through the soil and soybean. Physiol. Entomol. 31: 371–381.

Čokl, A., M. Zorović and J.G. Millar. 2007. Vibrational communication along plants by the stink bugs *Nezara viridula* and *Murgantia histrionica*. Behav. Processes 75: 40–54.

Čokl, A. 2014. Communication through plants in a narrow frequency window. pp. 171–195. *In*: Cocroft, R.B., M. Gogala, P.S.M. Hill and A. Wessel (eds.). Studying Vibrational Communication. Springer, Heidelberg New York Dordrecht London.

Čokl, A., R.A. Laumann, A. Žunič-Kosi, M.C. Blassioli-Moraes, M. Virant-Doberlet and M. Borges. 2015. Interference of overlapping insect vibratory communication signals: An *Euschistus heros* model. PloS One 10(6): 1–16.

De Groot, M., A. Čokl and M. Virant-Doberlet. 2010. Effects of heterospecific and conspecific vibrational signal overlap and signal-to-noise ratio on male responsiveness in *Nezara viridula* (L.). J. Exp. Biol. 213: 3213–3222.

De Groot, M., A. Čokl and M. Virant-Doberlet. 2011. Species identity cues: possibilities for errors during vibrational communication on plant stems. Beh. Ecol. 22: 1209–1217.

Eriksson, A., G. Anfora, A. Luchhi, M. Virant-Doberlet and V. Mazzoni. 2011. Inter-plant vibrational communication in a leafhopper insect. PloS ONE 6(5): 1–6.

Gogala, M., A. Čokl, K. Drašlar and A. Blaževič. 1974. Substrate-borne sound communication in Cydnidae (Heteroptera). J. Comp. Physiol. 94: 25–31.

Gogala, M. and R. Razpotnik. 1974. Metoda oscilografske sonagrafije za bioakustične raziskave (Method of oscillographic sonagraphy for bioacoustic research). Biološki Vestnik (Ljubljana) 22: 209–216.

Gogala, M. 2006. Vibratory signals produced by Heteroptera–Pentatomorpha and Cimicomorpha. pp. 275–295. *In*: Drosopoulos, S. and M.F. Claridge (eds.). Insect Sounds and Communication: Physiology, Behaviour, Ecology and Evolution. CRC Press Taylor and Francis Group, Boca Raton, FL.

Ichikawa, T. and S. Ishii. 1974. Mating signal of the brown planthopper, *Nilaparvus lugens* Ståhl (Homoptera: Delphacidae): vibration of the substrate. Appl. Ent. Zool. 9: 196–198.

Kavčič, A., A. Čokl, R.A. Laumann, M.C. Blassioli-Moraes and M. Borges. 2013. Tremulatory and abdomen vibration signals enable communication through air in the stink bug *Euschistus heros*. PloS One 8(2): 1–10.

Keuper, A. and R. Kühne. 1983. The acoustic behaviour of the bushcricket *Tettigonia cantans* II. Transmission of airborne sound and vibration signals in the biotope. Behav. Proc. 8: 125–145.

Keuper, A., C.W. Otto and A. Schatral. 1985. Airborne sound and vibration signals of bushcrickets and locusts. Their importance for the behaviour in the biotope. pp. 135–142. *In*: Kalmring, K. and N. Elsner (eds.). Acoustical and Vibrational Communication in Insects. Paul Parey, Berlin, D.

Latimer, W. and A. Schatral. 1983. The acoustic behaviour of the katydid *Tettigonia cantans* L. Behavioural responses to sound and vibration. Behav. Proc. 8: 113–124.

Magal, C., M. Schöller, J. Tautz and J. Casas. 2000. The role of leaf structure in vibration propagation. J. Acoust. Soc. Am. 108: 2412–2418.

Markl, H. 1983. Vibrational communication. pp. 332–353. *In*: Huber, F. and H. Markl (eds.). Neuroethology and Behavioural Physiology. Springer, Berlin Heidelberg New York.

Michelsen, A., F. Fink, M. Gogala and D. Traue. 1982. Plants as transmission channels for insect vibrational songs. Behav. Ecol. Sociobiol. 11: 269–281.

Michelsen, A. 2014. Physical aspects of vibrational communication. pp. 199–213. *In*: Cocroft, R.B., M. Gogala, P.S.M. Hill and A. Wessel (eds.). Studying Vibrational Communication. Springer, Heidelberg New York Dordrecht London.

Miklas, N., N. Stritih, A. Čokl and M. Virant-Doberlet. 2001. The influence of substrate on male responsiveness to the female calling song in *Nezara viridula*. J. Insect Beh. 14(3): 313–332.

Miklas, N., T. Lasnier and M. Renou. 2003. Male buigs modulate pheromone emission in response to vibratory signals of the conspecifics. J. Chem. Ecol. 29(3): 561–574.

Ossiannilsson, F. 1949. Insect drummers. Opusc. Entomol. (Suppl.) 10: 1–145.

Polajnar, J. and A. Čokl. 2008. The effect of vibratory disturbance on sexual behaviour of the southern green stink bug *Nezara viridula* (Heteroptera, Pentatomidae). Centr. Eur. J. Biol. 3(2): 189–197.

Polajnar, J., D. Svenšek and A. Čokl. 2012. Resonance in herbaceous plant stems as a factor in vibrational communication of pentatomid bugs (Heteroptera: Pentatomidae). J. Roy. Soc. Interface 9(73): 1898–1907.

Rössler, W., M. Jatho and K. Kalmring. 2006. The auditory-vibratory sensory system in bushcrickets. pp. 35–69. *In*: Drosopoulos, S. and M.F. Claridge (eds.). Insect Sounds and Communication: Physiology, Behaviour, Ecology and Evolution. CRC Press Taylor and Francis Group, Boca Raton, FL.

Schatral, A. and K. Kalmring. 1985. The role of the song for spatial dispersion and agonistic contacts in male bushcrickets. pp. 111–116. *In*: Kalmring, K. and N. Elsner (eds.). Acoustical and Vibrational Communication in Insects. Paul Parey, Berlin, D.

Shimozava, T., T. Mruakami and T. Kumagai. 2003. Cricket wing receptors: thermal noise for the highest sensitivity known. pp. 145–157. *In*: Barth, F.G., J. Humphrey and T. Secomb (eds.). Sensors and Sensing in Biology and Engineering. Springer Verlag Wien.

Spezia, S., L. Curcio, A. Fiasconaro, N. Pizzolato, D. Valenti, B. Spagnolo et al. 2008. Evidence of stochastic resonance in the mating behaviour of *Nezara viridula* (L.). Eur. Phys. J. B 65: 452–458.

Stritih, N. and A. Čokl. 2014. The role of frequency in vibrational communication of Orthoptera. pp. 375–393. *In*: Cocroft, R.B., M. Gogala, P.S.M. Hill and A. Wessel (eds.). Studying Vibrational Communication. Springer, Heidelberg New York Dordrecht London.

Traue, D. 1978. Vibrationskommunikation *bei Euides speciosa* Boh. (Homoptera-Cicadina: Delphacidae). Verh. Dtsch. Zool. Ges. 1978: 167.

Weidemann, S. and A. Keuper. 1987. Influence of the vibratory signals on the phonotaxis of the gryllid *Gryllus bimaculatus* De Geer (Ensifera: Gryllidae). Oecologia 74: 316–318.

Zgonik, V. and A. Čokl. 2014. The role of signals of different modalities in initiating vibratory communication in *Nezara viridula*. Central Eur. J. of Biol. 9(2): 200–2011.

Žunič, A., M. Virant-Doberlet and A. Čokl. 2008. Communication with signals produced by abdominal vibration, tremulation and percussion in *Podisus maculiventris* (Heteroptera: Pentatomidae). Ann. Rev. Soc. Amer. 101(6): 1169–1178.

Žunič, A., M. Virant-Doberlet and A. Čokl. 2011. Species recognition during substrate-borne communication in *Nezara viridula* (L.) (Pentatomidae: Heteroptera). J. Insect Behav. 24: 468–487.

CHAPTER 8

Plant and Stink Bug Interactions at Different Trophic Levels

Salvatore Guarino, Ezio Peri* and *Stefano Colazza*

Introduction

The interactions between plants and herbivores are among the most important ecological interactions in nature (Johnson 2011). Plants are sessile organisms and cannot run away from potential attackers, which are range from pathogenic microbes to grazing mammals. Among these attackers, arthropods, and in particular insects, comprise the largest and most diverse group of organisms with approximately 1–3 million species feeding on plants (Schoonhoven et al. 2005, Dicke 2009). Host-finding and acceptance or rejection of plants by herbivorous insects depend on their behavioral responses to physical and chemical plant features (Finch and Collier 2000). In this context, the insect-plant interactions have been metaphorically likened to warfare (Gonzalez and Nebert 1990, Berembaum and Zangerl 2008), and the process of reciprocating defense and counter-defense has been called a coevolutionary arms race (Whittaker and Feeny 1971). The concept of coevolution was first stated by Fraenknel (1953), who speculated that, as the majority of the green plants are essentially nutritionally equivalent, the so-called secondary metabolites are then likely to determine the patterns of host plant utilization. These concepts did not reach major scientific society until 1959 when Fraenkel published the famous article "The raison d'etre of secondary plant substances" in the journal Science. In this paper was cited the "first detailed description of a chemical insect-host plant relationship"

Dipartimento di Scienze Agrarie e Forestali, Università degli Studi di Palermo, Viale delle Scienze, Edificio 5, 90128 Palermo, Italy.
Emails: ezio.peri@unipa.it; stefano.colazza@unipa.it
* Corresponding author: salvatore.guarino@unipa.it

reported by Verschaffelt (1911) about the ability of sinigrin, a mustard oil glycoside from Brassicaceae, to stimulate feeding by pierid caterpillars. Among the plant secondary metabolites volatile organic compounds (VOC) can be distinguished, often constituted by the so-called green leaf volatiles, and non-volatile chemicals.

Volatile compounds leave the plant surface and upon arrival in new surroundings, air or soil, are transported away from their source of production; they can be exploited by plant feeding insects to find their plant hosts from a distance (Bernays and Chapman 2007), allowing the orientation of phytophagous insects towards the plants, and the ultimate recognition of host plants for feeding and oviposition. Host plant volatile or contact chemicals exploited by herbivores for plant location and selection are commonly defined as kairomones (Metcalf and Metcalf 1992, Ruther et al. 2002). Phytophagous insects have evolved specific olfactory receptor neuron (ORN) systems to perceive some of these plant volatiles, which then compile an odor that acts as a chemical message.

On the contrary front, plants have evolved an arsenal of defenses during the 410 million years that they have been consumed by herbivores, in response to the perennial threat and loss of productivity. These defenses play a critical role in shaping the interactions between plants and herbivores, such that understanding the evolution and ecology of plant defenses is tantamount to understanding the origin and functioning of extant ecosystems. In particular, plants have developed a multitude of chemical defense mechanisms that can either be constitutively present or are activated only after being attacked. Inducible defenses can be classified as direct defenses that affect the herbivore's biology directly (Howe and Schaller 2008) and indirect defenses that affect the herbivore by promoting the effectiveness of the natural enemies of herbivorous arthropods as "bodyguards" (Arimura et al. 2000, Dicke 2009, Heil 2015). In 1980, Price et al. highlighted that the theory on insect–plant interactions cannot progress realistically without consideration of the third trophic level as a part of a plant's set of defenses against herbivores. Plant responses to herbivore infestation can consist of the emission of mixtures of volatiles that not only differ in the total abundance of volatiles released, but more importantly, also in the composition of the volatile blend by producing specific herbivore-induced plant volatiles (HIPV) (Dicke et al. 2003). The emission of specific volatile chemicals released by plants is induced by insect feeding and also oviposition, in the latter case producing the so-called oviposition-induced plant volatiles (OIPV) (Hilker and Fatouros 2015). The emission of HIPV, induced by insect feeding activity, whether accompanied or not by oviposition is mostly influenced by the feeding mechanism of the phytophagous species itself: chewing or piercing-sucking (Paré and Tumlison 1996). Piercing-sucking insects use their stylets for feeding on the sap of xylem, phloem or from plant cells and, in contrast to chewing insects, cause no physical loss of plant tissues (Walling 2000). Consequently, piercing-sucking insects cause fewer injuries to the leaves as compared to chewing herbivores, and the overall impact on plant ecophysiology is also reduced (Hare and Elle 2002). Nevertheless, piercing-sucking insects are also responsible for serious crop losses and are a major threat for cultivations worldwide (Zvereva et al. 2010). In the last decades, the influence of piercing-sucking insects on plant VOC emission has been studied by several research groups (Du et al. 1998, Turlings et al. 1998, Guerrieri et al. 1999, Guerrieri 2016). In particular, VOC from the plants infested by the so-called true bugs (Hemiptera: Heteroptera) were investigated in some plant-pentatomid systems (Colazza et al. 2004a,b, Moraes et al. 2005, Conti et al. 2010, Moujahed et al. 2014).

The aim of this chapter is to address the role of plant traits in influencing the trophic webs including the host plant, phytophagous stink bug and their natural enemies. Here we focus on herbivorous stink bugs of terrestrial habitats as the majority of the studies on this subject have been conducted in this ecological context. In the first part we focus on the plant semiochemicals exploited by the stink bugs in host location and recognition of the host plant, site of feeding and/or oviposition. In the second part we address the role of host plant traits exploited by stink bugs' parasitoids. In particular are reviewed the volatile cues and the substrate-born cues exploited by egg parasitoids and the role of biotic and abiotic factors in influencing these cues.

Plant Traits Exploited by Stink Bugs

The host searching behavior of phytophagous insects consists of several consecutive steps that orientate the herbivore from a random dispersal movement to an orientate movement to the plant by walking, crawling or flying (Schoonhoven et al. 2005). This process permits the insects to choose the host plant before deciding to "select" it or not as a food source and/or egg-deposition site (Visser 1986, Bernays and Chapman 2007). In this process the behavior of the herbivore is influenced by visual and olfactory stimuli, with the latter having particular importance in long range host location. Among these cues the chemical volatiles emitted from the host plant play a crucial role in this process: acting like kairomones they can elicit a positive odor conditioned attraction (Schoonhoven et al. 2005). The herbivore's perception of host plant odor is a complex task in a natural environment, where insects are exposed to many different chemical volatiles, at different concentrations and in different combinations (Bruce and Pickett 2011). Obviously the response of insects is not unique; plant odors can trigger different responses in herbivorous insects, depending on the insect host-plant range (Martinez et al. 2013). The same plant volatile compound can act as an attractant for specialist phytophagous species and as a repellent for generalist species (Van der Meijden 1996). For herbivore heteropteran species the host plant is not just a place to feed but also a place to live, as a consequence, the host plant location is a critical process to satisfy not only their nutritional necessities but also to find suitable sites for mating and ovipositing. However, the role of the plant volatiles used for Heteroptera host plant location has been poorly investigated, while the research has been mainly focused on understanding the role of the plant volatiles in recruiting natural enemies of these pests (see section: Plant traits exploited by egg parasitoids of stink bugs).

Plant Volatiles

Herbivores. Herbivorous insects' host location is influenced by several intrinsic and extrinsic factors, such as insect polyphagy/monophagy or developmental stage, the insect's oviposition behavior, plant stage and environmental conditions. In general, polyphagous insects tend to exploit a great range of plant chemical volatiles, while specialist insect herbivores should show more efficient forms of adaptation, since they are expected to use specific cues. In the case of heteropteran herbivores the majority of them are polyphagous (Backus 1988), some species exhibit narrower host plant ranges and very few of them are monophagous; as a consequence, it is likely that they might respond to a blend of plant volatiles rather than a plant-specific compound. Other intrinsic factors influencing the response, and thus the relationships between phytophagous Heteroptera and their hosts are sex and the developmental stage of insects: for example, studies on the Miridae species, *Lygus lineolaris* (Palisot de Beauvois) and *Lygocoris pabulinus* (L.), indicate that females are more responsive to plant compounds than males; electroantennographic (EAG) analyses showed that the antennal responses of female bugs to green leaf volatiles and monoterpenes were far more marked than the antennal responses of male bugs (Chinta et al. 1994, Groot et al. 1999). Similarly, behavioral studies of *Lygus hesperus* Knight evidenced that females were more receptive to volatiles from alfalfa, *Medicago sativa* L., than males (Williams III et al. 2010). Analogously, different responses to plant volatiles can be ascribed to the development stage. In *L. hesperus*, nymphs show more responsiveness to odors emanating from a plant than adults (Blackmer et al. 2004), however, in this species, the response of these volatiles can be increased when supplemented by visual cues such as 530 nm green light-emitting diode [LED] (Blackmer and Canas 2005).

Ecological habits of herbivorous Heteroptera such as the sites of feeding, oviposition and mating also adapt the host selection process. Since the majority of the herbivorous heteropterans lay their eggs directly on plants (Martinez et al. 2013), many species are attracted to plants with growing shoots and developing seeds or fruits, which are the preferred sites for feeding and oviposition activity of these species (McPherson and McPherson 2000, Olson et al. 2011). For example, the western boxelder bug, *Boisea rubrolineata* Barber (Heteroptera: Rhopalidae) is attracted by phenylacetonitrile, a compound produced by pollen-bearing staminate trees and pistillate trees with maturing seeds of its host *Acer negundo* L., suggesting that this pest can track and exploit the availability of nutrient-rich food sources by adapting its reproductive ecology to the phenology of its host (Schwarz et al. 2009).

The attraction of herbivorous stink bugs (Heteroptera: Pentatomidae) to volatiles released by their host plants has been shown in a few cases (Table 8.1). The cabbage shield bug, *Eurydema pulchrum* Westwood, exploits plant volatiles for its host recognition. Rather et al. (2010) evidenced in olfactometer bioassays the positive response of the bug to volatiles from different plant species such as *Brassica oleracea* L., *Raphanus sativus* L. and *Solanum lycopersium* L. Studies on the painted bug, *Bagrada hilaris* Burmeister, a species native of the ancient world and invasive in North America (Palumbo et al. 2015), allowed in-depth examination of the compounds that can be involved in the host location process. Electroantennographic (EAG) experiments revealed antennal responses of *B. hilaris* to some of the main volatiles identified in the host plants *Lobularia maritima* L. and *B. oleracea* var botrytis (Palumbo et al. 2015, Guarino et al. in press). In particular, EAG responses were elicited by both host-specific compounds such as 3-butenyl isothiocyanate and 4-pentenyl isothiocyanate, typically emitted from brassicaceous crops, and from ubiquitary plant volatiles such as benzaldehyde, octanal, nonanal and acetic acid (Fig. 8.1) (Guarino et al. in press). Field bioassays carried out by testing various isothiocyanate compounds, molecules emitted from several *B. hilaris* host plants species, did not seem consistent (Palumbo et al. 2015), while laboratory bioassays testing a blend of benzaldehyde, octanal, nonanal and acetic acid elicited the attraction of the *B. hilaris* adults indicating that these chemicals can have a role in host location (Guarino et al. in press). This evidence suggests the possibility that for these insects host location is often mediated by ubiquitous volatile compounds perceived in a specific proportion (Bruce et al. 2005).

As previously reported, often phytophagous attacks induce the plants to modify their volatile emissions and this fact can modulate the relations between plants and phytophagous insects, including the host plant preference behavior towards conspecific insects. The females of *Tibraca limbativentris* (Stal.), the rice stalk stink bug, evidenced a preference for the volatile extracts of *Oryza sativa* L. from healthy plants rather than from plants damaged by conspecifics (Machado et al. 2014); on the contrary, in

Table 8.1. Behavioral responses of phytophagous stink bugs to host plant volatiles.

Species	Plant source		Behavioral response	References
	Latin name	Condition		
Bagrada hilaris	*Brassica oleracea* var botrytis	Healthy and/or bug damaged	attraction in olfactometer, preference for bug damaged plants	Guarino et al. personal obs.
Eurydema pulchrum	*Brassica oleracea* var botrytis	Healthy and macerated	attraction in olfactometer	Rather et al. 2010
	Brassica oleracea var capitata	Healthy and macerated	attraction in olfactometer, preference for macerated plants	Rather et al. 2010
	Raphanus sativus	Macerated	no attraction in olfactometer	Rather et al. 2010
	Solanum lycopersicum	Macerated	no attraction in olfactometer	Rather et al. 2010
Euschistus heros	*Glycine max*	Healthy or bug damaged	no preference in olfactometer	Michereff et al. 2011
Piezodorus guildinii	*Glycine max*	Healthy	no attraction in olfactometer	Molina and Trumper 2012
Nezara viridula	*Glycine max*			Panizzi et al. 2004
	Glycine max	Healthy	no attraction in olfactometer	Molina and Trumper 2012
Tibraca limbativentris	*Oryza sativa*	Extracts of healthy or bug damaged plant	attraction in olfactometer, preference of females for healthy plant extracts	Machado et al. 2014
Perillus bioculatus	*Solanum tuberosum*	Damaged by Colorado potato beetle larvae	attraction in walking track bioassay	Van Loon et al. 2000

B. hilaris the damage by conspecifics seemed to increase the attraction of the bugs to the plant volatiles (Guarino, personal observations).

Interestingly, the response of stink bugs to the volatiles emitted by plants as a consequence of the damage induced by conspecifics can depend on the insect's sex. For example, the preference showed by the rice stalk stink bug females towards volatiles from healthy rice plants was not evidenced by males (Machado et al. 2014). A possible explanation could be attributed to a different aim in the host plant searching behavior. The preference of females of rice stalk stink bug for healthy plants may be determined by the search of a place with reduced competition for oviposition and food.

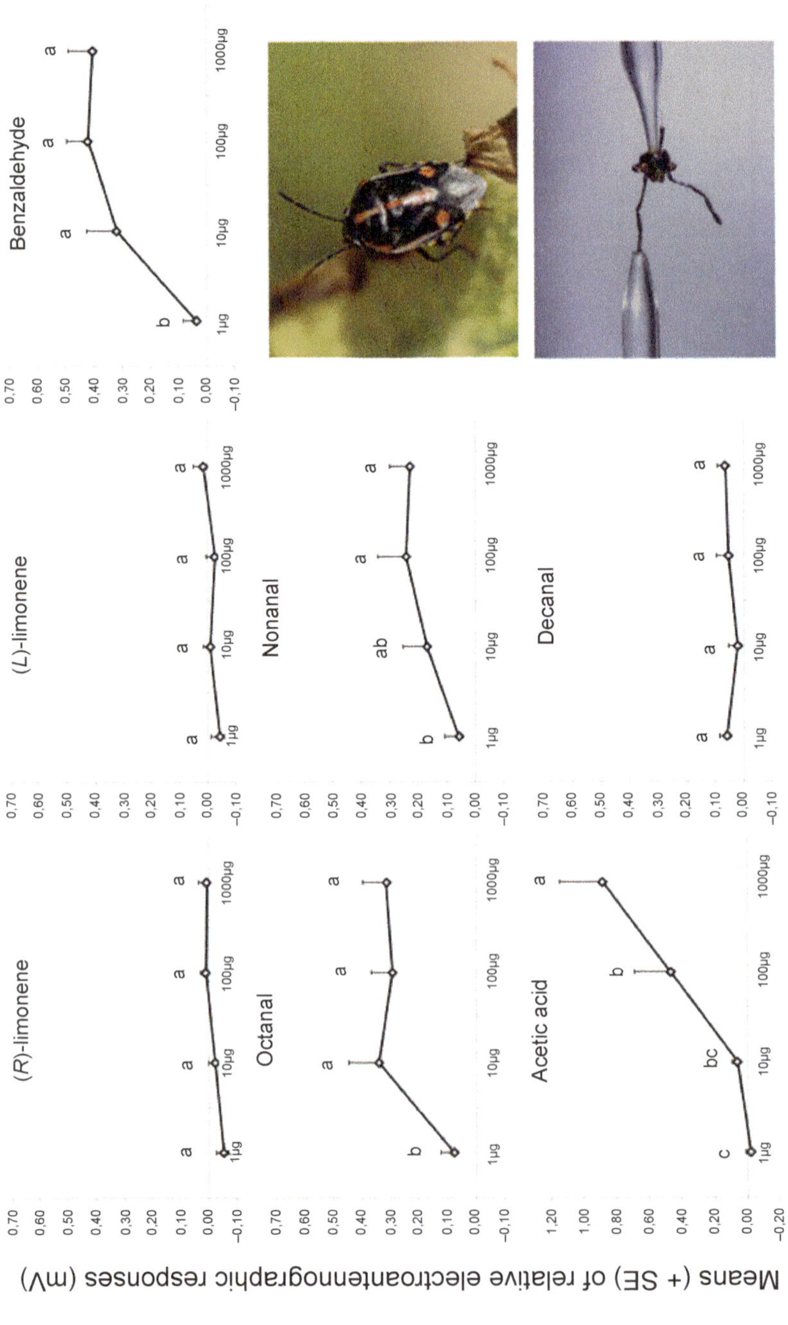

Figure 8.1. Electroantennographic (EAG) dose–response curves of *Bagrada hilaris* adults (A) to some volatiles produced from *Brassica oleracea* var botrytis leaves, EAG amplitudes were adjusted to a reference stimulus (hexane). (B) EAG preparation. Modified from Guarino et al. (in press).

Finally, in the case of *Euschistus heros* (F.) in olfactometer bioassays the adults did not discriminate between damaged and undamaged soybean plants nor between resistant and susceptible cultivars, albeit the VOC profile of damaged and undamaged host plants are different (Michereff et al. 2011).

Predators. Locating plants by orienting to VOC is a trait not only of phytophagous pentatomids: predaceous stink bugs are also attracted to volatiles emitted by the host plant of their prey. However, the few cases reported in the literature link the positive response of pentatomids to the presence of their prey on the plant where they feed. Therefore, in this case, the plant volatiles seem to be representing a cue in order to orient individuals towards potential prey and/or to facilitate aggregations (Sant'Ana et al. 1999). For example, *Podisus maculiventris* Say, the spined soldier bug, is sensitive to a number of plant volatiles that are released by the plant in response to feeding by one of its prey, the gipsy moth *Limantria dispar* L. (Sant'Ana et al. 1999). Similarly, the other predaceous stink bug *Perillus bioculatus* F., is attracted towards volatiles emitted by potato plants damaged by the Colorado Potato Beetle larvae, *Leptinotarsa decemlineata* (Say), whereas uninfested potato plants failed to elicit attraction (Weissbecker et al. 1999). In detail, this attraction seems to be mediated in particular by sesquiterpenes (Weissbecker et al. 2000, Van Loon et al. 2000).

Other Plant Traits

The external features of the plant influence several herbivorous species in mediating host finding processes: for several phytophagous insects, attraction to visual cues can result from responding to the color or shape of the host plant (Bernays and Chapman 2007), often in combination with olfactory cues that enhance the attraction to the host plant's odor (Prokopy and Owens 1983, Rull and Prokopy 2003). However, for Heteroptera species, the role of visual cues in host location has so far received less attention as compared to other plant traits. These plant characters may be either physical, such as tissue toughness and pubescence, or non-volatile chemical, such as toxins, digestibility-reducers, and nutrient balance, and can modify interactions between herbivores and their enemies by operating directly on the herbivore, the enemy, or both (Price et al. 1980). Once herbivores land on plants, in order to complete the plant selection behavior they examines the plant surface by contact testing, e.g., palpation of the leaf surface obtain additional information on plant quality that was not accessible during the previous phases of host selection (Schoonhoven et al. 2005). The importance of non-volatile chemicals in host selection for stink bugs has been described in a couple of studies. Silva et al. (2013) showed that in behavioral bioassays carried out on *E. heros,* non-volatile compounds of the resistant and susceptible cultivar of the same host plant might be involved in the choice of females to feed and oviposit. In particular the authors speculated about the possibility that the greater presence of reducing sugars and the lack of isoflavone forms induce the host preference of *E. heros*. The perception of these short range chemicals and their influence on oviposition behavior was also shown in Panizzi et al. (2004), who observed that *Nezara viridula* L. females prefer to oviposit in substrate treated with soybean plant extract in a two-choice arena bioassay.

Plant Traits Exploited by Egg Parasitoids of Stink Bugs

After the host location process is completed by phytophagous species and the activities of feeding and/or oviposition on the infested plant have started, changes in the ecophysiological traits and VOC emission of the plant itself usually occur (Kessler and Baldwin 2001) (Fig. 8.2). As plants cannot run away from attackers, they have developed different ways to respond to biotic factors such as herbivore infestation. Some activities of herbivores, such as feeding and/or oviposition, generally elicit an increase of the plant's emission of volatile compounds (Dicke and Van Loon 2000) or a qualitative change by producing compounds de novo (Dicke and Van Loon 2000, Martinez et al. 2013). These induced changes in the profiles of volatiles can occur locally or systemically, and may also provide chemical information about the status of attack of the emitting plant, which might be used by other herbivores or animals belonging to the higher trophic levels (Kessler and Baldwin 2001, van Poecke and Dicke 2004) or even by

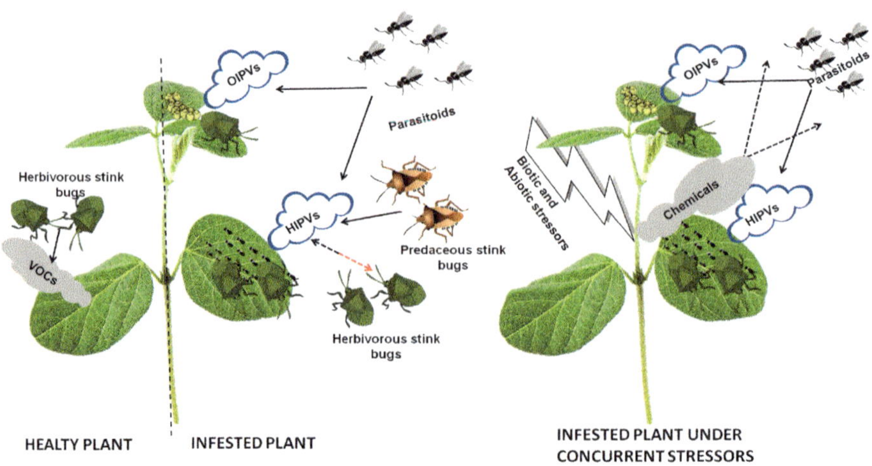

Figure 8.2. Scheme of the plant VOC exploited by stink bugs in host searching processes and the role of HIPV and OIPV in recruiting their parasitoids.

the neighboring plants of the same or another species (Baldwin and Schultz 1983, Arimura et al. 2000, Karban and Maron 2002).

Herbivore-induced plant volatiles may serve as reliable signals to other herbivores regarding the presence of competitors or natural enemies (Unsiker et al. 2009). Concerning the relations among competitors, the VOC changes induced by a phytophagous species can influence the behavior of the other herbivorous individuals at the intraspecific and interspecific level (Dicke and Baldwin 2010). At the interspecific level, HIPV can act as attractive cues for other species, in particular true bugs (Groot et al. 2003, Blackmer et al. 2004) or as oviposition repellents for several insect orders (Renwick and Chew 1994, Hilker and Meiners 2011). Concerning the response elicited at the intraspecific level, in the previous paragraph we reported the few records that demonstrate the influence of VOC emitted from plants damaged by oviposition and/or feeding of pentatomids on the host-finding behavior of conspecifics. However, as is extensively known, HIPV influence not only the other herbivores but also their natural enemies that are known to exploit these infochemicals during host location (Dicke 1994, Mattiacci et al. 2001, Soler et al. 2013, de Rijk et al. 2013, Fürstenberg-Hägg et al. 2013). Many parasitoids and predators are known to discriminate between the volatile chemicals produced by undamaged plants or plants infested by a particular herbivorous species (D'Alessandro and Turlings 2006, Hare 2011). The induced odors act as cues, leading female parasitoids to the habitat of their hosts (Turlings et al. 1991, 1993). The majority of the literature about herbivore-induced synomones is about chewing insects and mites, in comparison little is known about plant volatile production following herbivory by insects with piercing-sucking mouthparts.

Regarding the relationships among plants, pentatomids and their natural enemies, the role of chemical volatiles in mediating these interactions has been investigated only for some host plant—bug—egg parasitoid systems (Colazza et al. 2004a, Williams et al. 2005, Conti and Colazza 2012). Among the entomophagous insects associated with Heteroptera, egg parasitoids represent the largest group of species (Conti and Colazza 2012). These insects have to struggle with many difficult challenges concerning the nature of their hosts: bug eggs are in fact usually unapparent, especially when they are small, dispersed in the habitat, and concealed in plant tissue (Martinez et al. 2013). Moreover, the cues that are directly related to the presence of eggs may have high reliability but very low detectability (Vet and Dicke 1992, Hilker and McNeil 2007), as a consequence egg parasitoids have developed specialized strategies to overcome this reliability-detectability dilemma in order to accomplish a successful parasitism (Vinson 1998, Fatouros et al. 2008, Colazza et al. 2010), by optimizing exploitation of cues that are directly and indirectly related to host eggs and by discriminating those most reliable in indicating the presence of a suitable host (Hilker and McNeil 2007, Peñaflor et al. 2011). For example Manrique et al. (2005) reported that the volatiles released by several herbaceous plants, such as cotton, *Gossypium hirsutum* L., infested by adults of *L. hesperus* attract the egg parasitoid *Anaphes iole* Girault.

Feeding- and Oviposition-Induced Plant Volatile Cues

The exploitation of HIPV and OIPV by the natural enemies of herbivores, is well-known behavior (Mumm and Dicke 2010); these cues, defined synomones, are emitted in relatively great quantities and are, therefore, easily detectable by foraging parasitoids that, in this way, can get into the vicinity of the host to parasitize (Fatouros et al. 2008). Table 8.2 contains some examples of the attraction of egg parasitoids to HIPV and OIPV reported in tri-trophic systems involving stink bug species. Among pentatomids it has been observed that infestation activity of *E. heros* on two species of leguminous plants, soybean, *Glycine max* Merr., and pigeon pea, *Cajanus cajan* (L.), determined an increase in the attraction of this plant volatile for females of the egg parasitoid *Telenomus podisi* (Ashmead) (Moraes et al. 2005, 2009, Michereff et al. 2011). The VOC analysis indicates that the combination of feeding and oviposition damages induces the production of a volatile blend different from the individual types of damage, suggesting that *T. podisi* uses plant volatile cues associated with female *E. heros* damage in order to be present when the host lays its eggs, and thus ensures that its resource is

Table 8.2. Parasitoid behavioral responses to plant volatiles induced by the activity of feeding and/or oviposition of several stink bugs. From Conti and Colazza (2012) modified.

Parasitoid	Stink bug Host	Plant	Type of damage	Parasitoid response	Reference
Telenomus podisi	*Tibraca limbativentris*	*Oryza sativa*	feeding of females	attraction in olfactometer	Machado et al. 2014
Telenomus podisi	*Euschistus heros*	*Glycine max, Cajanus cajan*	feeding; cis-jasmone treatments	attraction in olfactometer	Moraes et al. 2005, Moraes et al. 2009, Mitchereff et al. 2011
Trissolcus basalis	*Tibraca limbativentris*	*Oryza sativa*	undamaged	attraction in olfactometer	Machado et al. 2014
Trissolcus basalis	*Nezara viridula*	*Vicia faba*	feeding + oviposition	attraction in olfactometer	Colazza et al. 2004a, Colazza et al. 2004b
Trissolcus brochymenae	*Murgantia histrionica*	*Brassica oleracea*	feeding; feeding + oviposition	increased searching behavior	Conti et al. 2008, Conti et al. 2010, Frati et al. 2013

optimal for parasitism (Moraes et al. 2005). In particular it has been observed that these changes in the emission of plant VOC are elicited in soybean by the naturally occurring plant activator *cis*-jasmone that plays a role in affecting different biosynthetic pathways related to defenses in soybean plants (Moraes et al. 2009, Vieira et al. 2013). Moraes et al. (2009) observed that plants treated with *cis*-jasmone attract *T. podisi* similarly to the plants damaged by the bug; the VOC responsible for the parasitoid attraction have not been identified yet, however it was observed that the emission of camphene, myrcene, (*E*)-ocimene, methyl salicylate, and (*E,E*)-4,8,12-trimethyltrideca-1,3,7,11-tetraene is induced by *cis*-jasmone treatment. In another study carried out by Michereff et al. (2011) it was observed that the compounds that contributed most to the difference in terms of VOC produced from healthy or bug-damaged soybean plants were (*E,E*)-*α*-farnesene, methyl salicylate, (*Z*)-3-hexenyl acetate, and (*E*)-2-octen-1-ol.

In the case of the egg parasitoid *Trissolcus basalis* Wollaston, the attraction to VOC emitted by host plants infested by the host *N. viridula* is elicited only when the contemporary activities of feeding and oviposition occur (Colazza et al. 2004a). In particular, it was observed that the combined activities of oviposition and feeding of *N. viridula* on *Vicia faba* L. and *Phaseolus vulgaris* L., induce an increase of (*E*)-*β*-caryophyllene as well as two sesquiterpenes (Colazza et al. 2004b). From an ecological point of view, the detection of volatile plant chemicals emitted as a result of herbivore egg deposition renders these cues highly detectable and also highly reliable (Hilker and Meiners 2010, 2011). Indeed, the combined activities of feeding and oviposition of *N. viridula* on the bean plants permit the emission of volatiles that attract the parasitoid only during a specific time window, i.e., 4–5 days, after egg deposition (Colazza et al. 2004a). Interestingly, this time interval is strictly linked to the acceptance behavior of *T. basalis*, since the egg parasitoid accepts host eggs up to 4–5 days old. Therefore, in this system, stimuli indirectly associated with host eggs can provide reliable information about the presence of suitable target hosts (Colazza et al. 2010).

Sometimes stink bug infestation activity can induce synomones perceived by the parasitoid only at a very short range. In the system *Murgantia histrionica* (Hahn)—*Trissolcus brochymenae* (Ashmead), female wasps did not seem to be attracted to volatiles from host-damaged plants, while they showed intense searching activity on the leaf surface, once they landed on plants damaged by *M. histrionica* feeding (Conti et al. 2010).

Substrate-Born Contact Cues

The plant surface shows an enormous variety of microtextures that are usually indiscernible to the unaided human eye, but are often of great importance in tri-trophic systems (Schoonhoven et al. 2005). The cuticles of most vascular plants are characterized by a thin layer of mostly hydrophobic components that defend the plant from desiccation or pathogen invasion (Schoonhoven et al. 2005). These microstructures of epicuticular waxes influence the interaction of physical and chemical cues among herbivores, plants and their natural enemies (Müller and Riederer 2005). In case of the tri-trophic systems involving pentatomids species, the function of epicuticular waxes was studied as they can absorb and release HIPV and OIPV allowing parasitoids to respond only after having contacted the plant or from a very short distance (Conti et al. 2008, 2010). Similarly, the physicochemical characteristics of the epicuticular waxes are involved in retaining contact kairomonal compounds released by the herbivore. Colazza et al. (2009) demonstrated that the leaf wax layer retains the chemical footprints from *N. viridula* adults walking on it. Even if the chemical composition of both the epicuticular wax layer and host footprints contains mainly linear hydrocarbons, *T. basalis* females are able to exploit the host chemical traces and differentiate between hosts of different sex. Indeed, the absence of *n*-nonadecane (*n*C19), a compound used by *T. basalis* to determine the host sex, as it is present in *N. viridula* males (Colazza et al. 2007), among the hydrocarbons of the epicuticular plant surfaces of several *N. viridula* host plants prevents any possible masking effects during *T. basalis* host searching activity (Lo Giudice et al. 2011). Moreover, leaf surface waxes by retaining these chemical traces permit females belonging to the *Trissolcus* genus to distinguish associated hosts from non-associated pentatomid species and to discriminate the host sex of associated hosts (Peri et al. 2013). These abilities developed by *Trissolcus* females allows them to modulate host search behavior, by spying cues strictly linked to host egg presence, and to avoid following 'false leads' that can cause wastage of time and energy in searching patches devoid of hosts.

Biotic and Abiotic Factors' Chemical Cues Exploited by Egg Parasitoids

In their natural environment plants are exposed simultaneously to several stress factors that can affect their VOC emission. Indeed, it was recently reported that the damage inflicted by non-host herbivores can vary in terms of intensity and duration, and that such factors can also affect plant responses as well as parasitoid foraging behavior (Ponzio et al. 2014). In this context, recent works have tried to establish the role of other biotic and abiotic factors that can influence the tri-trophic system *V. faba*—*N. viridula*—*T. basalis*.

A study carried out by Moujahed et al. (2014) evidenced how this system can be influenced by biotic factors such as the contemporary attack of another herbivorous species. In particular, the presence of *Sitona lineatus* L., a coleopteran species feeding on both above- and below-ground plant parts, alters the attraction of *T. basalis* to the plant that is also infested with its host *N. viridula* (Mouajhed et al. 2014). This disruption is probably linked to the fact that the infestation of *S. lineatus* determinates strong differences in the emission of the plant volatile blends produced. On the contrary, abiotic stress factors such as water stress can induce changes in plant VOC emission that in the same tropic system *V. faba —N. viridula—T. basalis,* seem to increase the parasitoid attraction to plants concurrently stressed by water reduction and phytophagous infestations (Salerno et al. 2017).

These aspects clearly show how difficult it is to predict the function of HIPV and OIPV in recruiting parasitoids in a multi-trophic context, such as the various natural environments, there are often several unpredictable variables that influence plant volatile emission and their consequential role in the so-called "cry for help" process. Indeed, in these multi-trophic environments, the parasitoid species can receive and base their decisions on the perceived chemical information from the different trophic levels which could be themselves influenced by biotic and abiotic factors (Vet and Dicke 1992).

Concluding Remarks

In this chapter we have summarized the current knowledge on the role of plant-produced semiochemicals in influencing the behavior of phytophagous stink bugs and their natural enemies. The role of the host plant volatiles used by stink bugs in host plant location has been poorly investigated, despite the notable weight of this taxon in terms of economic importance in agriculture. However, the role of plant volatiles in the host plant searching behavior of stink bugs has been observed, for several cases were reported here. Nevertheless the data obtained from the laboratory bioassays using olfactometer or electroantennographic techniques have not so far been often accompanied by adequate records from field bioassays. Different from the conditions in the laboratory, in the multi-trophic systems, such as the various natural environments, stink bugs need to pick out relevant host odor cues with ephemeral exposure and against high background noise. It has also to be pointed out that stink bugs have complex intraspecific aggregation behavior, often characterized by male-produced aggregation pheromones and intraspecific acoustic signals: possible interactions between these cues and plant semiochemicals cannot be excluded. Further research is needed in order to better clarify the ethology of these herbivorous species in order to understand in depth the chemicals cues involved in the interaction between stink bug species and their host plants. An understanding of the chemical ecology of herbivorous stink bugs could lead to the future exploitation of plant cues for monitoring and mass trapping these dangerous herbivorous species.

In the stink bug tri-trophic interactions, several studies have focused on understanding the role of the plant volatile in recruiting the stink bugs' natural enemies, in particular egg parasitoids. Interestingly, the egg parasitoids of stink bugs do not seem to exploit cues from healthy host plants to locate their host, unlike the egg parasitoids of other insect herbivores, such as some butterflies. The foraging strategies of several egg parasitoids of stink bugs have been reported here, showing in some cases similar patterns of exploitation of the semiochemicals induced by feeding and/or oviposition activity of the pest on the plant. These induced synomones are promptly detectable by foraging parasitoids that, in this way, can get into the vicinity of the host in order to parasitize. However, even if several chemical analysis of the VOC of healthy and bug-damaged plants have shown strong qualitative and or quantitative differences, the semiochemicals responsible for parasitoid attraction still remain to be identified. Among the other plant features playing an important function in these tri-trophic interactions, the cuticular leaf waxes seem to be of primary importance: their presence on plant surface permits the retention of the stink bugs' traces and can be exploited by parasitoids in their host searching behavior. Finally, recent studies have elucidated the importance of different biotic and abiotic factors in the searching behavior of stink bugs' parasitoids, such as the presence of a secondary below-ground herbivore or water stress conditions.

Acknowledgements

This work was supported by the Italian Ministry of Education, University and Research, within the Project "Assessing the impacts of invasive fungal pathogens and phytophagous insects on native plants, pathogens, phytophagous insects and symbionts" (FIRB 2012—Programma "Futuro in Ricerca" 2012—RBFR128ONN).

References

Arimura, G.I., R. Ozawa, T. Shimoda, T. Nishioka, W. Boland and J. Takabayashi. 2000. Herbivory-induced volatiles elicit defence genes in Lima bean leaves. Nature 406: 512–515.

Backus, E.A. 1988. Sensory systems and behaviours which mediate hemipteran plant-feeding: a taxonomic overview. J. Ins. Physiol. 34(3): 151–165.

Baldwin, I.T. and J.C. Schultz. 1983. Rapid changes in tree leaf chemistry induced by damage: evidence for communication between plants. Science 221: 277–279.

Berenbaum, M.R. and A.R. Zangerl. 2008. Facing the future of plant-insect interaction research: le retour à la "raison d'être". Plant Physiol. 146(3): 804–811.

Bernays, E.A. and R.F. Chapman. 2007. Host-plant Selection by Phytophagous Insects. Vol. 2. Springer Science and Business Media.

Blackmer, J.L., C. Rodriguez-Saona, J.A. Byers, K.L. Shope and J.P. Smith. 2004. Behavioral response of *Lygus hesperus* to conspecifics and headspace volatiles of alfalfa in a Y-tube olfactometer. J. Chem. Ecol. 30(8): 1547–1564.

Blackmer, J.L. and L.A. Canas. 2005. Visual cues enhance the response of *Lygus hesperus* (Heteroptera: Miridae) to volatiles from host plants. Environ. Entomol. 34(6): 1524–1533.

Bruce, T.J., L.J. Wadhams and C.M. Woodcock. 2005. Insect host location: a volatile situation. Trends Plant Sci. 10(6): 269–274.

Bruce, T.J. and J.A. Pickett. 2011. Perception of plant volatile blends by herbivorous insects–finding the right mix. Phytochemistry 72(13): 1605–1611.

Chinta, S., J.C. Dickens and J.R. Aldrich. 1994. Olfactory reception of potential pheromones and plant odors by tarnished plant bug, *Lygus lineolaris* (Hemiptera: Miridae). J. Chem. Ecol. 20(12): 3251–3267.

Colazza, S., A. Fucarino, E. Peri, G. Salerno, E. Conti and F. Bin. 2004a. Insect oviposition induces volatile emission in herbaceous plants that attracts egg parasitoids. J. Exp. Biol. 207(1): 47–53.

Colazza, S., J.S. McElfresh and J.G. Millar. 2004b. Identification of volatile synomones, induced by *Nezara viridula* feeding and oviposition on bean spp., that attract the egg parasitoid *Trissolcus basalis*. J. Chem. Ecol. 30(5): 945–964.

Colazza, S., G. Aquila, C. De Pasquale, E. Peri and J.G. Millar. 2007. The egg parasitoid *Trissolcus basalis* uses *n*-nonadecane, a cuticular hydrocarbon from its stink bug host *Nezara viridula*, to discriminate between female and male hosts. J. Chem. Ecol. 33(7): 1405–1420.

Colazza, S., M. Lo Bue, D. Lo Giudice and E. Peri. 2009. The response of *Trissolcus basalis* to footprint contact kairomones from *Nezara viridula* females is mediated by leaf epicuticular waxes. Naturwissenschaften 96(8): 975–981.

Colazza, S., E. Peri, G. Salerno and E. Conti. 2010. Host searching by egg parasitoids: exploitation of host chemical cues. pp. 97–147. *In*: Egg Parasitoids in Agroecosystems with Emphasis on *Trichogramma* Springer Netherlands.

Conti, E., C. Zadra, G. Salerno, B. Leombruni, D. Volpe, F. Frati et al. 2008. Changes in the volatile profile of *Brassica oleracea* due to feeding and oviposition by *Murgantia histrionica* (Heteroptera: Pentatomidae). Eur. J. Entomol. 105(5): 839–847.

Conti, E., G. Salerno, B. Leombruni, F. Frati and F. Bin. 2010. Short-range allelochemicals from a plant–herbivore association: a singular case of oviposition-induced synomone for an egg parasitoid. J. Exp. Biol. 213(22): 3911–3919.

Conti, E. and S. Colazza. 2012. Chemical ecology of egg parasitoids associated with true bugs. Psyche doi:10.1155/2012/651015.

D'Alessandro, M. and T.C. Turlings. 2006. Advances and challenges in the identification of volatiles that mediate interactions among plants and arthropods. Analyst 131(1): 24–32.

de Rijk, M., M. Dicke and E.H. Poelman. 2013. Foraging behaviour by parasitoids in multiherbivore communities. Animal Behav. 85(6): 1517–1528.

Dicke, M. 1994. Local and systemic production of volatile herbivore-induced terpenoids: their role in plant-carnivore mutualism. J. Plant Physiol. 143(4): 465–472.

Dicke, M. and J.J.A. Van Loon. 2000. Multitrophic effect of the herbivore-induced plant volatile in a evolutionary context. Ent. Exp. Appl. 97: 237–249.

Dicke, M., R.M. van Poecke and de J.G. Boer. 2003. Inducible indirect defence of plants: from mechanisms to ecological functions. Basic Appl. Ecol. 4(1): 27–42.

Dicke, M. 2009. Behavioural and community ecology of plants that cry for help. Plant Cell Environ. 32(6): 654–665.

Dicke, M. and I.T. Baldwin. 2010. The evolutionary context for herbivore-induced plant volatiles: beyond the 'cry for help'. Trends Plant Sci. 15(3): 167–175.

Du, Y., G.M. Poppy, W. Powell, J.A. Pickett, L.J. Wadhams and C.M. Woodcock. 1998. Identification of semiochemicals released during aphid feeding that attract parasitoid *Aphidius ervi*. J. Chem. Ecol. 24: 1355–1368.

Fatouros, N.E., M. Dicke, R. Mumm, T. Meiners and M. Hilker. 2008. Foraging behavior of egg parasitoids exploiting chemical information. Behav. Ecol. 19: 677–689.

Finch, S. and R.H. Collier. 2000. Host-plant selection by insects–a theory based on 'appropriate/inappropriate landings' by pest insects of cruciferous plants. Entomol. Exp. Appl. 96(2): 91–102.
Frati, F., G. Salerno and E. Conti. 2013. Cabbage waxes affect *Trissolcus brochymenae* response to short-range synomones. Insect Sci. 20(6): 753–762.
Fraenkel, G.S. 1953. The nutritional value of green plants for insects. pp. 90–100. *In*: Junk, W. (ed.). Transactions of the IXth International Congress of Entomology. The Hague. The Netherlands.
Fraenkel, G.S. 1959. The raison d'être of secondary plant substances. Science 129: 1466–1470.
Fürstenberg-Hägg, J., M. Zagrobelny and S. Bak. 2013. Plant defense against insect herbivores. Int. J. Mol. Sci. 14(5): 10242–10297.
Gonzalez, F.J. and D.W. Nebert. 1990. Evolution of the P450 gene superfamily: animal-plant "warfare", molecular drive and human genetic differences in drug oxidation. Trends Genet. 6: 182–186.
Groot, A.T., R. Timmer, G. Gort, G.P. Lelyveld, F.P. Drijfhout, T.A. Van Beek et al. 1999. Sex-related perception of insect and plant volatiles in *Lygocoris pabulinus*. J. Chem. Ecol. 25(10): 2357–2371.
Groot, A.T., A. Heijboer, J.H. Visser and M. Dicke. 2003. Oviposition preference of *Lygocoris pabulinus* (Het., Miridae) in relation to plants and conspecifics. J. Appl. Entomol. 127(2): 65–71.
Guarino, S., E. Peri, S. Colazza, N. Luchi, M. Michelozzi and F. Loreto. 2017. Impact of the invasive painted bug, *Bagrada hilaris* on physiological traits of its host *Brassica oleracea* var botrytis. Arthropod Plant Int. In press.
Guerrieri, E., G.M. Poppy, W. Powell, E. Tremblay and F. Pennacchio. 1999. Induction and systemic release of herbivore-induced plant volatiles mediating in-flight orientation of *Aphidius ervi*. J. Chem. Ecol. 25: 1247–1261.
Guerrieri, E. 2016. Who's Listening to Talking Plants? pp. 117–136. *In*: Blande, J.D. and R. Glinwood (eds.). Deciphering Chemical Language of Plant Communication, Signaling and Communication in Plant, Springer, Switzerland.
Hare, J.D. and E. Elle. 2002. Variable impact of diverse insect herbivores on dimorphic *Datura wrightii*. Ecology 83: 2711–2720.
Hare, J.D. 2011. Ecological role of volatiles produced by plants in response to damage by herbivorous insects. Annual Rev. Entomol. 56: 161–180.
Heil, M. 2015. Extrafloral nectar at the plant-insect interface: a spotlight on chemical ecology, phenotypic plasticity, and food webs. Annu. Rev. Entomol. 60: 213–232.
Hilker, M. and J. McNeil. 2007. Chemical and behavioral ecology in insect parasitoids: how to behave optimally in a complex odorous environment. pp. 693–705. *In*: Wajnberg, E., C. Bernstein and J. van Alphen (eds.). Behavioral Ecology of Insect Parasitoids, Blackwell Publishing.
Hilker, M. and T. Meiners. 2010. How do plants "notice" attack by herbivorous arthropods? Biol. Rev. Vol. 85: 267–280.
Hilker, M. and T. Meiners. 2011. Plants and insect eggs: how do they affect each other? Phytochemistry 72: 1612–1623.
Hilker, M. and N.E. Fatouros. 2015. Plant responses to insect egg deposition. Annu. Rev. Entomol. 60: 493–515.
Howe, G.A. and A. Schaller. 2008. Direct defenses in plants and their induction by wounding and insect herbivores. pp. 7–29. *In*: Induced Plant Resistance to Herbivory. Springer Netherlands.
Johnson, M.T. 2011. Evolutionary ecology of plant defences against herbivores. Funct. Ecol. 25(2): 305–311.
Karban, R. and J. Maron. 2002. The fitness consequences of interspecific eavesdropping between plants. Ecology 83: 1209–1213.
Kessler, A. and I. Baldwin. 2001. Defensive function of herbivore-induced plant volatile emissions in nature. Science 291: 2141–2144.

Lo Giudice, D., M. Riedel, M. Rostás, E. Peri and S. Colazza. 2011. Host sex discrimination by an egg parasitoid on brassica leaves. J. Chem. Ecol. 37(6): 622–628.
Manrique, V., W.A. Jones, L.H. Williams III and J.S. Bernal. 2005. Olfactory responses of *Anaphes iole* (Hymenoptera: Mymaridae) to volatile signals derived from host habitats. J. Insect Behav. 18(1): 89–104.
Martinez, G., R. Soler and M. Dicke. 2013. Behavioral ecology of oviposition-site selection in herbivorous true bugs. Advances in the Study of Behavior. No. 45. Elsevier, 175–207.
Machado, R.M., J. Sant'Ana, M.C. Blassioli-Moraes, R.A. Laumann and M. Borges. 2014. Herbivory-induced plant volatiles from *Oryza sativa* and their influence on chemotaxis behaviour of *Tibraca limbativentris* stal. (Hemiptera: Pentatomidae) and egg parasitoids. Bull. Entomol. Res. 104(03): 347–356.
Mattiacci, L., B.A. Rocca, N. Scascighini, M. D'Alessandro, A. Hern and S. Dorn. 2001. Systemically induced plant volatiles emitted at the time of "danger". J. Chem. Ecol. 27(11): 2233–2252.
McPherson, J.E. and R.M. McPherson. 2000. Stink Bugs of Economic Importance in America North of Mexico. CRC Press LLC.
Metcalf, R.L. and E.R. Metcalf. 1992. Plant Kairomones in Insect Ecology and Control (Vol. 1). Springer Science & Business Media.
Michereff, M.F.F., R.A. Laumann, M. Borges, M. Michereff-Filho, I.R. Diniz, A.L.F. Neto et al. 2011. Volatiles mediating a plant-herbivore-natural enemy interaction in resistant and susceptible soybean cultivars. J. Chem. Ecol. 37(3): 273–285.
Molina, G.A.R. and E.V. Trumper. 2012. Selection of soybean pods by the stink bugs, *Nezara viridula* and *Piezodorus guildinii*. J. Insect Sci. 12(1): 104.
Moraes, M.C.B., R.A. Laumann, E.R. Sujii, C. Pires and M. Borges. 2005. Induced volatiles in soybean and pigeon pea plants artificially infested with the neotropical brown stink bug, *Euschistus heros*, and their effect on the egg parasitoid, *Telenomus podisi*. Entomol. Exp. Appl. 115(1): 227–237.
Moraes, M.C.B., R.A. Laumann, M. Pareja, F.T. Sereno, M.F. Michereff, M.A. Birkett et al. 2009. Attraction of the stink bug egg parasitoid *Telenomus podisi* to defence signals from soybean activated by treatment with cis-jasmone. Entomol. Exp. Appl. 131(2): 178–188.
Moujahed, R., F. Frati, A. Cusumano, G. Salerno, E. Conti, E. Peri et al. 2014. Egg parasitoid attraction toward induced plant volatiles is disrupted by a non-host herbivore attacking above or belowground plant organs. Frontiers Plant Sci. 5: 601. DOI: 10.3389/fpls.2014.00601.
Müller, C. and M. Riederer. 2005. Plant surface properties in chemical ecology. J. Chem. Ecol. 31(11): 2621–2651.
Mumm, R. and M. Dicke. 2010. Variation in natural plant products and the attraction of bodyguards involved in indirect plant defense. Can. J. Zool. 88: 628–667.
Olson, D.M., J.R. Ruberson, A.R. Zeilinger and D.A. Andow. 2011. Colonization preference of *Euschistus servus* and *Nezara viridula* in transgenic cotton varieties, peanut, and soybean. Entomol. Exp. Appl. 139: 161–169.
Palumbo, J.C., T.M. Perring, J.G. Millar and D.A. Reed. 2015. Biology, ecology, and management of an invasive stink bug, *Bagrada hilaris*, in North America. Annu. Rev. Entomol. 61: 453–473.
Panizzi, A.R., M. Berhow and R.J. Bartelt. 2004. Artificial substrate bioassay for testing oviposition of southern green stink bug conditioned by soybean plant chemical extracts. Environ. Entomol. 33: 1217–1222.
Paré, P.W. and J.H. Tumlinson. 1996. Plant volatile signals in response to herbivore feeding. Florida Entomologist 79: 93–103.
Peñaflor, M.F.G.V., M. Erb, L.A. Miranda, A.G. Werneburg and J.M.S. Bento. 2011. Herbivore-induced plant volatiles can serve as host location cues for a generalist and a specialist egg parasitoid. J. Chem. Ecol. 37(12): 1304–1313.
Peri, E., F. Frati, G. Salerno, E. Conti and S. Colazza. 2013. Host chemical footprints induce host sex discrimination ability in egg parasitoids. PloS One 8(11): e79054.

Ponzio, C., R. Gols, B.T. Weldegergis and M. Dicke. 2014. Caterpillar-induced plant volatiles remain are liable signal for foraging wasps during dual attack with a plant pathogen or non-host insect herbivore. Plant Cell Environ. 37: 1924–1935.

Price, P.W., C.E. Bouton, P. Gross, B.A. McPheron, J.N. Thompson and A.E. Weis. 1980. Interactions among three trophic levels: influence of plants on interactions between insect herbivores and natural enemies. Annu. Rev. Ecol. Syst. 11: 41–65.

Prokopy, R.J. and E.D. Owens. 1983. Visual detection of plants by herbivorous insects. Annu. Rev. Entomol. 28(1): 337–364.

Rather, A.H., M.N. Azim and S. Maqsood. 2010. Host plant selection in a pentatomid bug *Eurydema pulchrum* Westwood. J. Plant Prot. Res. 50: 229–232.

Renwick, J.A.A. and F.S. Chew. 1994. Oviposition behavior in Lepidoptera. Annu. Rev. Entomol. 39(1): 377–400.

Rull, J. and R.J. Prokopy. 2003. Trap position and fruit presence affect visual responses of apple maggot flies (Dipt., Tephritidae) to different trap types. J. Appl. Entomol. 127(2): 85–90.

Ruther, J., T. Meiners and J.L. Steidle. 2002. Rich in phenomena-lacking in terms. A classification of kairomones. Chemoecology 12(4): 161–167.

Salerno, G., F. Frati, G. Marino, L. Ederli, S. Pasqualini, F. Loreto et al. 2017. Effect of water stress on emission of volatile organic compound by *Vicia faba* and consequences for attraction of the egg parasitoid *Trissolcus basalis*. J. Pest Sci. DOI: 10.1007/510340-016-0830-z.

Sant'Ana, J., R.F. Da Silva and J.C. Dickens. 1999. Olfactory reception of conspecific aggregation pheromone and plant odors by nymphs of the predator, *Podisus maculiventris*. J. Chem. Ecol. 25(8): 1813–1826.

Schoonhoven, L.M., J.J.A. Van Loon and M. Dicke. 2005. Insect-plant Biology. No. Ed. 2. Oxford University Press.

Schwarz, J., R. Gries, K. Hillier, N. Vickers and G. Gries. 2009. Phenology of semiochemical-mediated host foraging by the western boxelder bug, *Boisea rubrolineata*, an aposematic seed predator. J. Chem. Ecol. 35(1): 58–70.

Silva, F.A., M.C. Carrão-Panizzi, M.C. Blassioli-Moraes and A.R. Panizzi. 2013. Influence of volatile and nonvolatile secondary metabolites from soybean pods on feeding and on oviposition behavior of *Euschistus heros* (Hemiptera: Heteroptera: Pentatomidae). Environ. Entomol. 42(6): 1375–1382.

Soler, R., T.M. Bezemer and J.A. Harvey. 2013. Chemical ecology of insect parasitoids in a multitrophic above- and below-ground context. pp. 64–85. *In*: Chemical Ecology of Insects Parasitoids. Wiley United Kingdom.

Turlings, T.C.J., J.H. Tumlinson, F.J. Eller and W.J. Lewis. 1991. Larval-damaged plants: source of volatile synomones that guide the parasitoid *Cotesia marginiventris* to the micro-habitat of its hosts. Entomol. Exp. Appl. 58(1): 75–82.

Turlings, T.C., F.L. Wäckers, L.E. Vet, W.J. Lewis and J.H. Tumlinson. 1993. Learning of host-finding cues by hymenopterous parasitoids. pp. 51–78. *In*: Insect Learning. Springer US.

Turlings, T.C.J., M. Bernasconi, R. Bertossa, F. Bigler, G. Caloz and S. Dorn. 1998. The induction of volatile emissions in maize by three herbivore species with different feeding habits—possible consequences for their natural enemies. Biol. Control 11: 122–129.

Unsicker, S.B., G. Kunert and J. Gershenzon. 2009. Protective perfumes: the role of vegetative volatiles in plant defense against herbivores. Curr. Opin. Plant Biol. 12(4): 479–485.

Van Loon, J.J.A., E.W. Vos and M. Dicke. 2000. Orientation behaviour of the predatory hemipteran *Perillus bioculatus* to plant and prey odours. Entomol. Exp. Appl. 96(1): 51–58.

Van der Maijden, E. 1996. Plant defence, an evolutionary dilemma: contrasting effects of (specialist and generalist) herbivores and natural enemies. Entomol. Exp. Appl. 80: 307–310.

van Poecke, R.M.P. and M. Dicke. 2004. Indirect defense of plants against herbivores: using *Arabidopsis thaliana* as a model plant. Plant Biol. 6: 387–401.

Verschaffelt, E. 1911. The cause determining the selection of food in some herbivorous insects. K Akad Wetensch Amsterdam Proc. Sect. Sci. 13: 536–542.

Vet, L.E.M. and M. Dicke. 1992. Ecology of infochemical use by natural enemies in a tri-trophic context. Annu. Rev. Entomol. 37(1): 141–172.

Vieira, C.R., M.C.B. Moraes, M. Borges, E.R. Sujii and R.A. Laumann. 2013. *Cis*-Jasmone indirect action on egg parasitoids (Hymenoptera: Scelionidae) and its application in biological control of soybean stink bugs (Hemiptera: Pentatomidae). Biol. Control 64(1): 75–82.

Vinson, S.B. 1998. The general host selection behavior of parasitoid hymenoptera and a comparison of initial strategies utilized by larvaphagous and oophagous species. Biol. Control 11(2): 79–96.

Visser, J.H. 1986. Host odor perception in phytophagous insects. Annu. Rev. Entomol. 31(1): 121–144.

Walling, L. 2000. The myriad plant responses to herbivores. J. Plant Growth Regul. 19: 195–216.

Weissbecker, B., J.J.A. Van Loon and M. Dicke. 1999. Electroantennogram responses of a predator, *Perillus bioculatus*, and its prey, *Leptinotarsa decemlineata*, to plant volatiles. J. Chem. Ecol. 25(10): 2313–2325.

Weissbecker, B., J.J.A. Van Loon, M.A. Posthumus, H.J. Bouwmeester and M. Dicke. 2000. Identification of volatile potato sesquiterpenoids and their olfactory detection by the two-spotted stink bug *Perillus bioculatus*. J. Chem. Ecol. 26(6): 1433–1445.

Whittaker, R.H. and P. Feeny. 1971. Allelochemics: chemical interactions between species. Science 171: 757–770.

Williams, L., C. Rodriguez-Saona, P.W. Paré and S.J. Crafts-Brandner. 2005. The piercing-sucking herbivores *Lygus hesperus* and *Nezara viridula* induce volatile emissions in plants. Arch. Insect Biochem. Physiol. 58(2): 84–96.

Williams III, L., J.L. Blackmer, C. Rodriguez-Saona and S. Zhu. 2010. Plant volatiles influence electrophysiological and behavioral responses of *Lygus hesperus*. J. Chem. Ecol. 36(5): 467–478.

Zvereva, E., V. Lanta and M. Kozlov. 2010. Effects of sap-feeding insect herbivores on growth and reproduction of woody plants: a meta-analysis of experimental studies. Oecologia 163: 949–960.

CHAPTER 9

Use of Pheromones for Predatory Stink Bug Management

Diego Martins Magalhães

Introduction

The Pentatomidae are one of the largest heteropteran families, with more than 4,120 described species, however, a minority of them are predators (Panizzi et al. 2000, Rider 2011). The Asopinae (~ 300 spp.) mainly differ from other pentatomid subfamilies by having a crassate rostrum that is an adaptation for predaceous feeding habits (Gapud 1991, Bueno and van Lenteren 2012). They feed principally on the larval forms of soft-bodied insects such as Coleoptera, Hymenoptera and Lepidoptera (De Clercq 2008), but they can also feed on a wide variety of prey which can vary in quality (De Clerq et al. 1998a, Lemos et al. 2003, Zanuncio et al. 2005), including plant material (Ruberson et al. 1986). The Asopinae stink bugs are found throughout the world, but are most plentiful in the Nearctic and Neotropical regions (De Clercq 2000).

Predaceous stink bugs are abundant in a wide range of natural and agroecosystem habitats (Yeargan 1998) and are thought to play a great role in the biological control of economically important insect pests (DeBach and Rosen 1991, Hough-Goldstein and Whalen 1993, Torres et al. 2006). Nevertheless, the use of predatory stink bugs in pest management is still limited and a few species are used worldwide in commercial augmentative biological control: *Brontocoris tabidus* Signoret (for lepidopterans in Latin America), *Picromerus bidens* L. (for lepidopterans in Europe), *Podisus maculiventris* (Say) (for coleopterans and lepidopterans in Europe and

Universidade de Brasília, Instituto de Ciências Biológicas - Departamento de Zoologia, Campus Universitário Darcy Ribeiro, CEP 70910-900 Brasília-DF, Brazil.
Email: magalhaes.dmm@gmail.com

North America) and *Podisus nigrispinus* (Dallas) (for lepidopterans in Latin America) (van Lenteren 2012).

The increasing concern about the environment, human health and food insecurity has encouraged the implementation of Integrated Pest Management (IPM). In this context, pheromones and other semiochemicals might provide an efficient alternative for biorational control and can be implemented alongside other IPM strategies, such as augmentative biological control. In this chapter, the pheromones of predaceous stink bugs which have already been identified, their exploitation as kairomones and their application in pest control will be summarized.

Overview of Predaceous Stink Bug pheromones and Semiochemicals

In several asopine species, besides the metathoracic scent gland (MTG), males have two types of pheromone glands: dorsal abdominal scent glands (DAGs) and sternal glands (SGs) (Aldrich 1995, 1998). Polyphagous predators generally have dimorphic DAGs [e.g., *Brontocoris, Podisus* and *Supputius* spp.], meanwhile specialized predators have male-specific SGs [e.g., *Oplomus, Perillus* and *Stiretrus* spp.] (Thomas 1992, Aldrich 1998). The asopines with dimorphic DAGs are able to control their pheromone emission (Aldrich 1995) and those with male-specific SGs usually produce blends comprised of more non-volatile compounds (Aldrich et al. 1986a, Aldrich and Lubsy 1986). There are still some predator stink bugs that have neither DAG nor SG, but may produce pheromones from the cells present in the abdominal epidermis [e.g., *Apateticus, Dinorhynchus, Euthyrhynchus* and *Picromerus* spp.] (Aldrich 1988a,b, Aldrich 1995).

The first long-range attractant pheromone determined from a predacious stink bug was obtained from the spined soldier bug, *P. maculiventris*. The DAGs secretion of the male *P. maculiventris* is comprised of benzyl alcohol, (*E*)-2-hexenal, linalool, (*R*)-α-terpineol and terpinen-4-ol, and attracts both sexes and nymphs (Aldrich et al. 1984a,b, Sant'Ana et al. 1997). Starved spined soldier bugs respond most strongly to this aggregation pheromone in wind tunnel bioassays, supporting the hypothesis that this generalist predator uses the pheromone as a cue indicative of prey presence (Shetty and Hough-Goldstein 1998). *Podisus* species seem to have a pattern in the composition of male DAGs: usually (*E*)-2-hexenal and a major species-specific compound (Table 9.1). The males of *Podisus neglectus* (Westwood), *P. nigrispinus*, *Podisus placidus* Uhler and *Podisus mucronatus* Uhler, release compounds comprised of (*E*)-2-hexenal plus major species-specific components that include linalool, α-terpineol, 9-hydroxy-2-nonanone and (*E*)-2-hexenyl tiglate, respectively (Aldrich et al. 1986a, Aldrich and Borges 1991). For males of *Supputius cincticeps* Stal, the same pattern has

Table 9.1. Chemical composition of the male dorsal abdominal scent glands (DAGs) of *Podisus* spp. (Pentatomidae: Asopinae).

Podisus spp.	DAGs compounds	References
P. maculiventris	(*E*)-2-hexenal, benzyl alcohol, linalool, (*R*)-α-terpineol, terpinen-4-ol, traces of *cis* and *trans* piperitol	Aldrich et al. 1984a
P. neglectus	(*E*)-2-hexenal, benzyl alcohol, linalool, nerolidol, α-terpineol	Aldrich et al. 1986a
P. mucronatus	(*E*)-2-hexenal, benzyl alcohol, (*E*)-2-hexenol, (*E*)-2-hexenyl tiglate	Aldrich and Borges 1991
P. placidus	(*E*)-2-hexenal, (*E*)-2-hexenol, 9-hydroxy-2-nonanone, 2-(4-hydroxyphenyl)ethanol	Aldrich and Borges 1991
P. nigrispinus	(*E*)-2-hexenal, benzyl alcohol, linalool, *trans*-piperitol, α-terpineol, terpine-4-ol	Aldrich and Borges 1991
P. distinctus	(*E*)-2-hexenal, benzyl alcohol, linalool, α-terpineol	Aldrich et al. 1997
P. rostralis	(*E*)-2-hexenal, (*E*)-2-hexenoic acid, linalool, nonanal, nonanol, nerolidol	Aldrich et al. 1997
P. sagitta	(*E*)-2-hexenal, benzyl alcohol, linalool, (*E*)-2-hexenoic acid, α-terpineol	Aldrich et al. 1997

been observed: (*E*)-2-hexenal plus linalool as the major species-specific component (Aldrich et al. 1997). Interestingly, the males of *Oechalia schellenbergii* (Guérin-Méneville), a predator of several lepidopteran larvae in the Australian ecozone, produce a unique DAG secretion comprised of two major compounds: 3-methylenehexyl acetate and 9-hydroxygeranyl diacetate [2,6-dimethyl-2(*E*),6(*E*)-octadien-1,8-diol diacetate] (Aldrich et al. 1996).

The males of *Mineus, Oplomus, Perillus* and *Stiretrus* spp. possess pubescent patches on the abdomen with enlarged SGs that mainly release a single norterpene compound (Aldrich et al. 1986a, Aldrich 1998). The predator asopines *Mineus strigipes* (Herrich-Schaeffer), *Oplomus servus* (Spinola) and *Perillus bioculatus* (Fabricius) release 6,10,13-trimethyltetradecyl isovalerate (Aldrich et al. 1986a, Aldrich and Lubsy 1986), and *Stiretrus anchorago* (Fabricius) secretes the corresponding alcohol 6,10,13-trimethyltetradecanol (Aldrich et al. 1986a, Kochansky et al. 1989). Usually, the SGs have been reported in specialist predator stink bug species (Aldrich 1998), however, males of the generalist asopine *Eocanthecona furcellata* (Wolf) also have SGs and secrete 6,10,13-trimethyltetradecyl isovalerate (Ho et al. 2003). Nevertheless, the biological function of these male-specific SGs compounds remains unclear.

It is important to note that many of the compounds identified from scent glands have no supporting biological data, either from laboratory or field tests. The predatory asopines that have had their scent gland constituents biologically tested are *P. maculiventris* (Aldrich et al. 1984b, Aldrich et al. 1986b, Sant'Ana et al. 1997, Aldrich and Cantelo 1999, Kelly et al. 2014), *P. neglectus* (Aldrich et al. 1986b), *P. nigrispinus* (Aldrich et al. 1997), *Podisus distinctus* (Stal) (Aldrich et al. 1997) and *S. cincticeps* (Aldrich et al. 1997). Several other compounds have been isolated and identified from scent glands but their biological activity has not been reported in the general literature.

Field Test

There has been an effort to develop sustainable approaches to augment and conserve predators, especially stink bug predators, for biological control purposes. Several works have focused on the predatory spined soldier bug because this stink bug is the most widespread asopine predator in North America (McPherson 1982, De Clercq 2008), and its adults and nymphs are voracious predators of a wide variety of eggs and larvae of many phytophagous species (McPherson 1982). Field-testing using synthetic blends mimicking male DAG secretions have demonstrated that this secretion constitutes the long-range attractant pheromone of *P. maculiventris* (Aldrich et al. 1984a, Sant'Ana et al. 1997, Aldrich and Cantelo 1999). Pheromone-baited and live-insect-baited traps were tested and there were no significant differences in the number of insects captured in both traps (Aldrich et al. 1984b), indicating that the pheromone formulation contains the compounds responsible for *P. maculiventris* attraction. Moreover, Aldrich et al. (1986b) demonstrated that an effective artificial pheromone for the spined soldier bug can be made with a racemic α-terpineol and (*E*)-2-hexenal (2:1 volume ratio). This discovery was very important in the context of pest management because a synthetic blend with a lower number of components is more advisable if we consider aspects such as applicability and production costs (Collatz and Dorne 2012).

A commercial *P. maculiventris* pheromone formulation is currently available (Soldier Bug Attractors®) and field tests have shown its efficacy (Sant'Ana et al. 1997, Aldrich and Cantelo 1999). An example of *P. maculiventris* pheromone efficacy is the field collection described by Aldrich and Cantelo (1999) showing that for three consecutive years, by using 30 pheromone-baited traps (rebaited every 2–4 days for ~ 3 weeks in early spring), ca. 4,600 spined soldier bug adults were captured each year. It is noteworthy that not only male and female spined soldier bugs are attracted by male-produced pheromones; nymphs are also attracted by these semiochemicals (Sant'Ana et al. 1997, Aldrich and Cantelo 1999). This is crucial information for augmentative biological control, meaning that the pheromone can be used to disperse nymphs from the points of augmentation, reducing the probability of emigration from the field because they cannot fly away from the release site (Kelly et al. 2014). For the management of the Colorado potato beetle, *Leptinotarsa decemlineata* (Say) (Coleoptera: Chrysomelidae), potato field plots treated with *P. maculiventris* pheromone resulted in significantly better yields than the untreated control plots (Aldrich and Cantelo 1999). Recently in tomato fields, tests with *P. maculiventris* pheromone showed that in optimal weather conditions the use of pheromone further reduced spined soldier bug emigrations, at the same time attracting wild conspecifics to the tested plots (Kelly et al. 2014). This study also provides evidence that *P. maculiventris* long-range attractant pheromone manipulates its behaviour in order to enhance pest consumption.

Besides *P. maculiventris*, few other predatory stink bug pheromones have been field tested. An artificial pheromone for *P. neglectus* can be prepared from (*E*)-2-hexenal and racemic linalool (1:2 volume ratio), successfully attracting male and female adults, but in contrast to *P. maculiventris* more females than males are captured in pheromone-baited traps (Aldrich et al. 1986b). In Brazil, pheromone field tests have been done with two Neotropical species, *P. nigrispinus* and *S. cincticeps*. Although no conspecific was captured in the traps baited with *P. nigrispinus* and *S. cincticeps* pheromones, a few adults of *P. distinctus* were captured in the traps (Aldrich et al. 1997). Chemical analysis of male *P. distinctus* DAGs indicated that this species produces a blend of compounds containing the major components of *P. nigrispinus* and *S. distinctus* putative pheromones (see Table 9.1 and previous item). Very few pheromones have been identified for predator stink bugs and the only case where the long-range attractant pheromone has been used effectively in biological control is that of *P. maculiventris*.

Exploitation of Pheromones as Kairomones

A complex of parasitoids and predators exploit stink bugs' pheromones as host finding kairomones. There are two main dipteran parasitoids of *Podisus* spp., the specialist *Hemyda aurata* Robineau-Desvoidy and the generalist *Euclytia flava* (Townsend) (Diptera: Tachinidae) that also parasitizes several phytophagous stink bugs (Aldrich 1985). Both species are reliant on the male pheromones produced by *P. maculiventris* and *P. neglectus* and the males usually have more tachinid eggs on them than the females (Aldrich et al. 1984b). *Euclytia flava* seems to be more abundant than *H. aurata* in the U.S. In three consecutive seasons, parasitoids were attracted five times more to pheromone-baited traps than *P. maculiventris* individuals (Aldrich 1985). Moreover, females of the ectoparasitoid *Forcipomyia crinite* Saunders (Diptera: Ceratopogondae) are also attracted by *P. maculiventris*' pheromone and were observed sucking the blood of male spined soldier bugs (Aldrich et al. 1984b).

The egg parasitoid *Telenomus calvus* Johnson (Hymenoptera: Plastygastridae) uses the aggregation pheromone of *P. maculiventris* in order to facilitate its phoratic behaviour (Aldrich et al. 1984b). Female parasitoids come into the vicinity of *Podisus* males releasing pheromones and wait for the female stink bugs to arrive. As soon as the inseminated female stink bugs oviposit, the wasps parasitize their eggs (Aldrich 1985). Surprisingly the eastern yellow jacket, *Vespula maculifrons* (Hymenoptera: Vespidae), is also attracted to *P. neglectus* and *P. maculiventris* pheromones (Aldrich et al. 1986b), although there are no records of *V. maculifrons* feeding on these stink bugs. There are some hypotheses as to why the eastern yellow jackets are attracted to these pheromones: (I) they are opportunistic, stealing prey from the stink bugs; (II) they prey on the tachinid flies that are attracted by the stink bugs' pheromones; (III) they identify the pheromone constituents such as the volatiles from damaged leaves in order to find phytophagous larvae (Aldrich et al. 1986b). Whichever the hypothesis, the number of yellow jackets captured in *Podisus* pheromone-baited traps are impressive: from among all the captured insects, including *P. neglectus* and *V. maculifrons,* in a two-year experiment, more than 70% corresponded to yellow jackets (Aldrich et al. 1986b).

Parasitic insects may affect host distribution, behaviour, breeding sites and emergence time (Price 1975), and they seem to be important selective agents affecting the life strategy of *Podisus* stink bugs. Males spined soldier bugs release large amounts of pheromones becoming vulnerable to parasitism, some males probably remain silent and go to calling males and try to ambush the attracted females (Aldrich 1985), thus minimizing their own risk of being parasitized (Aldrich 1995).

Concluding Remarks

The use of semiochemical attractants derived from either insects or plants has incredible potential to increase the recruitment and retention of natural enemies in agroecosystems. In this scenario the identification, synthesis and field testing of these compounds are crucial steps for the development of pest management strategies. Nevertheless, the spined soldier bug aggregation pheromone is one of the few examples of predator's pheromones that can be found commercially in order to be used to enhance biological control. And despite being used mainly in home gardens, it can be used to suppress the Colorado potato beetle in augmentative biological control (Sant'Ana et al. 1997, Aldrich and Cantelo 1999).

Several studies have shown that for augmentative biological control a large number of predators are necessary in order to suppress the pest prey (Hough-Goldstein and Whalen 1993, Aldrich 1998). For the spined soldier bug and other *Podisus* spp., large amounts of individuals can be captured by using pheromone-baited traps, enabling the production of young predators. Aldrich and Cantelo (1999) have described that using this technology, it was possible to capture over 1,700 *P. maculiventris* females during 2 ± 3 weeks in early spring, and that this total represents the potential to produce large amounts of young predators based on biological parameters such as fecundity and survival (De Clercq et al. 1998b). In field conditions, porous nursery cages and pheromone dispensers can be used for the nymphs in order to promote the dispersal of these young predators (Sant'Ana et al. 1997, Aldrich and Cantelo 1999). Besides the introduced predators, native predators can also be attracted from the surrounding vegetation and retained in agroecosystems using the pheromone traps (Kelly et al. 2014). Synthetic pheromones can also be used for monitoring in order to know whether these natural enemies have become established in the crops and to direct predators to the infestation points or out of the infested fields prior to insecticide application (Aldrich and Lubsy 1986).

The downside of this strategy is that predatory stink bugs subject themselves to intense parasitic pressure from tachinid and scelionid wasps that use their pheromones as host-finding kairomones (Aldrich et al. 1984b, Aldrich 1995). Consequently, the high parasitoid attraction can cancel out the benefits associated with the retention of predatory stink bugs in the area. Besides that, augmentative biological control studies have shown low efficiency in reducing pest densities due to the excessive emigration of predators (Heimpel and Asplen 2011) and due to the unfavourable climate conditions (Collier and van Steenwyk 2004). Poor retention considering loss by emigration can make augmentative biological control less cost-effective as far more predators would be purchased than are actually needed. An alternative strategy to reduce emigration is to use behaviour-modifying semiochemicals (pheromones and/or plant volatiles) for the retention of predatory stink bugs in the crop and at the same time to recruit natural populations from the adjacent habitats (Kelly et al. 2014, Kaplan and Lewis 2015). Kelly et al. (2014) showed that under favourable climate conditions (dry days), pheromones could be effective in retaining predatory stink bugs in the field and also that semiochemicals (pheromones and plant volatiles) could direct stink bugs from the central release site in order to enhance pest consumption and move the predators towards infestation points. Using nymphs can also reduce the likelihood of loss because they cannot fly away from the release site (Sant'Ana et al. 1997, Aldrich and Cantelo 1999). Especially needed are more studies on field evaluation on the effectiveness of putative pheromones for their assessment of biological activity. The lack of appropriate bioassays has left us in the middle of the identification process with hardly any supporting biological data for DAG secretions (Millar 2005), meaning that their biological roles remain unknown for many asopine species.

References

Aldrich, J.R., W.R. Lubsy, J.P. Kochansky and C.B. Abrams. 1984a. Volatile compounds from the predatory insect *Podisus maculiventris* (Hemiptera: Pentatomidae): male and female metathoracic scent gland and female dorsal abdominal gland secretions. J. Chem. Ecol. 10: 561–568.

Aldrich, J.R., J.P. Kochansky and C.B. Abrams. 1984b. Attractant for a beneficial insect and its parasitoids: pheromone of the predatory spined soldier bug, *Podisus maculiventris* (Hemiptera: Pentatomidae). Environ. Entomol. 13: 1031–1036.

Aldrich, J.R. 1985. Pheromone of a true bug (Hemiptera-Heteroptera): attractant for the predator, *Podisus maculiventris*, and kairomonal effects. pp. 95–119. *In*: Acree, T.E. and D.M. Soderlund (eds.). Semiochemistry: Flavour and Pheromones. Walter de Gruyter, Berlin, Germany.

Aldrich, J.R. and W.R. Lusby. 1986. Exocrine chemistry of beneficial insects: male-specific secretions from predatory stink bugs (Hemiptera: Pentatomidae). Comp. Biochem. Physiol. 85: 639–642.

Aldrich, J.R., J.E. Oliver, W.R. Lubsy and J.P. Kochansky. 1986a. Identification of male-specific exocrine secretions from predatory stink bugs (Hemiptera: Pentatomidae). Arch. Insect Biochem. Physiol. 3: 1–12.

Aldrich, J.R., W.R. Lubsy and J.P. Kochansky. 1986b. Identification of a new predaceous stink bug pheromone and its attractiveness to the eastern yellow jacket. Experientia 42: 583–85.

Aldrich, J.R. 1988a. Chemistry and biological activity of Pentatomoid sex pheromones. pp. 417–431. *In*: Cutler, H.G. (ed.). Biologically Active Natural Products: Potential use in Agriculture. ACS Symposium Series, No. 380, Washington, DC, USA.

Aldrich, J.R. 1988b. Chemical ecology of the Heteroptera. Ann. Rev. Entomol. 33: 211–238.

Aldrich, J.R. and M. Borges. 1991. Pheromone blends of predaceous bugs (Heteroptera: Pentatomidae: *Podisus* spp.) Z. Naturforsch. 46: 264–269.

Aldrich, J.R. 1995. Chemical communication in the true bugs and parasitoid exploitation. pp. 318–363. *In*: Cardé, R.T. and W.J. Bell (eds.). Chemical Ecology of Insects 2. Chapman & Hall, Springer, New York, NY, USA.

Aldrich, J.R., J.E. Oliver, G.K. Waite, C. Moore and R.M. Waters. 1996. Identification of presumed pheromone blend from Australasian predaceous bug, *Oechalia schellenbergii* (Heteroptera: Pentatomidae). J. Chem. Ecol. 22: 729–738.

Aldrich, J.R., J.C. Zanuncio, E.F. Vilela, J.B. Torres and R.D. Cave. 1997. Field tests of predaceous pentatomid pheromones and semiochemistry of *Podisus* and *Supputius* species (Heteroptera: Pentatomidae: Asopinae). An. Soc. Entomol. Brasil 26: 1–14.

Aldrich, J.R. 1998. Status of semiochemicals research on predatory Heteroptera. pp. 33–48. *In*: Coll, M. and J.R. Ruberson (eds.). Predatory Heteroptera in Agroecosystems: Their Ecology and Use in Biological Control. Thomas Say Publications in Entomology, Entomol. Soc. of Amer., Lanham, MD, USA.

Aldrich, J.R. and W. Cantelo. 1999. Suppression of Colorado potato beetle infestation by pheromone-mediated augmentation of the predatory spined soldier bug, *Podisus maculiventris* (Say) (Heteroptera: Pentatomidae). Agric. and Forest Entomol. 1: 209–217.

Bueno, V.H.P. and J.C. van Lenteren. 2012. Predatory bugs (Heteroptera). pp. 539–570. *In*: Panizzi, A.R. and J.R.P. Parra (eds.). Insect Bioecology and Nutrition for Integrated Pest Management. CRC Press, Boca Raton, FL, USA.

Collatz, J. and S. Dorn. 2012. A host-plant-derived volatile blend to attract the apple blossom weevil *Anthonomus pomorum*–the essential volatiles include a repellent constituent. Pest Manag. Sci. 69: 1092–1098.

Collier, T. and R. van Steenwyk. 2004. A critical evaluation of augmentative biological control. Biol. Control 31: 245–256.

De Bach, P. and D. Rosen. 1991. Biological Control by Natural Enemies. Cambridge University Press, Cambridge, UK.

De Clercq, P., F. Merleved and L. Tirry. 1998a. Unnatural prey and artificial diets for rearing *Podisus maculiventris* (Heteroptera: Pentatomidae). Biol. Contr. 12: 137–142.
De Clercq, P., M. Vandewalle and L. Tirry. 1998b. Impact of inbreeding on performance of the predator *Podisus maculiventris*. Biocontrol 43: 299–310.
De Clercq, P. 2000. Predaceous stinkbugs (Pentatomidae: Asopinae). pp. 737–790. *In*: Schafer, C.W. and A.R. Panizzi (eds.). Heteroptera of Economic Importance. CRC Press, Boca Raton, FL, USA.
De Clercq, P. 2008. Predatory stink bugs (Hemiptera: Pentatomidae, Asopinae). pp. 3042–3045. *In*: Capinera, L.J. (ed.). Encyclopaedia of Entomology. Springer Science, Business Media B.V., Netherlands.
Gapud, V. 1991. A generic revision of the subfamily Asopinae, with consideration of its phylogenetic position in the family Pentatomidae and superfamily Pentatomoidea (Hemiptera, Heteroptera) Pts. 1 & 11. Phillip. Entomol. 8: 865–961.
Heimpel, G.E. and M.K. Asplen. 2011. A 'goldilocks' hypothesis for dispersal of biological control agents. BioControl 56: 441–450.
Ho, H.Y., R. Kou and H.H. Tseng. 2003. Semiochemicals from the predatory stink bug *Eocanthecona furcellata* (Wolf): components of metathoracic gland, dorsal abdominal gland, and sternal gland secretions. J. Chem. Ecol. 29: 2101–2114.
Hough-Goldstein, J. and J. Whalen. 1993. Inundative release of predatory stink bugs for control of Colorado potato beetle. Biol. Control 3: 343–347.
Kaplan, I. and D. Lewis. 2015. What happens when crops are turned on? Simulating constitutive volatiles for tritrophic pest suppression across an agricultural landscape. Pest Manag. Sci. 71: 139–150.
Kelly, J.L., J.R. Hagler and I. Kaplan. 2014. Semiochemical lures reduce emigration and enhance pest control services in open-field predator augmentation. Biol. Control 71: 70–77.
Kochansky, J.P., J.R. Aldrich and W.R. Lubsy. 1989. Synthesis and pheromonal activity of 6,10,13-trimethyltetradecanol for predatory stink bug, *Stiretrus anchorago* (Heteroptera: Pentatomidae). J. Chem. Ecol. 15: 1717–1728.
Lemos, W.P., F.S. Ramalho, J.E. Serrão and J.C. Zanuncio. 2003. Effects of diet on development of *Podisus nigrispinus* (Dallas) (Het., Pentatomidae), a predator of cotton leafworm. J. Appl. Entomol. 127: 389–395.
McPherson, J.E. 1982. The Pentatomidae (Hemiptera) of Northeastern North America with Emphasis on the Fauna of Illinois. Southern Ill. Univ. Press, Carbonadale, USA.
Millar, J.G. 2005. Pheromones of true bugs. Top. Curr. Chem. 240: 37–84.
Panizzi, A.R., J.E. McPherson, D.G. James, M. Javhery and R.M. McPherson. 2000. Stink bugs (Pentatomidae). pp. 421–474. *In*: Schafer, C.W. and A.R. Panizzi (eds.). Heteroptera of Economic Importance. CRC Press, Boca Raton, FL, USA.
Price, P.W. 1975. Evolutionary Strategies of Parasitic Insects and Mites. Plenum Press, New York, NY, USA.
Rider, D. 2011. Number of genera and species of Pentatomidae. Pentatomoidea Home Page. https://www.ndsu.edu/ndsu/rider/Pentatomoidea/index.htm. Accessed 26 August 2015.
Ruberson, J., M.J. Tauber and C.A. Tauber. 1986. Plant feeding by *Podisus maculiventris* (Heteroptera: Pentatomidae): Effect on survival, development, and pre oviposition period. Environ. Entomol. 15: 894–897.
Thomas, D.B. 1992. Taxonomic Synopsis of the Asopine Pentatomidae (Heteroptera) of the Western Hemisphere. The Thomas Say Foundation, Entomological Society of America, Lanham, USA.
Torres, J.B., J. Zanuncio and M.A. Moura. 2006. The predatory stinkbug *Podisus nigrispinus*: biology, ecology and augmentative releases for lepidopteran larval control in *Eucalyptus* forests in Brazil. CAB Rev. 15: 1–18.
Sant'Ana, J., R. Bruni, A.A. Abdul-Baki and J.R. Aldrich. 1997. Pheromone-induced movement of nymhs of the predator, *Podisus maculiventris* (Heteroptera: Pentatomidae). Biol. Control 10: 123–128.

Shetty, P.N. and J.A. Hough-Goldstein. 1998. Behavioral response of *Podisus maculiventris* (Hemiptera: Pentatomidae) to its synthetic pheromone. J. Entom. Sci. 331: 72–81.
van Lenteren, J.C. 2012. The state of commercial augmentative biological control: plenty of natural enemies, but frustrating lack of uptake. BioControl 57: 1–20.
Yeargan, K.V. 1998. Predatory Heteroptera in North American agroecosystems: an overview. pp. 7–19. *In*: Coll, M. and J.R. Ruberson (eds.). Predatory Heteroptera in Agroecosystems: Their Ecology and use in Biological Control. Thomas Say Publications in Entomology, Entomological Society of America, Lanham, USA.
Zanuncio, J.C., E.B. Beserra, A.J. Molina-Rugama, T.V. Zanuncio, T.B.M. Pinon and V.P. Maffia. 2005. Reproduction and longevity of *Supputius cincticeps* (Het.: Pentatomidae) fed with larvae of *Zophobas confusa, Tenebrio molitor* (Col.: Tenebrionidae) or *Musca domestica* (Dip.: Muscidae). Braz. Arch. Biol. Technol. 48: 771–777.

CHAPTER 10

Use of Pheromones for Monitoring Phytophagous Stink Bugs (Hemiptera: Pentatomidae)

P. Glynn Tillman[1,*] and *Ted E. Cottrell*[2]

Introduction

Phytophagous stink bugs (Hemiptera: Pentatomidae) are primary pests responsible for millions of dollars in losses and cost of control in most fruit, vegetable, grain, and row crops worldwide (McPherson and McPherson 2000, Schaefer and Panizzi 2000). Pheromones have been identified and synthesized for several species of economically important stink bug pests. When stink bug traps are baited with lures containing one or more of these pheromones, stink bugs are captured in the field. Thus, pheromone-baited stink bug traps can be used as monitoring tools to assess the presence and seasonal activity of certain stink bug pest species.

Stink Bug Pheromones and Pheromone Dispensers

In the late 1980's, researchers began identifying compounds attractive to stink bugs. The two major components of the male-produced pheromone of *Nezara viridula* (L.) were identified approximately simultaneously by Aldrich et al. (1987) and Baker et al. (1987) as *trans*-(*Z*)-(1*S*,2*R*,4*S*)-epoxybisabolene and the corresponding *cis*-(*Z*)-(1*R*,2*S*,4*S*)-epoxybisabolene in a 3:1 ratio.

[1] United States Department of Agriculture, Agricultural Research Service, Crop Protection & Management Research Laboratory, PO Box 748, Tifton, GA 31793, USA.
[2] United States Department of Agriculture, Agricultural Research Service, Southeastern Fruit & Tree Nut Research Laboratory, 21 Dunbar Road, Byron, GA 31008, USA.
Email: Ted.Cottrell@ars.usda.gov
* Corresponding author: Glynn.Tillman@ars.usda.gov

These two components also are produced by the male *Chinavia hilaris* (Say) but in a 19:1 ratio (Aldrich et al. 1989, McBrien et al. 2001). Since then, aggregation pheromones have been identified for many species of stink bug pests. For example, Aldrich et al. (1991) identified the major component of the male-specific aggregation pheromone, methyl (*E*,*Z*)-2,4-decadienoate (MDD), of the *Euschistus* spp., including *E. servus* (Say), *E. tristigmus* (Say), *E. politus* Uhler, *E. conspersus* (Uhler), and *E. ictericus* (L.). Sugie et al. (1996) identified the aggregation pheromone, methyl (*E*,*E*,*Z*)-2,4,6-decatrienoate (MDT), produced by the male *Plautia stali* Scott. In Asia, *Halyomorpha halys* (Stål) is cross-attracted to this pheromone (Tada et al. 2001, Lee et al. 2002). In the U.S., both *H. halys* and *C. hilaris* are cross-attracted to MDT (Aldrich et al. 2007, Khrimian et al. 2008, Tillman et al. 2010). Recently, the male-produced aggregation pheromone of *H. halys* was identified as a mixture of (3*S*,6*S*,7*R*,10*S*)-10,11-epoxy-1-bisabolen-3-ol and (3*R*,6*S*,7*R*,10*S*)-10,11-epoxy-1-bisabolen-3-ol (Khrimian et al. 2014). The male-produced aggregation pheromones of *Chlorochroa uhleri* (Stål) and *C. ligata* (Say) consist of the same major compound, (*R*)-3-(*E*)-6-2,3-dihydrofarnesoate, along with small amounts of the analogs methyl (*E*,*E*)-2,6-farnesoate and methyl (*E*)-5-2,6,10-trimethyl-5,9-undecadienoate, whereas the pheromone of *C. sayi* (Stål) consists primarily of methyl geranate with traces of methyl (*E*)-6-2,3-dihydrofarnesoate and methyl citronellate (Ho and Millar 2001a,b). These aggregation pheromones elicit responses from all mobile stages; nymphs and adults of both sexes (Harris and Todd 1980, Aldrich et al. 1991, Tillman et al. 2010, Weber et al. 2014). In contrast, stink bug sex pheromones primarily attract adult females. Sex pheromones for *Thyanta perditor* (F.), methyl (*E*,*Z*,*Z*)-2,4,6-decatrienoate, *Thyanta pallidovirens* (Stål), a blend of methyl (*E*,*Z*,*Z*)-2,4,6-decatrienoate and the sesquiterpenes (+)-α-curcumene, (–)-zingiberene, and (–)-β-sesquiphellandrene, and *Euschistus heros* (F.), methyl 2,6,10-trimethyltridecanoate, have been identified and synthesized (Borges et al. 1998, McBrien et al. 2002, Moraes et al. 2005). In fact, synthetic pheromones for *Euschistus* spp., *P. stali*, *H. halys*, and *N. viridula* are commercially available (see Chapter 5 for more details about the chemistry of stink bug pheromones).

Pheromone-baited traps offer an opportunity to monitor stink bugs in various cropping systems. However, the appropriate pheromone(s) for capturing stink bugs in the field need to be utilized. Across various cropping systems and locations, traps with MDD lures consistently capture more Nearctic *Euschistus* spp., including *E. servus*, *E. tristigmus*, *E. conspersus*, *E. politus*, and *E. quadrator*, than non-baited traps (Aldrich et al. 1991, Leskey and Hogmire 2005, Tillman et al. 2010, Cottrell and Horton 2011). The sex pheromone of the Neotropical *E. heros* is distinct from the MDD produced by the Nearctic *Euschistus* spp. (Borges et al. 1998), likely explaining why MDD failed to capture *E. heros* in Brazil (Aldrich et al. 1991). Traps baited with the pheromone of *E. heros*, though, effectively capture females in soybean (Borges et al. 2011). Only one study has been published on the response of *Euschistus* nymphs to MDD in the field. In that study, more early instars of *E. conspersus* were captured in traps baited with this pheromone than in control traps (Aldrich et al. 1991). In a recent field study, a significantly higher number of *E. servus* and *E. tristigmus* nymphs were captured in pyramid traps with MDD lures than in unbaited traps in peanut fields and alongside pecan and peach orchards (Tillman and Cottrell 2016a). This further confirms that nymphs are not randomly entering traps but are actually attracted to the pheromones. Nymphal attraction to pheromones has also been demonstrated for *N. viridula*, *C. hilaris*, *Podisus maculiventris* (Say), and *H. halys* in the U.S. (Sant'Ana et al. 1997, Tillman et al. 2010, Weber et al. 2014, P.G.T., unpublished data). In Japan, *P. stali* nymphs and adults were captured in traps with MDT lures in the forests where *P. stali* reproduces (Mishiro and Ohira 2002). Beyond stink bugs, a synthetic blend of the active components of the male-produced aggregation pheromone of *Riptortus pedestris* (= *clavatus*) (Heteroptera: Alydidae) attracts nymphs as well as adults (Leal et al. 1995).

Field trails demonstrated that both pheromone components of *H. halys* were important for the optimal attraction of this pest, but that the presence of additional stereoisomers does not hinder attraction (Khrimian et al. 2014, Weber et al. 2014, Leskey et al. 2015a). Thus, relatively inexpensive mixtures can be used for monitoring. Recently, Weber et al. (2014) showed that when MDT is deployed in combination with the *H. halys* aggregation pheromone in black pyramid traps, a synergistic response is observed. In season-long totals, combined lures caught between 1.9 to 3.2 times the number of adults and between 1.4 to 2.5 times the number of nymphs that were expected from an additive effect of the lures alone. However, for the 22 biweekly periods in which significant differences were detected, nymphal capture was similar for traps with both combined lures and MDT alone, for 10 of those periods, mainly late-season. Adult *H. halys* have not been found to be responsive to the MDT pheromone alone until mid-August (Leskey et al. 2012a). Thus, with the combination of the economical mixed-isomer aggregation pheromone and the MDT lure, a practical season-long attractant combination for all mobile life stages of *H. halys* is available.

The pheromone dosage in lures can also impact trap capture. In one study, more *E. servus* and *E. tristigmus* were captured with a higher dose of MDD (Cottrell and Horton 2011). In another study, four times as many *E. servus* and *E. tristigmus* were captured in traps with MDD lures purchased from IPM Technologies, Inc. as compared to the traps with lures purchased from Suterra (Leskey and Hogmire 2005). The difference may have been due to the higher amount of pheromones loaded into the IPM Technologies, Inc. lures over the Suterra ones. Traps baited with the aggregation pheromone and Rescue MDT lures (Sterling International, Inc.) captured more *H. halys* adults and nymphs late-season and more nymphs mid-season than those baited with the aggregation pheromone plus MDT lures from AgBio Inc., probably because the amount of active ingredient in the Rescue lures was nearly twice the amount of the AgBio lures (Leskey et al. 2015a). Adult and nymphal captures of *H. halys* were dose-dependent regardless of whether the lure contained pheromonal or non-pheromonal components.

Pheromone-baited Stink Bug Traps

Pheromone-baited traps capture stink bugs in crop and non-crop habitats. Exploiting stink bug behavior can increase the efficiency of capture in traps. Generally, stink bugs exhibit negative geotaxis, moving upwards on objects, and even on a flat surface such as the ground, they walk to the nearby vertical objects and climb (Millar et al. 2010). Consequently, traps should be made so as to facilitate either flying or walking stink bugs to walk up and into the insect-collecting part of the trap. Pyramidal traps (Fig. 10.1) can capture various stink bug species in a variety of crops and habitats.

Figure 10.1. Yellow corrugated plastic pyramid stink bug trap with insect-collecting device; 1.26 m in height and 0.49 wide at the base (A) and the lid of the insect-collecting device with eyelets to allow the escape of stink bug adult parasitoids (B).

Mizell and Tedders (1995) modified a pyramidal trap, originally designed to monitor the pecan weevil, *Curculio caryae* (Coleoptera: Curculionidae) (Tedders and Wood 1994), to capture stink bugs. The pyramid trap combines visual and chemical (i.e., pheromone) stimuli in order to capture stink bugs. The color of a pyramid trap can impact trap capture. Prokopy and Owens (1983) noted that a large number of phytophagous insects respond to yellow. Mizell and Tedders (1995) found that unbaited yellow pyramid traps captured more native stink bug species in the U.S. than traps painted light and dark green and black, or covered with aluminum foil. However, Hogmire and Leskey (2006) detemined that the captures of native stink bugs were greater in clear and white (no particular visual stimulus) and green (foliar mimic) pyramid traps than in black ones, with the captures in yellow (foliar stimulus) traps being intermediate among the first three "colors" and black. Regardless of the trap color, baited traps captured more stink bugs than unbaited ones. Black pyramid traps capture more *H. halys* adults and nymphs than yellow, white, and clear traps (Leskey et al. 2012b). This invasive stink bug species commonly reproduces on arboreal host plants (Hoebeke and Carter 2003); thus the black pyramid trap may mimic a tree trunk. The lighter, less expensive corrugated plastic pyramid trap with a single piece of rebar in the center for securing the trap (Fig. 10.1) is more convenient to use and is just as durable as masonite, whereas using a plywood base is problematic because it is heavier and can rot under field conditions unless pressure treated for exterior usage (P.G.T., personal observation). A black, corrugated plastic pyramid trap was the most effective trap for *H. halys*, capturing more adults than six other designs including the wooden pyramid experimental standard (Morrison et al. 2015).

Recently, trap capture was examined for two types of trap bases, a pyramid base and a bamboo pole base the same height as the pyramid one (Tillman and Cottrell 2015). The insect-collecting device (Fig. 10.1) as described by Cottrell et al. (2000) was used with both base types. Traps with a pyramid base captured a higher number of *E. servus* and *E. tristigmus* nymphs and adults than traps with a bamboo pole base suggesting that the pyramid base provides a broader platform for the nymphs to crawl into the insect-collecting device at the top of the trap and directs adults into the top of the trap exploiting their innate behavior. We note, though, that pheromone-baited traps other than a pyramid trap can effectively capture stink bugs. A pheromone-baited sticky trap captures *P. stali* adults and nymphs (Mishiro and Ohira 2002, Toyama et al. 2015). A double-cone cylinder screen trap efficiently catches *C. uhleri*, *C. sayi*, and *E. conspersus* (Millar et al. 2010). The specific trap chosen for monitoring stink bugs may depend on the stink bug species, host plant, or habitat.

Because stink bugs move into an insect-collecting device, it must be easy for them to enter, but difficult for them to escape. Trap capture was greater when pyramid traps were topped with an insect-collecting jar with a 1.6 cm diameter opening than with a jar with a 5 cm diameter opening (Hogmire and Leskey 2006). For the insect-collecting jar used by Cottrell et al. (2000), the tip of the pyramid trap fits snugly into the 5 cm diameter opening with ≈ 3 cm of the tip protruding past the opening, so the crawling space for stink bug entry is a pie-shaped quarter of the circle. Also, because stink bugs tend to crawl upwards, the extension of the tip of the trap into the jar may help the bugs move higher into the jar than case of when it does not extend into the device. Jar traps without air vents can become very hot when exposed to sunlight, and stink bugs avoid crawling into them (T.E.C., personal observation). Stink bugs detect the location of the cool air entering the device and head that way. Thus, if the trap is adequately vented, fewer of them will escape before succumbing to the insecticidal ear tag. The trap marketed by Sterling International, Inc. for *H. halys* captured fewer nymphs and adults than six other traps (Morrison et al. 2015) perhaps because the plastic jar top was not well-ventilated. Stink bugs do not like to enter darkened insect collecting devices (Mizell and Tedders 1995), so these devices should be transparent. Trap capture was numerically lower for the collecting jars painted yellow (dark inside) than for clear jars (Leskey and Hogmire 2005). The addition of a killing agent, either a Saber Extra insecticidal ear tag or a piece of Hercon Vaportape II, should be placed in a collecting device in order to help prevent the escape of insects (Cottrell 2001, Leskey et al. 2012b). Parasitoids of adult stink bugs also use their host's pheromone as a host-finding kairomone (Aldrich et al. 2006). The capture of natural enemies in pheromone-baited traps can be decreased by inserting 6 mm diameter eyelets in the screen of the top of the insect-collecting device (Fig. 10.1). This allows the parasitoids *Trichopoda pennipes* (F.), *Euthera tentatrix* Loew, and *Cylindromyia binotata* (Bigot) (Diptera: Tachinidae), but not adult stink bugs, to escape, thus conserving these natural enemies (Tillman et al. 2015). This could also be important for conserving the predatory digger wasp *Astata occidentalis* (Hymenoptera: Sphecidae) which is attracted to the pheromone produced by *T. pallidovirens*, as well as to MDT (Cottrell et al. 2014).

In case of trees, the trap position can impact trap capture. More *E. servus* adults were captured in traps on the ground than in traps placed in the canopy of pecan and apple trees, and more *E. tristigmus* and *C. hilaris* adults were captured in the canopy of pecan and/or apple trees (Cottrell et al. 2000, Hogmire and Leskey 2006). Ground and hanging traps performed equally well in capturing *H. halys* adults, though hanging traps yielded lower captures of nymphs relative to ground traps (Morrison et al. 2015). Similarly, fewer *E. servus*, *E. tristigmus*, and *C. hilaris* nymphs in pecan trees are captured in hanging pyramid traps than ground-deployed ones (P.G.T., unpublished data); this is likely due to the fact stink bug nymphs tend to crawl up the pyramid base into the collection device.

The ecology of stink bugs should be taken into consideration when positioning traps within an agroecosystem. Stink bugs move between closely associated host plant habitats throughout the growing season in response to the deteriorating suitability of their current host plants (Velasco and Walter 1992, Bundy and McPherson 2000, Ehler 2000, Reeves et al. 2010). In case of *H. halys*, injury to apples and peaches in commercial orchards was usually greater at the exterior of orchards relative to the interior, suggesting that adults emigrating from overwintering sites during the early season and from woodlands or cultivated hosts such as corn and soybean later in the season, constantly invade orchards (Leskey et al. 2012a). Similar patterns of movement have been observed for stink bug species in other cropping systems (Tillman et al. 2009, Tillman et al. 2014, Tillman and Cottrell 2015). Pheromone-baited traps used for monitoring *H. halys* have been positioned between agricultural production and unmanaged areas, particularly woodland habitats (Khrimian et al. 2014, Weber et al. 2014, Morrison et al. 2015, Leskey et al. 2015b), for each of the unmanaged areas were either known to harbor host plants of *H. halys* or likely harbored these hosts. These traps effectively monitor *H. halys* in crops. Traps with MDD lures successfully captured *Euschistus* spp. dispersing from peanut into cotton when positioned at peanut-cotton interfaces (Tillman et al. 2015). Nielsen et al. (2011) reported that pyramid traps positioned at the field edges, near a hedgerow consisting of host plants of *H. halys*, successfully monitored this pest in soybean. High densities of adult stink bugs were captured for an extended period thus indicating that the traps detected individuals dispersing into the crop. Borges et al. (2011) determined that the placement of traps around the borders of the soybean crop was as effective as placement inside the crop, suggesting that *E. heros* density surveys could even be carried out only along the crop edges in order to minimize the effort required to monitor large areas of the crop. In pecan orchards, except during October, ground traps along the hedgerow, orchard edge, and orchard center captured a similar number of *E. servus* and *E. tristigmus* possibly due to the open space in the orchard between the ground and lower limbs of the pecan tree (Cottrell et al. 2000).

Plant height can also influence trap capture. In soybean, a large pyramid trap, 142 cm tall, captured more *H. halys* bugs than a smaller trap, 81.3 cm tall (Nielsen et al. 2011), likely because of the similarity between soybean canopy height and the height of the smaller trap. Higher canopy height can obscure the visual stimuli (i.e., yellow) and decrease the likelihood of stink bug adults alighting on the pyramid base. For *H. halys*, a smaller pyramid trap (0.29 m tall) performed equally well in capturing adults as compared to the experimental standard (1.22 m tall) in apple orchards, likely because ground cover was low in the orchard. Therefore, for ground-deployed traps, decisions on trap height need to take into consideration the height of the crop in which stink bugs will be monitored.

Aggregation of Stink Bugs on Plants Associated with Pheromone Lures

Aggregation pheromones draw respondents to the plants where both feeding and mating occur. Aldrich et al. (1999) proposed that the male-produced pheromones of migratory Heteroptera are an integral part of a resource-based mating system in which a male colonizes a new host and guides potential mates to the host with his pheromone. Generally, stink bug males develop before females and thus newly-emerged males are the first to migrate to a new host when the current host begins to senesce. These males feed and produce the aggregation pheromone which attracts females to the host. Additionally, nymphs of many stink bug species are attracted to aggregation pheromones (Aldrich et al. 1991, Millar et al. 2010, Khrimian et al. 2014) supporting the idea that pheromones are associated with food because it is not expected that nymphs are seeking mates. Indeed, starving *P. stali* nymphs are attracted to the male-produced pheromone of this species (Toyama et al. 2015). Mating is necessary for oviposition by *N. viridula* (Fortes et al. 2011), and likely other stink bug species. The male-produced aggregation pheromone concentrates both sexes of adults on plants in a limited area. Aggregated adults then use an extensive repertoire of substrate-borne vibrational signals to locate and recognize each other for mating (Cokl and Millar 2009). For both *E. conspersus* and *E. servus*, aggregation formations and mating occurred on mullein plants either baited with pheromone lures or in plots containing lures (Krupke et al. 2001, Leskey and Hogmire 2007). In farmscapes where peanut is closely associated with cotton, newly-emerged adults disperse from peanut into nearby cotton, where they aggregate and begin feeding and mating on cotton (Tillman et al. 2009). Because aggregation pheromones concentrate or "arrest" individuals in an area, stink bugs are sometimes present on plants near pheromone-baited traps instead of inside the collecting devices. In some cases, more stink bugs have been found on plants near the traps than were actually caught in the traps (Aldrich et al. 1991, James et al. 1996, Krupke et al. 2001). In a recent study, a significantly higher number of stink bugs were sampled from the two cotton rows (1.8-m in width and length) adjacent to a pheromone-baited trap than were found on cotton unassociated with a trap (P.G.T., unpublished data). Adult *E. servus* aggregated on mullein plants over a zone of 3.14 m^2 around a *Euschistus* spp. baited trap (Leskey and Hogmire 2007). *Halyomorpha halys*, *Euschistus* spp., *C. hilaris*, and *N. viridula* were attracted by the aggregation pheromone of *H. halys* and the pheromone synergist and arrested on unbaited traps within a 2.5-m radius of a baited trap (Morrison et al. 2016). As a result, the number of stink bugs captured in traps are relative measures of stink bug population abundance.

Monitoring Stink Bugs Using Pheromone-baited Traps

Once the appropriate pheromone(s) and trap combination for effectively capturing a stink bug species has been determined, traps can be deployed for monitoring. Monitoring stink bugs using pheromone-baited traps has been reported to (1) determine seasonal variation in the occurrence of stink bug nymphs and adults, (2) detect or predict the timing of dispersal into and/or out of a crop or overwintering habitat, and (3) assess the relative abundance of stink bugs in crops or habitats which are difficult to sample using the standard sampling techniques. In addition, the capture of stink bugs in traps in a grid across an agricultural landscape can be analyzed in order to discover both spatial and temporal distribution in addition to dispersal within and between the various crop commodities and non-crop habitats. Also, traps can aid in the recapture of marked insects in mark-recapture studies, providing new insights into the dispersal behavior of stink bugs.

Season-long monitoring of stink bugs in agroecosystems can be useful in determining relative abundance over time and detecting emergence from overwintering sites, subsequent immigration to a host plant(s), and finally movement to overwintering sites. In fact, the sensitivity of detection by traps is quite high as shown by the pheromone-baited traps capturing *H. halys* adults and nymphs earlier in the season than sweep net samples and blacklight traps (Nielsen et al. 2011). Recently, traps baited with the *H. halys* aggregation pheromone in association with MDT were used to monitor the season-long activity around the areas of crop production, including fruit orchards, vegetables, ornamentals, and row crops, across many parts of the U.S. (Leskey et al. 2015a). Adults dispersing from overwintering sites were first detected in April with peak captures in mid-to-late May. Nymphs were first detected in late May indicating intial reproduction on a crop. Adult captures typically peaked in early September. Captures sharply declined in autumn indicating that the adults were moving into the overwintering sites. In an earlier study, year-round adult captures of *E. servus* and *E. tristigmus* in traps baited with MDD lures were very low from December through February (Cottrell et al. 2000). This likely indicated that the adults were overwintering. Although both stink bug species were present in the orchard, *E. servus* was more abundant earlier than later in the season. Pecan was a mid-to-late season host for *E. tristigmus* indicating that *E. tristigmus* probably reproduced on another host early in the season and then dispersed into pecan. In a recent study, traps with MDD lures were used to monitor stink bugs along the field borders between peanut or cotton and woodland habitats (Tillman and Cottrell 2016c). The presence of a relatively high number of adult *E. servus* in the early-season traps, well before the crops were susceptible to stink bug attack, strongly indicated that

E. servus initially was present on the non-crop hosts. Then, in the mid-to-late season, the adults apparently dispersed from the non-crop hosts because they entered both crops. Traps with MDD lures elicited a highly reliable and statistically significant response by flight capable, sexually mature, ovipositing *E. conspersus* females migrating into tomato fields prior to fruit development (Cullen and Zalom 2005, 2006). These examples illustrate the usefulness of traps as a tool for early detection. The University of California Pest Management Guidelines for tomato recommend to begin monitoring *E. conspersus* with pheromone-baited cone traps in tomato fields at flowering (University of California 2015). In the areas of the field where *E. conspersus* is consistently found in traps, they suggest taking plant samples beginning when the fruits reach a diameter of 2.54 cm to determine when economic thresholds are reached. This further emphasizes the utility of pheromone-baited traps and how they may be used to increase the precision and efficiency of sampling a crop for stink bug damage.

Because pheromone traps baited with aggregation pheromones capture stink bug nymphs, they can be used to monitor nymphal development over time in order to discern host plant suitablity and detect the development of a new generation of stink bugs on a host plant. For example, monitoring *H. halys* nymphs season-long in traps in various crops showed they were present in crops in late May and peaked in late August (Leskey et al. 2015a). For *H. halys* in soybean, positive population growth was indicated during the R3 and R4 stages by the capture of early instars in traps (Nielson et al. 2011). Ground-deployed traps with MDD and MDT lures captured *E. servus*, *E. tristigmus*, and *C. hilaris* nymphs in pecan trees in woodland habitats strongly indicting that these stink bug species were reproducing on these trees (P.G.T., unpublished data).

Detecting the nymphal populations of stink bugs on a host plant could be useful in predicting infestations in the successive host crops. In Japan, *P. stali* largely reproduces in the plantation forests of Japanese cypress and uses their cones as a food source during the summer and autumn (Yamada and Noda 1985). Although this stink bug does not develop and reproduce on fruit crops (Shiga and Moriya 1984), depletion of cypress cones causes starving bugs to depart from this food source into cultivated fields (Tsutsumi 2001). Toyama et al. (2015) used sticky traps with MDT lures to trap *P. stali* nymphs in cypress. A morphological indicator of nutritional status showed that the nymphs attracted to the pheromone were starving. The first attraction of 3rd to 5th instars to pheromone-baited traps coincided each year with the times when the mean number of stylet sheaths from *P. stali* feeding exceeded 25, indicating that the cones were depleted. They concluded that the pheromone-baited sticky trap could be a useful tool for predicting *P. stali* orchard infestations. Similarly, imminent infestations of native stink bugs and *H. halys* in crops could be predicted by monitoring nymphal populations on non-crop hosts in woodlands. For example, early-to-late instars of *C. hilaris* were captured in early July in traps with MDD lures near black cherry trees (*Prunus serotina* Ehrh.) bordering a peanut-cotton field (Tillman and Cottrell 2015). By mid-July, *C. hilaris* adults were aggregated in cotton at the boundary of the peanut and cotton field.

Pheromone-baited traps or plants can useful in examining the spatial and temporal distribution and dispersal of stink bugs, not only within a crop, but also across an agricultural landscape. In a landscape-scale study, traps with MDD lures were used to monitor the season-long abundance patterns of *E. servus* in several commodities; one of the significant findings was detecting the dispersal of *E. servus* from pecan into cotton (T.E.C., unpublished data). In a recent study, the release-recapture tests of marked individuals revealed that *Murgantia histrionica* (Hahn) was not only highly attracted to collard plants baited with lures of the synthetic pheromone of this stink bug, but also that emigration from the trap plants seemed to be unrelated to the stink bug density on the plants (Walsh et al. 2016). In another study, the recapture of marked individuals in traps with MDT and MDD lures in cotton has demonstrated the dispersal of *C. hilaris*, *E. tristigmus*, *E. servus*, and *Thyanta custator custator* (F.) from elderberry (*Sambucus nigra* subsp. *canadensis* [L.] R. Bolli) into cotton (Tillman and Cottrell 2016b).

If pheromone-baited traps are to be useful for making pest management decisions, trap capture should reflect the relative size of the stink bug population being monitored. However, certain factors sometimes make it difficult to relate trap capture with the abundance in the field when using standard monitoring tools. In the mid-Atlantic region, apple and peach orchards were monitored for *H. halys* using four sampling techniques: pyramid traps baited with MDT lures, sweep net sampling of ground flora, limb jarring sampling, and visual inspection of trees (Leskey et al. 2012a). Numerically, more stink bugs were found using traps than with the other monitoring techniques. This may be due to the difficulty of sampling trees using the typical sampling techniques. Therefore, it may be easier to effectively obtain seasonal data for stink bugs in trees using pheromone-baited traps. This is particularly important for woodland habitats where the seasonal abundance of stink bugs remains basically unexplored. Trap captures of *E. servus*, *E. tristigmus*, and *C. hilaris*, though, were correlated with tree beating samples in apple orchards, reflecting the overall relative size of the population (Leskey and Hogmire 2005). However, this relationship was not significant in peach orchards possibly because the traps also captured the stink bugs dispersing from the nearby sources. When traps baited with sex pheromone and shake cloth plant samples were used to monitor *E. heros* females in soybean, trap captures showed a positive relationship with population density in soybean during the initial to medium reproductive stages (R1–R5) of the crop (Borges et al. 2011). However, trap captures during the R7-R8 stages did not reflect the stink bug abundance in soybean. This was due to two factors: nymphal populations increased on soybean at the end of the season, but the sex pheromone was only attractive to females and the capture of *E. heros* females was correlated to the females' reproductive status (measured as the mean number of eggs in the reproductive tract). During the R1–R5 stages, there was an increase in trap captures and female reproductive maturity, but during the R6-R7 stages, trap captures decreased with a decrease in female reproductive maturity. Even though the traps may not be as effective in capturing females in the final stages of soybean development, this may not negatively impact pest management decisions. Insecticides are generally not applied to soybean during these stages in Brazil, for the crop is less susceptible to stink bug damage at these stages (Corrêa-Ferreira and Azevedo 2002). The seasonality of *H. halys* in soybean fields was investigated by comparing the monitoring efficiency of sweep net sampling and pyramid traps baited with MDT pheromone (Nielsen et al. 2011). The traps caught higher densities of *H. halys* than sweep net samples. This may have been due to the fact that the traps caught a higher number of small nymphs than sweep net samples which recovered mainly 5th instars.

Potential for Using Pheromones to Manage Stink Bugs

Two management strategies—attract and kill and trap cropping in combination with stink bug traps baited with aggregation pheromones-take advantage of the propensity of stink bugs to aggregate on plants near synthetic pheromone attractants. The attract and kill strategy is used to manage stink bugs on select, baited plants at field or orchard borders in order to attract and aggregate stink bugs in spatially precise locations (Yamanaka et al. 2011, Leskey et al. 2015a, Morrison et al. 2016). By only treating these plants or a limited number of crop rows with insecticide, the overall amount of material applied against the stink bugs is reduced. Baiting non-crop plants could also lower crop injury. At peanut-cotton interfaces, pheromone-baited traps alone are not effective in deterring the entry of stink bugs into cotton because these pests still tend to aggregate on nearby cotton (Tillman et al. 2015). However, combining an attractive food source, i.e., soybean, with pheromone-baited traps was an effective management tactic which trapped and killed stink bugs within the trap crop throughout the growing season. We note that caution should be taken not to position the traps near cotton. Because stink bugs tend to aggregate near synthetic pheromones, traps baited with a pheromone probably should not be used for managing stink bugs in cash crops. For example, Sargent et al. (2014) determined that tomatoes grown in gardens with traps baited with MDT sustained significantly more injury by *H. halys* than tomatoes grown in gardens without traps.

References

Aldrich, J.R., J.E. Oliver, W.R. Lusby, J.P. Kochansky and J.A. Lockwood. 1987. Pheromone strains of the cosmopolitan pest, *Nezara viridula* (Heteroptera: Pentatomidae). J. Exper. Zool. 244: 171–176.

Aldrich, J.R., W.R. Lusby, B.E. Marron, K.C. Nicolaou, M.P. Hoffmann and L.T. Wilson. 1989. Pheromone blends of green stink bugs and possible parasitoid selection. Naturwissenschaften 76: 173–175.

Aldrich, J.R., M.P. Hoffmann, J.P. Kochansky, W.R. Lusby, J.E. Eger and J.A. Payne. 1991. Identification and attractiveness of a major component for Nearctic *Euschistus* spp. stink bugs (Heteroptera: Pentatomidae). Environ. Entomol. 20: 477–483.

Aldrich, J.R., J.E. Oliver, T. Taghizadeh, J.T.B. Ferreira and D. Liewehr. 1999. Pheromones and colonization: reassessment of the milkweed bug migration model (Heteroptera: Lygaeidae: Lygaeinae). Chemoecol. 9: 63–71.

Aldrich, J.R., A. Khrimian, A. Zhang and P.W. Shearer. 2006. Bug pheromones (Hemiptera, Heteroptera) and tachinid fly host-finding. Denisia 19: 1015–1031.

Aldrich, J.R., A. Khrimian and M.J. Camp. 2007. Methyl 2,4,6-decatrienoates attract stink bugs (Hemiptera: Heteroptera: Pentatomidae) and tachinid parasitoids. J. Chem. Ecol. 33: 801–815.

Baker, R., M. Borges, N.G. Cooke and R.H. Herbert. 1987. Identification and synthesis of (Z)-(1′S,3′R,4′S)(−)-2-(3′,4′-epoxy-4′-methylcyclohexyl)-6-methylhepta-2,5-diene, the sex pheromone of the southern green stink bug, *Nezara viridula* (L.). J. Chem. Soc., Chemical Communications 1987: 414–116.

Borges, M., K. Mori, M.L.M. Costa and E.R. Sujii. 1998. Behavioural evidence of methyl-2,6,10-trimethyltridecanoate as a sex pheromone of *Euschistus heros* (Heteroptera: Pentatomidae). J. Appl. Entomol. 122: 335–338.

Borges, M., M.C.B. Moraes, M.F. Peixoto, C.S.S. Pires, E.R. Sujii and R.A. Laumann. 2011. Monitoring the neotropical brown stink bug *Euschistus heros* (F.) (Hemiptera: Pentatomidae) with pheromone-baited traps in soybean fields. J. Appl. Entomol. 135: 68–80.

Bundy, C.S. and R.M. McPherson. 2000. Dynamics and seasonal abundance of stink bugs (Heteroptera: Pentatomidae) in a cotton-soybean ecosystem. J. Econ. Entomol. 93: 697–706.

Cokl, A. and J.G. Millar. 2009. Manipulation of insect signaling for monitoring and control of pest insects. pp. 279–316. *In*: Ishaaya, I. and A.R. Horowitz (eds.). Biorational Control of Arthropod Pests: Application and Resistance Management. Springer, New York, NY.

Corrêa-Ferreira, B.S. and J. Azevedo. 2002. Soybean seed damage by different species of stink bugs. Agric. For. Entomol. 4: 145–150.

Cottrell, T.E., C.E. Yonce and B.W. Wood. 2000. Seasonal occurrence and vertical distribution of *Euschistus servus* (Say) and *Euschistus tristigmus* (Say) (Hemiptera: Pentatomidae) in pecan orchards. J. Entomol. Sci. 35: 421–431.

Cottrell, T.E. 2001. Improved trap capture of *Euschistus servus* and *Euschistus tristigmus* (Hemiptera: Pentatomidae) in pecan orchards. Fla. Entomol. 84: 731–732.

Cottrell, T.E. and D. Horton. 2011. Trap capture of brown and dusky stink bugs (Hemiptera: Pentatomidae) as affected by pheromone dosage in dispensers and dispenser source. J. Entomol. Sci. 46: 135–147.

Cottrell, T.E., P.J. Landolt, Q.H. Zhang and R.S. Zack. 2014. A chemical lure for stink bugs (Hemiptera: Pentatomidae) is used as a kairomone by *Astata occidentalis* (Hymenoptera: Sphecidae). Fla. Entomol. 97: 233–237.

Cullen, E.M. and F.G. Zalom. 2005. Relationship between *Euschistus conspersus* (Hem., Pentatomidae) pheromone trap catch and canopy samples in processing tomatoes. J. Appl. Entomol. 129: 505–514.

Cullen, E.M. and F.G. Zalom. 2006. *Euschistus conspersus* female morphology and attraction to methyl (2E,4Z)-decadienoate pheromone-baited traps in processing tomatoes. Entomol. Exp. Appl. 119: 163–173.

Ehler, L.E. 2000. Farmscape ecology of stink bugs in northern California. Memorial Thomas Say Publications of Entomology, Entomological Society of America Press, Lanham, MD, 59 pp.

Fortes, P., G. Salvador and F.L. Cônsoli. 2011. Ovary development and maturation in *Nezara viridula* (L.) (Hemiptera: Pentatomidae). Neotrop. Entomol. 40: 89–96.

Harris, V.E. and J.W. Todd. 1980. Male-mediated aggregation of male, female and 5th-instar southern green stink bugs and concomitant attraction of a tachinid parasite, *Trichopoda pennipes*. Entomol. Exp. Appl. 27: 117–126.

Hoebeke, E.R. and M.E. Carter. 2003. *Halyomorpha halys* (Stål) (Heteroptera: Pentatomidae): A polyphagous plant pest from Asia newly detected in North America. Proc. Entomol. Soc. Wash. 105: 225–237.

Ho, H.-Y. and J.G. Millar. 2001a. Identification and synthesis of a male-produced sex pheromone from the stink bug *Chlorochroa sayi*. J. Chem. Ecol. 27: 1177–1201.

Ho, H.-Y. and J.G. Millar. 2001b. Identification and synthesis of male-produced sex pheromone components of the stink bugs *Chlorochroa ligata* and *Chlorochroa uhleri*. J. Chem. Ecol. 27: 2067–2095.

Hogmire, H.W. and T.C. Leskey. 2006. An improved trap for monitoring stink bugs in apple and peach orchards. J. Entomol. Sci. 41: 9–21.

James, D.G., R. Heffer and M. Amaike. 1996. Field attraction of *Biprorulus bibax* (Hemiptera: Pentatomidae) to synthetic aggregation pheromone and (E)-2-hexenal, a pentatomid defense chemical. J. Chem. Ecol. 22: 1697–1708.

Khrimian, A., P.W. Shearer, A. Zhang, G.C. Hamilton and J.R. Aldrich. 2008. Field trapping of the invasive brown marmorated stink bug, *Halyomorpha halys* (Stål.), with geometric isomers of methyl 2,4,6-decatrienoate. J. Agric. Food Chem. 56: 197–203.

Khrimian, A., A. Zhang, D.C. Weber, H.-Y. Ho, J.R. Aldrich, K.E. Vermillion et al. 2014. Discovery of the aggregation pheromone of the brown marmorated stink bug (*Halyomorpha halys*) through the creation of stereoisomeric libraries of 1-bisabolen-2-ols. J. Natural Products 77: 1708–1717.

Krupke, C.H., J.F. Brunner, M.D. Doerr and A.D. Kahn. 2001. Field attraction of the stink bug *Euschistus conspersus* (Hemiptera: Pentatomidae) to synthetic pheromone-baited host plants. J. Econ. Entomol. 94: 1500–1505.

Leal, W.S., H. Higuchi, N. Mizutani, H. Nakamori, T. Kado-sawa and M. Ono. 1995. Multifunctional communication in *Riptortus clavatus* (Heteroptera: Alydidae): conspecific nymphs and egg parasitoid *Ooencyrtus nezarae* use the same adult attractant pheromone as chemical cue. J. Chem. Ecol. 21: 973–985.

Lee, K.C., C.H. Kang, D.W. Lee, S.M. Lee, C.G. Park and H.Y. Choo. 2002. Seasonal occurrence trends of hemipteran bug pests monitored by mercury light and aggregation pheromone traps in sweet persimmon orchards. Korean J. Appl. Entomol. 41: 233–238.

Leskey, T.C. and H.W. Hogmire. 2005. Monitoring stink bugs (Hemiptera: Pentatomidae) in mid-Atlantic apple and peach orchards. J. Econ. Entomol. 98: 143–153.

Leskey, T.C. and H.W. Hogmire. 2007. Response of the brown stink bug (Hemiptera: Pentatomidae) to the aggregation pheromone, methyl (2E,4Z)-decadienoate. J. Entomol. Sci. 42: 548–557.

Leskey, T.C., B.D. Short, B.R. Butler and S.E. Wright. 2012a. Impact of the invasive brown stink bug, *Halyomorpha halys* (Stål), in mid-Atlantic fruit orchards in the United States: case studies of commercial management. Psyche. DOI: 10.1155/2012/535062.

Leskey, T.C., S.E. Wright, B.D. Short and A. Khrimian. 2012b. Development of behaviorally based monitoring tools for the brown marmorated stink bug, *Halyomorpha halys* (Stål) (Heteroptera: Pentatomidae) in commercial tree fruit orchards. J. Entomol. Sci. 47: 76–85.

Leskey, T.C., A. Khrimian, D.C. Weber, J.R. Aldrich, B.D. Short, D.-H. Lee et al. 2015a. Behavioral responses of the invasive *Halyomorpha halys* (Stål) to traps baited with stereoisomeric mixtures of 10,11-epoxy-1-bisabolen-3-OL J. Chem. Ecol. 41: 418–429.

Leskey, T.C., A. Agnello, J.C. Bergh, G.P. Dively, G.C. Hamilton, P. Jentsch et al. 2015b. Attraction of the invasive *Halyomorpha halys* (Hemiptera: Pentatomidae) to traps baited with semiochemical stimuli across the United States. Environ. Entomol. 1–11. Doi: 10.1093/ee/nvv049.

McBrien, H.L., J.G. Millar, L. Gottlieb, X. Chen and R.E. Rice. 2001. Male-produced sex attractant pheromone of the green stink bug, *Acrosternum hilare* (Say). J. Chem. Ecol. 27: 1821–1839.
McBrien, H.L., J.G. Millar, R.E. Rice, J.S. McElfresh, E. Cullen and F.G. Zalom. 2002. Sex attractant pheromone of the red-shouldered stink bug *Thyanta pallidovirens*: A pheromone blend with multiple redundant components. J. Chem. Ecol. 28: 1797–1818.
McPherson, J.E. and R.M. McPherson. 2000. Stink Bugs of Economic Importance in America North of Mexico. CRS Press LLC, Boca Raton, FL.
Millar, J.G., H.M. McBrien and J.S. McElfresh. 2010. Field trial of aggregation pheromones for the stink bugs *Chlorochroa uhleri* and *Chlorochroa sayi* (Hemiptera: Pentatomidae). J. Econ. Entomol. 103: 1603–1612.
Mishiro, K. and Y. Ohira. 2002. Attraction of a synthetic aggregation pheromone of the brown-winged green bug, *Plautia crossota stali* Scott to it its parasitoid, *Gymnosoma rotundata* and *Trissolcus plautiae*. Proc. Assoc. Plant Prot. Kyushu. 48: 76–80.
Mizell, R.F. and W.L. Tedders. 1995. A new monitoring method for detection of the stink bug complex in pecan orchards. Proc. Southeast. Pecan Growers Assoc. 88: 36–40.
Moraes, M.C.B., J.G. Millar, R.A. Laumann, E.R. Sujii, C.S.S. Pires and M. Borges. 2005. Sex attractant from the neotropical red-shouldered stink bug, *Thyanta perditor* (F.). J. Chem. Ecol. 31: 1415–1427.
Morrison, W.R. III, J.P. Cullum and T.C. Leskey. 2015. Evaluation of trap designs and deployment strategies for capturing *Halyomorpha halys* (Hemiptera: Pentatomidae). J. Econ. Entomol. 108: 1683–1692.
Morrison, W.R. III, D.-H. Lee, B.D. Short, A. Khrimian and T.C. Leskey. 2016. Establishing the behavioral basis for an attract-and-kill strategy to manage the invasive *Halyomorpha halys* in apple orchards. J. Pest. Sci. 89: 81–96.
Nielsen, A.L., G.C. Hamilton and P.W. Shearer. 2011. Seasonal phenology and monitoring of the non-native *Halyomorpha halys* (Hemiptera: Pentatomidae) in soybean. Environ. Entomol. 40: 231–238.
Prokopy, R.J. and E.D. Owens. 1983. Visual detection of plants by herbivorous insects. Ann. Rev. Entomol. 28: 337–364.
Reeves, R.B., J.K. Greene, F.O.F. Reay-Jones, M.D. Toews and P.D. Gerard. 2010. Effects of adjacent habitat on populations of stink bugs (Heteroptera: Pentatomidae) in cotton as part of a variable agricultural landscape in South Carolina. Environ. Entomol. 39: 1420–1427.
Sant'Ana, J., R. Bruni, A.A. Abdul-Baki and J.R. Aldrich. 1997. Pheromone-induced movement of nymphs of the predator, *Podisus maculiventris* (Heteroptera: Pentatomidae). Biol. Control 10: 123–128.
Sargent, C., H.M. Martinson and M.J. Raupp. 2014. Traps and trap placement may affect location of brown marmorated stink bug (Hemiptera: Pentatomidae) and increase injury to tomato fruits in home gardens. Environ. Entomol. 43: 432–438.
Schaefer, C.W. and A.R. Panizzi. 2000. Heteroptera of Economic Importance. CRS Press, NY. 828 pp.
Shiga, M. and S. Moriya. 1984. Utilization of food plants by *Plautia stali* Scott (Hemiptera, Heteroptera, Pentatomidae), an experimental approach. Bull. Fruit Tree Res. Stn. A 11: 107–121.
Sugie, H., M. Yoshida, K. Kawasaki, H. Noguchi, S. Moriya, K. Takagi et al. 1996. Identification of the aggregation pheromone of the brown-winged green bug, *Plautia stali* Scott (Heteroptera: Pentatomidae). Appl. Entomol. Zool. 31: 427–431.
Tada, N., M. Yoshida and Y. Sato. 2001. Monitoring of forecasting for stink bugs in apple. 2. The possibility of forecasting with aggregation pheromone. Ann. Rept. Plant Prot. North Japan 52: 227–229.
Tedders, W.L. and B.W. Wood. 1994. A new technique for monitoring pecan weevil emergence (Coleoptera: Curculionidae). J. Entomol. Sci. 29: 18–30.
Tillman, P.G., T.D. Northfield, R.F. Mizell and T.C. Riddle. 2009. Spatiotemporal patterns and dispersal of stink bugs (Heteroptera: Pentatomidae) in peanut-cotton farmscapes. Environ. Entomol. 38: 1038–1052.

Tillman, P.G., J.R. Aldrich, A. Khrimian and T.E. Cottrell. 2010. Pheromone attraction and cross-attraction of *Nezara*, *Acrosternum*, and *Euschistus* spp. stink bugs (Heteroptera: Pentatomidae) in the field. Environ. Entomol. 39: 610–617.

Tillman, P.G., T.E. Cottrell, R.F. Mizell III and E. Kramer. 2014. Effect of field edges on dispersal and distribution of colonizing stink bugs across farmscapes of the Southeast US. Bull. Entomol. Res. 104: 56–64.

Tillman, P.G., A. Khrimian, T.E. Cottrell, X. Luo, R.F. Mizell III and J. Johnson. 2015. Trap cropping systems and a physical barrier for suppression of stink bug (Hemiptera: Pentatomidae) in cotton. J. Econ. Entomol. 1–11. Doi: 10.1093/jee/tov217.

Tillman, P.G. and T.E. Cottrell. 2015. Spatiotemporal distribution of *Chinavia hilaris* (Hemiptera: Pentatomidae) in peanut-cotton farmscapes. J. Insect Sci. 15: 101. DOI: 10.1093/jisesa/iev081.

Tillman, P.G. and T.E. Cottrell. 2016a. Attraction of stink bug (Hemiptera: Pentatomidae) nymphs to Euschistus aggregation pheromone in the field. Fla. Entomol. 99: 678–682.

Tillman, P.G. and T.E. Cottrell. 2016b. Density and egg parasitism of stink bugs (Hemiptera: Pentatomidae) in elderberry and dispersal into crops. J. Insect Sci. 16: 106. DOI: 10.1093/jisesa/iew091.

Tillman, P.G. and T.E. Cottrell. 2016c. Stink bugs (Hemiptera: Pentatomidae) in pheromone-baited traps near field crops in Georgia, USA. Fla. Entomol. 99: 363–370.

Toyama, M., H. Kishimoto, K. Mishiro, R. Nakano and F. Ihara. 2015. Sticky traps baited with synthetic pheromone predict fruit orchard infestations of *Plautia stali* (Hemiptera: Pentatomidae). J. Econ. Entomol. 108: 2366–2372.

Tsutsumi, T. 2001. A method for inspecting stylet sheaths of stink bugs on the cone of Japanese cypress. Plant Prot. 55: 500–502.

University of California. 2015. UC IPM Pest Management Guidelines: Tomato, UC ANR Publication 3470. University of California Cooperative Extension, Davis, CA. http://ipm.ucdavis.edu/PMG/r783300211.html. (Last accessed 14 Oct 2015).

Velasco, L.R.I. and G.H. Walter. 1992. Availability of different host plant species and changing abundance of the polyphagous bug *Nezara viridula* (Hemiptera: Pentatomidae). Environ. Entomol. 21: 751–759.

Walsh, G.C., A.S. Dimeglio and A. Khrimian. 2016. Marking and retention of harlequin bug, *Murgantia histrionica* (Hahn) (Hemiptera: Pentatomidae), on pheromone-baited and unbaited plants. J. Pest Sci. 89: 21–29.

Weber, D.C., T.C. Leskey, G.C. Walsh and A. Khrimian. 2014. Synergy of aggregation pheromone with methyl (E,E,Z)-2,4,6-decatrienoate in attraction of *Halyomorpha halys* (Hemiptera: Pentatomidae). J. Econ. Entomol. 107: 1061–1065.

Yamada, K. and M. Noda. 1985. Studies on the forecasting of the occurrence for the stink bugs infesting fruits. Bull. Fukuoka Agric. Res. Cent. B-4: 17–24.

Yamanaka, T., M. Teshiba, M. Tuda and T. Tsutsumi. 2011. Possible use of synthetic aggregation pheromones to control stink bug *Plautia stali* in kaki persimmon. Agric. Forest Entomol. 13: 321–331.

CHAPTER 11

Use of Vibratory Signals for Stink Bug Monitoring and Control

Raul Alberto Laumann,[1,*] *Douglas Henrique Bottura Maccagnan*[2] and *Andrej Čokl*[3]

Introduction

Of the eight recognized subfamilies of stink bugs, only species of two subfamilies, Edessinae and Pentatominae, are serious crop pests. Because of their polyphagous feeding habits, stink bugs are serious pests in crops such as legumes, grasses, grains, vegetables and fruit and nut trees (Panizzi et al. 2000). In recent years, the incorporation of *Bacillus thuringiensis* genetically modified-based crops (*Bt* crops), primarily against lepidopteran pests, has led to a reduction in insecticide use. As a consequence, some stink bugs and mirids have increased their population levels, reaching outbreak levels (Green et al. 2001, Kennedy 2008, Lu et al. 2010). Additionally, extended non-tillage practices appear to contribute to the pervasive population rise in these insects (Chocorosqui and Panizzi 2004, Seffrin et al. 2006). Another recent phenomenon, which is related to climate change, is the invasion of temperate climate regions by tropical or subtropical species, such as *Halyomorpha halys* (Stål), *Bagrada hilaris* (Burmeister) and *Piezodorus guildinii* (Westwood) (Panizzi 2015). These situations challenge entomologists to develop efficient pest management strategies for stink bugs.

---

[1] Laboratório de Semioquímicos, Embrapa Recursos Genéticos e Biotecnologia, Avda W5 Norte (Final), 71070-917, Brasilia, DF, Brazil.

[2] Laboratory of Entomology, Universidade Estadual de Goiás, Campus Iporá, Avda R2, 76200-000, Iporá, GO, Brazil.
Email: douglas.hbm@ueg.br

[3] Department of Organisms and Ecosystems Research, National Institute of Biology, Večna pot 111, SI-1000 Ljubljana, Slovenia.
Email: andrej.cokl@nib.si

* Corresponding author: raul.laumann@embrapa.br

In general, the management of stink bugs has been based on population control with insecticide applications, which have increased dramatically in recent years (Panizzi 2013). This practice is unsustainable over the long-term because of the known side-effects of synthetic insecticides.

Many authors recommend the management of stink bugs based on the Integrated Pest Management (IPM) practices with efficient monitoring systems and alternatives for population control (Borges et al. 2011a,b, Bueno et al. 2011, Panizzi 2013). In this way, the incorporation of information related to sexual communication and vibratory signals may contribute to some biorational control alternatives.

Vibratory communication is important in stink bug reproductive behaviour (complete description in Chapters 6 and 7 of this book). During the calling and courtship phases of reproduction, vibratory signals interact with signals of other modalities such as chemical or visual stimuli that together create opportunities to be used in stink bug management alone or combined (Borges et al. 2011b, see also Chapter 4).

In this chapter, we review some applications of acoustic, and more specifically, substrate-borne vibrations for pest management. We also introduce some ideas regarding how these strategies at the current state knowledge of the vibratory communication systems of stink bugs can be applied in pest management systems. We focus our attention on three principal areas of application: (1) the use of vibratory signals in monitoring systems, (2) the disruption of stink bug mating by the interference of vibratory communication and (3) applications in biological control. Wherever possible, we consider the technical aspects and solutions.

Use of Vibratory Signals for Insect Monitoring

Monitoring is the central point of any pest management strategy. In general, monitoring in agriculture systems can be defined as any system or technique used to determine the identity, number, density, spatial location and/or level of pest damage in the field (Mankin et al. 2011, Lampson et al. 2013).

The detection and monitoring of insects using acoustic devices has been reported in the literature since the early decades of the past century. After the most apparent and easily detectable air-borne sounds, the first reports on the use of substrate vibrations in insect monitoring were published on the models of the termites and insects present in stored products (Mankin et al. 2011).

Insects produce substrate-borne vibrations in order to communicate through different media such as plants, water or soil or incidentally during feeding, moving or other activities (Hill 2009). For monitoring purposes, different devices are used to confirm the presence of an insect by detection, or to measure and analyse species signals directly from substrates, or else to playback these signals on substrates in order to attract insects to specific places, where they can be caught or captured by a trap.

Most of the successful monitoring systems that are based on vibratory signals are used for insects that live in cryptic conditions on plants, in wood, on soil, in stored products and in human buildings (Scheffrahn et al. 1993, Mankin 2012, Mankin et al. 2000a, 2011, 2016, Johnson et al. 2007, Potamitis et al. 2009, Zorovič and Čokl 2015). In most of these situations, insects are detected by recording the substrate vibrations produced. Records may be obtained using a wide variety of sensors such as condenser microphones, geophones, tachometers, stethoscopes, magnetic cartridges and piezoelectric transducers, including piezoelectric accelerometers and, more recently, Doppler laser vibrometers (reviewed in Mankin et al. 2011).

In most cases, the monitoring techniques are related to pest detection, but some examples of their use in estimating population density (Hagstrum et al. 1991, 1996) and spatial distribution (Brandhorst-Hubbard et al. 2001, Mankin et al. 2007) have been reported. In these situations, the principal challenges are to extract insect-produced vibrations from environmental noise (Mankin et al. 2011) and to classify the signals for the identification of taxa (Lampson et al. 2013). In this sense, laser vibrometers have an advantage in measuring vibrations without the necessity of direct contact with the substrate surface. Laser vibrometry is especially useful for the measurement of vibrations on substrates such as green plants that cannot be loaded with contact sensors whose weight changes the substrate's mechanical properties.

The sensitivity and accuracy of measurement depend highly on the sensor-substrate interface. For example, when microphones are used as sensors, the waves are highly attenuated by the transmission from a solid (the substrate) through air on the microphone membrane. To avoid masking the signals by environmental noise, it is necessary to increase the signal-to-noise ratio, so amplifiers with appropriate filters should be used, taking care to avoid filtering in the frequency range characteristic of the target species' signals (Zorovič and Čokl 2015).

The other way to use vibratory signals to monitor insect populations is by incorporating them into devices such as traps to attract insects. There are many examples in the literature describing the attraction and monitoring of insects and even some vertebrates into traps using incorporated airborne acoustic signals. Such examples include monitoring and controlling mosquitoes through their attraction to sound traps (Belton 1994, Stone et al. 2013). Other insects such as members of the Orthoptera (Gryllidae) and Diptera (Tachinidae) orders have also been successfully monitored with acoustic traps (Walker 1996, Mankin 2012). An interesting example is the technology used to attract and kill the cicada *Quesada gigas* (Olivier) (Hemiptera: Cicadidae) in coffee crops in Brazil. A system was developed that incorporates the emission of a selected sequence from a previously recorded cicada song. This species emits an acoustic signal characterized by a sequence of pulses with two characteristic parts; the first is a sequence of pulses having a mean duration of ~ 12 s and the second is a continuous pulse of ~ 8 s. The dominant frequency of this song is 1705 Hz. Propagation of this signal attracts the cicadas to a trap that sprays insecticide (Maccagnan 2008, Maccagnan et al. 2008).

To date, no traps with incorporated vibratory signals are in commercial use. However, some progress has been made in the development of an autonomous system (AS) to monitor and trap the Asian citrus psyllid *Diaphorina citri* Kuwayama (Hemiptera: Psyllidae). The responses of females and males to signals 0.3 to 0.5 s in length with spectra that contain a sequence of harmonics (ranging from 200 to 1400 Hz) that mimic insect signals had been described previously (Rhode et al. 2013). These signals were incorporated into a microcontroller platform that reproduces vibratory signals and monitors insect responses. The system was programmed to induce vibrations on plants using different *D. citri* signals with a Piezo buzzer. Responses to these devices were tested with *D. citri* females, and the results showed that the playback of mimicked signals elicits similar responses as natural songs, showing that this system has the potential to be used as a trap for psyllids (Mankin et al. 2013).

A similar system was developed to detect and attract a species of leafhopper from the genus *Aphrodes* (Hemiptera: Cicadellidae) (Korinšec et al. 2016), as females of this species respond to calling males with a vibratory signal. In this case, an AS was developed using a laser vibrometer and software to detect male signals via a recognition algorithm based on linear prediction cepstral coefficient (LPCC) feature vectors and a multilayer perceptron classifier (MLP). The potential of this AS to recognize male call signals and to emit female signals, thereby mimicking the natural duets of this species, was tested with live insects. Female playback signals (with a 44 ms duration, 26 ms pulse repetition time and a dominant frequency of 1062 Hz in a 2031 Hz frequency range) attract males to the emission point, demonstrating the potential of this AS to recognize and attract males (Korinšec et al. 2016).

Application in Stink Bug Monitoring

Traditionally, stink bug monitoring has been performed using sweep nets or beat sheet sampling procedures (Kogan and Pitre 1980). In many crops, and especially in soybean, one of the principal cultivated host plants attacked by stink bugs, the monitoring is usually conducted using the beat sheet technique (Côrrea-Ferreira 2012). This monitoring technique is very useful in estimating stink bug population density (Tood and Herzog 1980). However, due to changes in agricultural systems and the expansion of cultured areas, including increases in plant populations caused by a reduction in row spacing and the use of soybean genotypes that produce phenotypes of taller plants, this technique has become cumbersome. For example, for a 100 ha area it is necessary to take 10 samples distributed at regular points. Considering that growing areas in Brazil, the U.S. and Argentina may reach 500 to 1000 ha in size, this technique becomes very expensive and time-consuming. For these reasons, the control of stink bugs usually proceeds by using insecticide application based on fixed dates or the phenological stage of the crops, without specific information related to population densities (Sosa-Gomez and Silva 2010).

Indirect monitoring

Indirect monitoring, i.e., the estimation of population size without the direct counting of insects, has the advantage of not requiring the capture or direct observation of insects. Such monitoring systems can be based on sensors that register communicative vibratory signals in the field and process them in order to classify their origin. To achieve this goal, it is necessary to have specific knowledge of the target species' ecology and behaviour as well as engineering technical support. For example, an accurate knowledge of the spatial and temporal distribution of stink bugs in crops may be necessary to determine the best locations to obtain samples. Sensors used to record and identify signals produced by other insect groups may also be applied in stink bug monitoring. In addition, the contact or non-contact recording devices used to record the vibrations produced by wood-boring insects may also be used to record the signals of stink bugs living on woody plants. The recording of vibratory signals from herbaceous plants in the field demands adjustments to increase the signal-to-noise ratio. Noise produced by wind, rain and others environmental factors can be removed using specific sound filters. Several mathematical algorithms have been developed to record, identify and recognize acoustic signals emitted by different animals such as bats, whales, dolphins, amphibians and insects (Aide et al. 2013).

For stink bugs, a Gaussian mixture model (GMM) and a probabilistic neural network (PNN) were used to discriminate the vibratory signals from the insect incidental sounds (ex. locomotion) and to identify two different species, the brown stink bug, *Euschistus servus* (Say) and the green stink bug, *Nezara viridula* (L.) (Lampson et al. 2013). Vibratory signals emitted by the insects on cotton plants were recorded using a piezoelectric accelerometer. Recordings were conducted for single pairs and from 30 individuals of each species per plant (low and high incidental noise environments). In order to identify the signals, the system processed vibratory signal characteristics such as the dominant frequency, pulse duration, and 1st through 6th linear order frequency cepstral coefficients. With both classification models (GMM and PNN) and in low and high incidental noise environments, the authors could discriminate the signals from the incidental sounds and between species with an accuracy of above 70% (in most cases above 90%) (Lampson et al. 2013).

In addition to the potential of these models, the authors also noted the limitations of the study because it was developed under laboratory conditions. Interference was restricted only to the incidental noise produced by the tested insects and the excluded abiotic noises produced by wind or water drops, as well as by the vibrations produced by other animals (see Chapter 7). However, this study shows that stink bug monitoring is possible using this approach. The potential of GMM and PNN has been previously demonstrated for the identification and classification of the airborne sounds of cicadas as well as the vibratory signals produced by the red palm weevil *Rhynchophorus ferrugineus* (Olivier) (Coleoptera: Curculionidae) (Pinhas et al. 2008). Classification systems based on the other methods previously mentioned (i.e., LPCC and MLP) have also shown their potential for vibratory signal discrimination (Korinšec et al. 2016).

Direct monitoring

Direct monitoring is based on counting insects previously captured in devices or by directly observing them. Pheromone-baited traps represent a useful tool for the direct monitoring of stink bugs. Many of the examples described in Chapters 5 and 10 show that with an appropriate trap design and pheromone blend, population densities can be efficiently estimated. This technique is based on the knowledge of the calling phase of the stink bug mating behaviour, which includes the attraction of stink bugs to the same plant through the emission of male pheromone (Borges et al. 1987). The second step of sexual communication includes locating the mate on the plant through directional movement of the male mediated by female vibratory calling song signals (Ota and Čokl 1991, Čokl et al. 1999) (For details see Chapters 4, 5 and 6).

This particular bimodal communication system that employs both chemical and vibratory signals is most likely the reason for some reports of reduced catch in pheromone traps with the observation of bugs clustered in their vicinity (Aldrich et al. 1991, James et al. 1996). For a more precise estimate of population densities, one also needs to count the insects occurring around the traps, usually in a radius of approximately 1 to 3 m (Cullen and Zalom 2005, Aldrich et al. 2007). As a result of these observations, the incorporation of vibratory signals in pheromone traps has been suggested (Millar et al. 2002, Borges et al. 2011b), but until now devices including both pheromone and vibratory signals have not been developed.

The female calling song of the target stink bug species is the most promising vibratory signal for incorporation in pheromone traps. In the green stink bug, *Nezara viridula,* and many other pentatomine species, the female calling song triggers the directional movement of males to the signal source (Čokl et al. 1999, Virant-Doberlet et al. 2006). Male directional movement and vibratory responses can also be elicited by computer-synthesized pure tone signals with temporal and spectral characteristics that resemble those of the natural female calling song (Žunič et al. 2011). Directional movements mediated by the female calling songs have also been reported in many Neotropical species of stink bugs (Blassioli-Moraes et al. 2005, 2014, Silva et al. 2012, Laumann et al., in preparation). This opens up the possibility of using synthetic vibratory stimuli to attract stink bugs, which is especially important for the development of devices to be incorporated in traps.

In a preliminary study, the directional responses of the males of the Neotropical brown stink bug, *Euschistus heros* (Fabricius), towards the female calling song and towards synthesized pure tone signals of the frequency and temporal characteristics resembling those of the natural song, were studied. In laboratory experiments it was observed that males moved selectively towards the vibrating bean plant leaves. The potential of these signals to improve the efficacy of pheromone traps was tested in artificial (cages with potted soybean plants) and natural (soybean fields) environments. In cages, vibrated traps attracted significantly more insects than non-vibrated traps (Fig. 11.1a), and in field tests a similar pattern of capture was observed (Fig. 11.1b) (Laumann et al., unpublished data).

This initial study shows the potential of a bimodal bait, composed of pheromone and vibratory signals, to increase the efficacy of stink bug monitoring traps. In order to improve this method, it will be necessary to determine the periods when the species are reproductively active and the periods of the day when the insects are more responsive. Another relevant consideration is establishing how vibratory signals interact with chemical signals (pheromones) in traps. The most plausible hypothesis is that both signals have a synergistic effect on insect captures, especially when sex pheromones are used in traps. Furthermore, it should be established whether the interactions of the insects inside the traps (ex. copulation) and the emission of other chemicals or vibratory signals could lead to increased or reduced additional captures. It is known that the vibratory signals of the female *Nezara viridula* increase the amount of pheromones emitted by the males (Miklas et al. 2003). In this way, the males caught in a vibrated trap emit a higher amount of pheromone, thereby acting as a natural source of pheromone. Thus, the use of synthetic pheromones could be reduced or eliminated, decreasing the cost of the traps. Zgonik and Čokl (2014) showed that *N. viridula* females stimulated by male pheromones emit calling songs more often, representing an additional favourable aspect of the interaction between pheromones and vibratory signals.

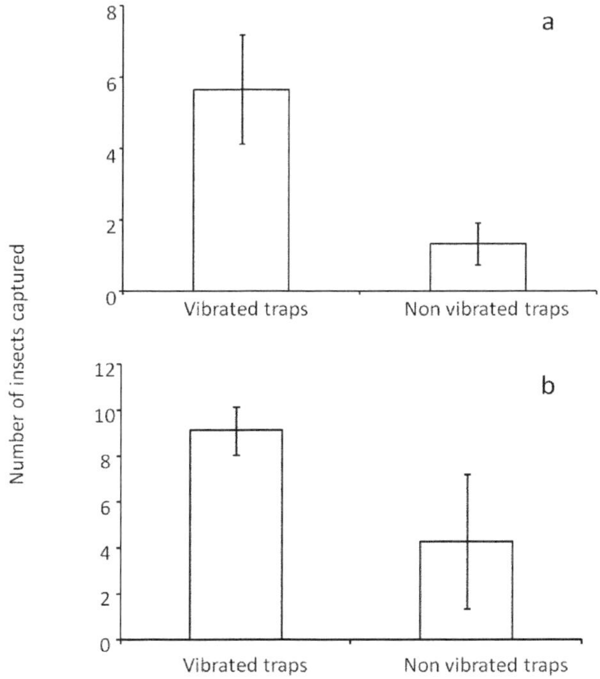

Figure 11.1. The number of insects captured in vibrated or non-vibrated traps in cage (a) and field (b) experiments. In the cage experiments a total of 30 *E. heros* males were released into the cages and in field experiments a total of 50 females and 50 males were released into a 50 m^2 soybean field. The number of insects captured in cages was recorded after 10 hours, and bars represent the results of 3 replicates. The number of insects captured in field experiments represents the number observed per hour during a 7-hour experiment. In both cases, the number of insects captured in vibrated traps was significantly higher than the number captured in non-vibrated traps ($z = 2.60$, $gl = 4$, $p = 0.009$ and $z = 3.09$, $df = 12$, $p = 0.002$ for cage and field experiments, respectively, GLM analyses).

The synergistic effects of both types of signals on behaviour have also been shown in other insect taxa. *Anastrepha suspensa* (Loew) (Diptera: Tephritidae) females exposed to sex pheromones increase their responses to male calling songs (Mankin et al. 2000b). Recently, it was demonstrated in the noctuid moth *Spodoptera littoralis* (Boisduval) that the sensitivity of the olfactory neurons could be modulated by exposure to a pulsed stimulus mimicking the echolocation sounds of bats (Anton et al. 2011). Interactions with stimuli of other modalities such as visual signals can also change the responses to acoustic signals (Mankin et al. 2004). However, the efficacy of such traps may also be reduced because the emission of pheromones by males inside the traps could induce rivalry and suppress the capture of additional males that move away to avoid competition. All of these aspects need to be addressed in future research.

Use of Vibratory Signals in Disrupting Behaviour or Interference in Communication

The use of acoustic signals to interfere in insect communication or as a deterrent of insect behaviour was proposed fifty years ago (Walker 1996). Most of the work done so far has focused on the interference of airborne signals, most likely because the best known examples are those of the escape behaviour of moths from the ultrasonic signals of bats (Connor and Corcoran 2012). Many works have attempted to develop ultrasonic devices to interfere with the communication of pest moth species and that of other insects such as mosquitoes, lacewings, field crickets, locusts, cockroaches, and praying mantids (Walker 1996, Mankin 2012). However, this strategy has been shown to be ineffective in moth control because it does not reduce oviposition or the economic damage to crops, most likely due to the habituation of insects to disrupting signals (Bomford and O'Brien 1990, Mankin 2012).

Other possible applications of airborne and substrate-borne signals include interfering with the physiology of insects and disrupting their normal development. Most of these studies were conducted to control stored product insects. The effects of these signals have been demonstrated in terms of direct mortality and the interference of neuroendocrine processes that have negative effects on the duration of development and the subsequent gains in weight and size (Kight and Swaddle 2011, Adamo and Baker 2011, Polajnar et al. 2015).

Many works have addressed the effects of biotic (conspecific and heterospecific singles, rivals, and predators or parasitoids) and abiotic (environmental, primarily rain and wind for insects that feed on plants) noise on insect vibratory communication (for a review see Chapter 7). The application of this knowledge in pest management is still incipient. The first report on the use of vibratory signals for pest control was related to the tests of the effects of vibrations in disrupting sexual communication of the cotton leafhopper, *Amrasca devastans* (Dist.) (Hemiptera: Cicadellidae), and the brown rice planthopper, *Nilaparvata lugens* (Stål) (Hemiptera: Delphacidae) (Saxena and Kumar 1980). In both species, stimulation by airborne signals picked up by their host plants interrupted their sexual behaviour for at least 4 hours. Even after the stimulation ended, the effect was maintained for another 4-hour period, with only 26% of the leafhopper couples observed mating (Saxena and Kumar 1980).

Over the past 20 years, new perspectives in pest control and management have arisen through the development of transducers that propagate vibrations at longer distances with high accuracy and by direct contact with solid substrates, including plants. An instructive example of this method, which was used to disrupt mating, was reported for the leafhopper *Scaphoideus titanus* Ball (Hemiptera: Cicadellidae) (Eriksson et al. 2012), a pest of vineyards. The authors disrupted mating behaviour using disturbance signals pre-recorded from S. *titanus* during male rivalry (Mazzoni et al. 2009a,b), and these signals were reproduced on plants using electromagnetic shakers. In a semi-field test, potted vineyard plants were vibrated by the shakers in direct contact with the leaves. Field tests were conducted in a mature rooted grapevine plantation with plants (1.5 m in height) cultivated in rows with stems tied to a supporting metal wire. Signals were transmitted using an electromagnetic shaker attached to the wire. In these semi-field and field tests, more than 80% of the females placed on vibrated plants remained virgins as compared to 20% under control conditions (non-vibrated plants) (Eriksson et al. 2012). In a subsequent study it was determined that the velocity threshold for a disruptive signal effect is approximately 0.015 mm/s, and that in the above described system the effect can be achieved at up to a 10 m distance from the source, within which 90% of the tested females remained virgins after 24 hours (Polajnar et al. 2016).

Another interesting example is the use of the vibratory signals of bark beetles (Coleoptera; Scolytidae) to reduce reproductive success. Bark beetles use vibratory signals produced by stridulation or friction with the substrate in different behaviours such as mate recognition and communication, territoriality, species recognition, predator escape and possibly host selection (Hofstetter et al. 2014). A combination of the vibratory alarm signals of three bark beetle species applied to ponderosa pine wood pieces significantly reduced gallery construction and oviposition and caused moderate mortality (Hofstetter et al. 2014).

The studies described above show the potential use of vibratory signals as disruption strategies for pest control. However, habituation to noise is a well-documented response in insects (Mankin et al. 2011) and should therefore be considered. Insects have been shown to develop several mechanisms to either increase the signal-to-noise ratio or to avoid interference (Čokl et al. 2015, Polajnar et al. 2016). Some well-known and relevant strategies to avoid habituation include the emission of disruptive signals using non-continuous and random programs, the emission of disruptive signals with irregular variations in pulse duration, changes in the position of the signal source or the development of devices that simulate motion (Agee 1969) and the use of natural signals, to which the probability of habituation is reduced over synthetic or environmental signals (Polajnar et al. 2015).

Studies of the efficacy of disruptive signals in pest control should be conducted in populations of varying densities. As observed in the pheromone-based mating disruption systems, the efficacy of this approach may be reduced at high population levels because of the higher probability that the insects will find mates by chance (Welter et al. 2005, Epstein et al. 2006). An effective disruption signal also needs to cover the signal temporal and spectral variability of each species, including those changes induced in response to noise. Another important consideration is related to the amplitude of disruptive signals. Vibratory signals in rod-like plant structures propagate by bending waves with readily repeated amplitude minima and maxima as a consequence of the resonance of plant tissues (Polajnar et al. 2012) (for details see Chapter 6). Low amplitude signals have little or no disruptive effects and those with high amplitude repel insects from the plant. Both reduce the efficiency of the method based on the application of disruptive signals.

Application in Stink Bug Management

The characteristics of stink bug signals and the impact of biotic and abiotic noise on communication in stink bugs are described in Chapters 6 and 7.

All examined stink bugs communicate using narrow band low frequency signals with the main energy emitted at frequencies ranging between 50 and 200 Hz. This generally excludes the possibility of communication interference by lower frequency vibrations typical of abiotic noises such as wind and water drops (see Chapter 7).

The efficiency of interfering with or disrupting the communication of stink bugs using pure tones with frequencies in the stink bug-specific range needs to be investigated in detail. The results of one study of *Nezara viridula* showed little effect of a 100 Hz disturbance on female calling (Polajnar and Čokl 2008). Females change the dominant frequency of the calling song signals in order to increase their difference from the 100 Hz background vibrations and thus improve song discrimination. Similar responses have been recently described in the individuals of *Euschistus heros,* which avoid interference by modifying the temporal and spectral characteristics of their signals to avoid overlapping (Čokl et al. 2015).

More promising for the disruption of communication is the use of the male rival song; the female repellent song and/or male signalling prior to copulation (see Chapter 6). The male rival song silences other males and females, the female repellent song halts male courting and male copulation signals stop female and male emission of different songs. The ability of these signals to disrupt communication in the field needs to be investigated.

The effects of interference vibrations on the reproductive behaviour of the stink bug species under study were tested with artificial pure tone signals in the range of stink bug emissions that were used to vibrate soybean plants. Interference using frequencies close to the dominant frequency characteristics of the calling and courtship signals of *E. heros*, as reported by Blassioli-Moraes et al. (2005) and Čokl et al. (2015), was disruptive of mating behaviour because most of the couples observed (> 80%) did not mate in the presence of these signals (Laumann et al., unpublished data). Insect stimulation with pure tone vibrations emitted vibratory signals having similar changes to those found by Čokl et al. (2015), in which the signals produced by brown stink bugs overlapped with those from conspecifics. Future research should consider alternatives to reduce habituation and the possible side-effects on parasitoids and predators.

Buzzing signals produced by wing vibrations have not been studied as extensively as abdomen vibration signals in stink bugs (for details see Chapter 6). These high amplitude signals could be transmitted between plants with airborne components (Kavčič et al. 2013). Their function in stink bug communication has not yet been described, but observations in the brown stink bugs suggest that they might have advertisement functions (Čokl and Laumann, unpublished data). Therefore, buzzing signals could also have the potential to be incorporated into a mating disruption strategy. Finally, the effects of vibratory signals on the immature stages of stink bugs are unknown, and the possible use of these signals to interfere with the physiology or behaviour of stink bug nymphs could offer novel opportunities to apply vibratory interference in pest control.

Substrate-borne Vibrations and Biological Control

Natural enemies, predators and parasitoids use substrate vibration in both prey/host search and localization (Cocroft and Rodrigues 2005). Such examples of interspecific vibratory interactions have been described in many arthropod species. Scorpions use seismic vibrations produced by movement to recognize and localize their prey (Brownell and Farley 1979, Brownell and van Hemmen 2001). Spiders detect prey by sensing vibrations induced by movement in the web or on a continuous substrate (Barth 1998). Male vibratory signals of the leafhopper *Aphrodes makarovi* attract tangle-web spiders *Enoplognatha ovata* (Clerck) (Araneae: Theridiidae), which then feed on them (Viran-Doberlet et al. 2011). Parasitoid wasps (Braconidae, Eulophidae and Pteromalidae) locate their hosts using vibrations produced incidentally by host movement, feeding or other activities (Meyhöfer et al. 1997, Meyhöfer and Casas 1999, Broad and Quicke 2000, Vilhelmesen et al. 2001). Additionally, some Orussidae (Vilhelmsen et al. 2001) and Ichneumonidae (Wäckers et al. 1998, Broad and Quicke 2000) wasps use echolocation to find their hosts.

The most promising biological technique for the control of stink bug populations is based on the use of egg parasitoids (Platygastridae) (Corrêa-Ferreira and Moscardi 1996, Côrrea-Ferreira 2002). Some efforts have been undertaken to manipulate their recruitment in crops from natural or adjacent cultivated areas using semiochemicals (Vieira et al. 2013, 2014, Michereff et al. 2015, also see details in Chapter 8). *Telenomus podisi* Ashmead, is one of the most relevant platygastrid egg parasitoids of stink bugs in Brazil.

Females of this species can recognize and use the vibratory signals of stink bugs as cues for spatial orientation. This orientation response (taxis) is highly selective in parasitoid females responding preferentially to stink bug female calling songs and to songs of their preferred host, *E. heros* (Laumann et al. 2007, 2011). This opens up the possibility of using species-specific vibratory signals to attract (recruit) their natural enemies. The ultimate goal of this strategy is to increase the efficiency of the natural enemies in the region by recruiting them to the infested areas and thus increasing the current conservative levels of biological control. This strategy may be used in combination or complementarily with the use of semiochemicals. To develop this strategy, additional studies investigating the interactions and synergisms among different modalities are needed.

Predatory stink bugs (subfamily Asopinae) also use vibratory signals during reproductive behaviour (for details see Chapter 3). Most of these insects prey on lepidopteran caterpillars and are important biological control agents. It may therefore be possible to use these vibratory signals in two ways. One way is to recruit predators in a similar way as described for parasitoids and in combination with semiochemicals (see Chapter 9). In addition, it has been demonstrated that the predatory stink bug *Podisus maculiventris* (Say) can locate prey using the vibratory signals produced by chewing when feeding on plants (Pfannenstiel et al. 1995); thus, the vibratory feeding cues of insect herbivores could also be used in predator recruitment. The second possibility is to disrupt herbivores by transmitting predatory stink bug signals to plants. This strategy is based on the fact that many herbivorous insects respond to vibratory noises, showing escape behaviour that usually entails dropping from the plant (Gross 1993, Losey and Denno 1998, Castellanos and Barbosa 2006).

Concluding Remarks

Information regarding substrate vibration by insects can be used as an important tool in IPM and biorational insect control. The most likely immediate application of the current knowledge on stink bug substrate-borne vibratory communication may be focused on monitoring the population densities of species with higher impacts as pests. Direct monitoring using pheromone traps has been developed in recent years (Borges et al. 2011a,b; for details see Chapter 10), and the incorporation of vibratory calling signals appears to have the potential to improve the efficiency of insect attraction and capture. This measure requires technical expertise to incorporate vibrations into traps and to enlarge the coverage area to neighbouring plants.

The other promising potential use of the information on stink bug vibratory communication is indirect monitoring through identification of species- and gender-specific stink bug signals from plants. Among the different types of sensors available for this purpose, laser vibrometers have proved to be most effective, as shown by the detection of wood-boring insects in trees (Zorovič and Čokl 2015). A disadvantage of the use of this method in the field is the movement of lightweight green plants by wind, disrupting the focus of the laser beams and disabling continuous long-term monitoring (Čokl and Laumann, observation in soybean fields). The use of contact sensors such as accelerometers, piezoelectric transducers, piezoelectric membranes or similar instruments is also possible for this purpose, but limited by substrate mechanical properties and weight, which must be at least three times greater than the weight of the sensor. The incorporation of sensors into unmanned aerial vehicles (UAVs) could also extend the indirect monitoring system over larger areas without human displacement.

Automated systems for pest monitoring with sensors to detect signals without direct human intervention have been developed in recent years (Hagstrum et al. 1996, Ganchev and Potamitis 2007, Raman et al. 2007, Jianga et al. 2008, Aide et al. 2013, Korinšek et al. 2016). Vibrations detected in the field by different sensors need to be analysed and identified. Various types of software are available to improve the detected signal-to-noise ratio, analyse its temporal and spectral characteristics and determine its nature and origin by species or gender. These data may be transmitted to remote data processing stations by various methods, even using cell phones (Jianga et al. 2008). Long-term automated monitoring also gives growers valuable information regarding trends in pest population densities in the field to predict insect migrations, estimate insect infestations at the regional scale and to determine their spatial distributions.

The use of vibration in the disruption of mating appears to be a control method that demands deeper inquiry into the target species' biology, ecology, mating behaviour and communication. Research on major stink bug pest species communication and general information regarding their biology provide a good basis for the efficient application of different biorational control measures. Well-equipped laboratories with high levels of expertise in communication using signals of different modalities offer great possibilities in providing fast and relevant knowledge on pest species communication specificities that can be used in the field for pest population control.

Acknowledgments

We thank the National Council for Scientific and Technological Development (CNPq), the Brazilian Corporation of Agricultural Research (EMBRAPA), the Research Support Foundation of the Federal District (FAP-DF) and the Slovenian Research Agency.

References

Agee, H.R. 1969. Response of bollworm moths to pulsed ultrasound while resting, feeding, courting, mating, and ovipositing. Ann. Entomol. Soc. Am. 62: 1122–1128.

Aide, T.M., C. Corrada-Bravo, M. Campos-Cerqueira, C. Milan, G. Vega and R. Alvarez. 2013. Real-time bioacoustics monitoring and automated species identification. PeerJ 1: e103; DOI 10.7717/peerj.103.

Adamo, S.A. and J.L. Baker. 2011. Conserved features of chronic stress across phyla: the effects of long-term stress on behavior and the concentration of the neurohormone octopamine in the cricket, *Gryllus texensis*. Horm. Behav. 60: 478–483.

Aldrich, J.R., M.P. Hoffmann, J.P. Kochansky, W.R. Lusby, J.E. Eger and J.A. Payne. 1991. Identification and attractiveness of a major pheromone component for Nearctic *Euschistus* spp. stink bugs (Heteroptera: Pentatomidae). Environ. Entomol. 20: 477–483.

Aldrich, J.R., A. Khrimian and M.J. Camp. 2007. Methyl 2,4,6-decatrienoates attract stink bugs and tachinid parasitoids. J. Chem. Ecol. 33: 801–815.

Anton, S., K. Evengaard, R.B. Barrozo, P. Anderson and N. Skals. 2011. Brief predator sound exposure elicits behavioral and neuronal long-term sensitization in the olfactory system of an insect. Proc. Natl. Acad. Sci. USA 108: 3401–3405.

Barth, F.G. 1998. The vibrations sense in spiders. *In*: Hoy, R.R., A.N. Popper and R.R. Fay (eds.). Comparative Hearing: Insects. Springer, New York, USA.

Belton, P. 1994. Attraction of male mosquitoes to sound. J. Am. Mosq. Control Assoc. 10: 297–301.

Blassioli-Moraes, M.C., R.A. Laumann, A. Čokl and M. Borges. 2005. Vibratory signals of four Neotropical stink bug species. Physiol. Entomol. 30: 175–188.

Blassioli-Moraes, M.C., D.M. Magalhaes, A. Čokl, R.A. Laumann, J.P. Da Silva, C.C.A. Silva and M. Borges. 2014. Vibrational communication and mating behaviour of *Dichelops melacanthus* (Hemiptera: Pentatomidae) recorded from the loudspeaker membrane and plants. Physiol. Entomol. 39: 1–11.

Bomford, M. and P. O'Brien. 1990. Sonic deterrents in animal damage control: a review of device tests and effectiveness. Wild Soc. Bull. 148: 411–422.

Borges, M., P.C. Jepson and P.E. Howse. 1987. Long-range mate location and close-range courtship behaviour in the green stink bug, *Nezara viridula* and its mediation by sex pheromone. Entomol. Exper. Et Appl. 44: 205–212.

Borges, M., M.C.B. Moraes, M.F. Peixoto, C.S.S. Pires, E.R. Sujii and R.A. Laumann. 2011a. Monitoring the neotropical brown stink bug *Euschistus heros* (F.) (Hemiptera: Pentatomidae) with pheromone-baited traps in soybean fields. J. Appl. Entomol. 135: 68–80.

Borges, M., M.C.B. Moraes, R.A. Laumann, M. Pareja, C.C. Silva, M.F. Michereff et al. 2011b. Chemical ecology studies in soybean crop in Brazil and their application to pest management. pp. 31–66. *In*: Tzi-Bun Ng. [Org.]. Soybean—Biochemistry, Chemistry and Physiology, 1st ed. Intech Open Access Publisher, Rijeka, Croatia.

Brandhorst-Hubbard, J.L., K.I. Flanders, R.W. Mankin, E.A. Guertal and R.I. Crocker. 2001. Mapping of soil insect infestations sampled by excavation and acoustic methods. J. Econ. Entomol. 94: 1452–1458.

Broad, G.R. and D.L.J. Quicke. 2000. The adaptive significance of host location by vibrational sounding in parasitoid wasps. Proc. Roy. Soc. Lon. Series B 267: 2103–2109.

Brownell, P. and R.D. Farley. 1979. Orientation to vibration in sand by the nocturnal scorpion *Paurocfonus mesaensis*: mechanism of target localization. J. Comp. Phys. 131: 31–38.

Brownell, P.H. and J.L. van Hemmen. 2001. Vibration sensitivity and a computational theory for prey-localizing behavior in sand scorpions. Amer. Zool. 41: 1229–1240.

Bueno, A.F., M.J. Batistela, R.C.O.F. Bueno, J.B. França-Neto, M.A.N. Nishikawa and A. Libério Filho. 2011. Effects of integrated pest management, biological control and prophylactic use of insecticides on the management and sustainability of soybean. Crop Prot. 30: 937–945.

Castellanos, I. and P. Barbosa. 2006. Evaluation of predation risk by a caterpillar using substrate-borne vibrations. An. Behav. 72: 461–469.
Chocorosqui, V.R. and A.R. Panizzi. 2004. Impact of cultivation systems on *Dichelops melacanthus* (Dallas) (Heteroptera: Pentatomidae) population and damage and its chemical control on wheat. Neotrop. Entomol. 33: 487–492.
Cocroft, R.B. and R.L. Rodriguez. 2005. The behavioral ecology of insect vibrational communication. Bioscience 55: 323–334.
Connor, W.E. and A.J. Corcoran. 2012. Sound strategies: the 65-million year-old battle between bats and insects. Ann. Rev. Entomol. 57: 21–39.
Čokl, A., M. Virant-Doberlet and A. McDowell. 1999. Vibrational directionality in the southern green stink bug, *Nezara viridula* (L.), is mediated by female song. An. Behav. 58: 1277–1283.
Čokl, A., R.A. Laumann, A. Žunič-Kosi, M.C. Blassioli-Moraes, M. Virant-Doberlet and M. Borges. 2015. Interference of overlapping insect vibratory communication signals: an *Euschistus heros* model. PloS One 10(6): 1–16.
Corrêa-Ferreira, B.S. and F. Moscardi. 1996. Biological control of soybean stink bugs by inocluative releases of *Trissolcus basalis*. Ent. Exp. Et Appl. 79: 1–7.
Corrêa-Ferreira, B.S. 2002. *Trissolcus basalis* para o controle de percevejos da soja. pp. 449–476. *In*: Parra, J.R.P., P.S. Botelho, B. Corrêa-Ferreira and J.M.S. Bento (eds.). Controle Biológico no Brasil, Parasitóides e Preadores. Manole Ltda.
Côrrea-Ferreira, B.S. 2012. Amostragem de pragas da soja. *In*: Hoffmann-Campo, C.B., B.S. Corrêa-Ferreira and F. Moscardi (eds.). Soja: Manejo Integrado de Insetos e Outros Artrópodes-praga. Embrapa, Brasília, DF. Embrapa, Tecnologias de produção de soja–região central do Brasil 2012 e 2013. Londrina: Embrapa Soja, 2011. 261 p. (Embrapa Soja. Sistemas de Produção, 15).
Cullen, E.M. and F.G. Zalom. 2005. Relationship between *Euschistus conspersus* (Hem., Pentatomidade) pheromone trap catch and canopy samples in processing tomatoes. J. Appl. Entomol. 129: 505–514.
Epstein, D.L., L.L. Stelinski, T.P. Reed, J.R. Miller and L.J. Gut. 2006. Higher densities of distributed pheromone sources provide disruption of codling moth (Lepidoptera: Tortricidae) superior to that of lower densities of clumped sources. J. Econ. Entomol. 99: 1327–1333.
Eriksson, A., G. Anfora, A. Lucchi, F. Lanzo, M. Virant-Doberlet and V. Mazzoni. 2012. Exploitation of insect vibrational signals reveals a new method of pest management. PLoS One 7(3): 1–5.
Ganchev, T. and I. Potamitis. 2007. Automatic acoustic identification of singing insects. Bioacoustics 16: 281–328.
Green, J.K., S.G. Turnipseed, M.J. Sullivan and O.L. May. 2001. Treatment thresholds for stink bugs (Hemiptera: Pentatomidae) in cotton. J. Econ. Entomol. 94: 403–409.
Gross, P. 1993. Insect behavioural and morphological defenses against parasitoids. Annu. Rev. Entomol. 38: 251–273.
Hagstrum, D.W., K.W. Vick and P.W. Flinn. 1991. Automated acoustical monitoring of *Tribolium castaneum* (Coleoptera: Tenebrionidae) populations in stored wheat. J. Econ. Entomol. 84: 1604–1608.
Hagstrum, D.W., P.W. Flinn and D. Shuman. 1996. Automated monitoring using acoustical sensors for insects in farm-stored wheat. J. Econ. Entomol. 89: 211–217.
Hill, P.S.M. 2009. How do animals use substrate-borne vibrations as an information source? Nature 96: 1355–1371.
Hofstetter, R.W., D.D. Dunn, R. McGuire and K.A. Potter. 2014. Using acoustic technology to reduce bark beetle reproduction. Pest Manag. Sci. 70: 24–27.
James, D.G., R. Heffer and M. Amaike. 1996. Field attraction of *Biprorulus bibax* Breddin (Hemiptera: Pentatomidae) to synthetic aggregation pheromone and (*E*)-2-hexenal, a pentatomid defense chemical. J. Chem. Ecol. 22: 1697–1708.
Jianga, J.A., L.T. Chwan, F.M. Lua, E.Ch. Yang, Z.S. Wua, Ch.P. Chena et al. 2008. A GSM-based remote wireless automatic monitoring system for field information: A case study for

ecological monitoring of the oriental fruit fly, *Bactrocera dorsalis* (Hendel). Comp. Elect. Agr. 62: 243–259.
Johnson, S.N., J.W. Crawford, P.J. Gregory, D.V. Grinev, R.W. Mankin, G.J. Masters et al. 2007. Non-invasive techniques for investigating and modelling root-feeding insects in managed and natural systems. Agric. Forest Entomol. 9: 39–46.
Kavčič, A., A. Čokl, R.A. Laumann, M.C. Blassioli-Moraes and M. Borges. 2013. Tremulatory and abdomen vibration signals enable communication through air in the stink bug *Euschistus heros*. PloS One 8(2): 1–10.
Kennedy, G.G. 2008. Integration of insect-resistant genetically modified crops with IPM programs. p. 17. *In*: Romeis, J., A.M. Shelton and G.G. Kennedy (eds.). Integration of Insect-Resistant Genetically Modified Crops with IPM Programs. Springer, New York, USA.
Kight, C.R. and J.P. Swaddle. 2011. How and why environmental noise impacts animals: an integrative, mechanistic review. Ecol. Lett. 14: 1052–1061.
Kogan, M. and H.N. Pitre Junior. 1980. General sampling methods for above-ground populations of soybean arthropods. pp. 30–60. *In*: Kogan, M. and D.C. Herzog (eds.). Sampling Methods in Soybean Entomology. Springer-Verlag, New York, USA.
Korinšek, G., M. Derlink, M. Virant-Doberlet and T. Tuma. 2016. An autonomous system of detecting and attracting leafhopper males using species- and sex-specific substrate borne vibrational signals. Comp. Elect. Agr. 23: 29–39.
Lampson, B.D., Y.J. Han, A. Khalilian, J. Greene, R.W. Mankin and E.G. Foreman. 2013. Automatic detection and identification of brown stink bug, *Euschistus servus*, and southern green stink bug, *Nezara viridula*, (Heteroptera: Pentatomidae) using intraspecific substrate-borne vibrational signals. Comp. Elect. Agr. 91: 154–159.
Laumann, R.L., M.C. Blassioli-Moraes, A. Čokl and M. Borges. 2007. Eavesdropping on sexual vibratory signals of stink bugs (Hemiptera: Pentatomidae) by the egg parasitoid *Telenomus podisi*. Anim. Behav. 73: 637–649.
Laumann, R.A., A. Čokl, A.P.S. Lopes, J.B.C. Fereira, M.C. Blassioli-Moraes and M. Borges. 2011. Silent singers are not safe: selective response of a parasitoid to substrate-borne vibratory signals of stink bugs. Anim. Behav. 82: 1175–1183.
Losey, J.E. and R.F. Denno. 1998. The escape response of pea aphids to foliar-foraging predators: factors affecting dropping behaviour. Ecol. Entomol. 23: 53–61.
Lu, Y., K. Wu, Y. Jiang, B. Xia, P. Li, H. Feng et al. 2010. Mirid bug outbreaks in multiple crops correlated with wide-scale adoption of *Bt* cotton in China. Science 328: 1151–1154.
Maccagnan, D.H.B. 2008. Cigarra (Hemiptera: Cicadidae): emergência, comportamento acústico e desenvolvimento de armadilha sonora. 2008. f. 90. Tese (Doutorado em Ciências: Entomologia Agrícola)—Faculdade de Filosofia, Ciências e Letras, Universidade de São Paulo, Ribeirão Preto, Brasil.
Maccagnan, D.H.B., N. Martinelli, T.K. Matuo and T. Matuo. 2008. Cigarras do café: biologia, ecologia e manejo. pp. 147–155. *In*: Núcleo de Estudos em Fitopatologia da UFLA [Org.]. Manejo fitossanitário da cultura do cafeeiro. Sociedade Brasileira de Fitopatologia, Brasília, Brasil.
Mankin, R.W., J. Brandhorst-Hubbard, K.L. Flanders, M. Zhang, R.L. Crocker, S.L. Lapointe et al. 2000a. Eavesdropping on insects hidden in soil and interior structures of plants. J. Econ. Entomol. 93: 1173–1182.
Mankin, R.W., E. Petersson, N. Epsky, R.R. Heath and J. Sivinski. 2000b. Exposure to male pheromones enhances *Anastrepha suspensa* (Diptera: Tephritidae) female response to male calling song. Fl. Entomol. 83: 411–21.
Mankin, R.W., J.B. Anderson, A. Mizrach, N.D. Epsky, D. Shuman, R.R. Heath et al. 2004. Broadcasts of wing-fanning vibrations recorded from calling male Ceratitis capitata (Wiedemann) (Diptera: Tephritidae) increase captures of females in traps. J. Econ. Entomol. 97: 1299–1309.

Mankin, R.W., J.L. Hubbard and K.L. Flanders. 2007. Acoustic indicators for mapping infestation probabilities of soil invertebrates. J. Econ. Entomol. 100: 790–800.

Mankin, R.W., D.W. Hagstrum, M.T. Smith, A.L. Roda and M.T.K. Kairo. 2011. Perspective and promise: a century of insect acoustic detection and monitoring. Am. Entomol. 57: 30–44.

Mankin, R.W. 2012. Applications of acoustics in insect pest management. CAB Rev. 7: 1–7.

Mankin, R.W., B.B. Rohde, S.A. Mcneill, T.M. Paris, N.I. Zagvazdina and S. Greenfeder. 2013. *Diaphorina citri* (Hemiptera: Liviidae) responses to microcontroller-buzzer communication signals off potential use in vibration traps. Fl. Entomol. 96: 1546–1555.

Mankin, R.W., H.Y. Al-Ayedh, Y. Aldryhim and B. Rohde. 2016. Acoustic detection of *Rhynchophorus ferrugineus* (Coleoptera: Dryophthoridae) and *Oryctes elegans* (Coleoptera: Scarabaeidae) in *Phoenix dactylifera* (Arecales: Arecacae) trees and offshoots in Saudi Arabian orchards. J. Econ. Entomol. 16: 1–7.

Mazzoni, V., J. Prešern, A. Lucchi and M. Virant-Doberlet. 2009a. Reproductive strategy of the Nearctic leafhopper *Scaphoideus titanus* Ball (Hemiptera: Cicadellidae). Bull. Entomol. Res. 99: 401–413.

Mazzoni, V., A. Lucchi, A. Čokl, J. Prešern and M. Virant-Doberlet. 2009b. Disruption of the reproductive behavior of *Scaphoideus titanus* by playback of vibrational signals. Entomol. Exp. Appl. 133: 174–185.

Meyhöfer, R., J. Casas and S. Orn. 1997. Vibration-mediated interactions in a host–parasitoid system. Proc. Roy. Soc. London, Series B 264: 261–266.

Meyhöfer, R. and J. Casas. 1999. Vibratory stimuli in host location by parasitic wasps. J. Ins. Physiol. 45: 967–971.

Michereff, M.M.F., M. Michereff Filho, M.C. Blassioli-Moraes, R.A. Laumann, I.R. Diniz and M. Borges. 2015. Effect of resistant and susceptible soybean cultivars on the attraction of egg parasitoids under field conditions. J. Appl. Entomol. 139: 207–216.

Miklas, N., T. Lasnier and M. Renou. 2003. Male bugs modulate pheromone emission in response to vibratory signals of the conspecifics. J. Chem. Ecol. 29: 561–574.

Millar, J.G., H.L. McBrien, H.Y. Ho, R.E. Rice, E. Cullen, F.G. Zalom et al. 2002. Pentatomid bug pheromone in IPM: possible applications and limitations. IOBC WPRS Bull. 25: 1–11.

Ota, D. and A. Čokl. 1991. Male location in the southern green stink bug *Nezara viridula* (Heteroptera: Pentatomidae) mediated through substrate-borne signals on ivy. J. Insect Behav. 4: 441–447.

Panizzi, A.R., J.E. McPherson, D.G. James, M. Javahery and R.M. McPherson. 2000. Economic importance of stink bugs (Pentatomidae). pp. 421–474. *In*: Schaefer, C.W. and A.R. Panizzi (eds.). Heteroptera of Economic Importance. CRC Press, Boca Raton, Florida, USA.

Panizzi, A.R. 2013. History and contemporary perspectives of the integrated pest management of soybean in Brazil. Neotrop. Entomol. 42: 119–127.

Panizzi, A.R. 2015. Growing problems with stink bugs (Hemiptera: Heteroptera: Pentatomidae): species invasive to the U.S. and potential neotropical invaders. Am. Entomol. 61: 223–233.

Pfannenstiel, R.S., R.E. Hunt and K.V. Yeargan. 1995. Orientation of a hemipteran predator to vibrations produced by feeding caterpillars. J. Ins. Behav. 8: 1–9.

Pinhas, J., V. Soroker, A. Hetzroni, A. Mizrach, M. Teicher and J. Goldberger. 2008. Automatic acoustic detection of the red palm weevil. Comp. Elect. Agr. 63: 131–139.

Polajnar, J. and A. Čokl. 2008. The effect of vibratory disturbance on sexual behaviour of the southern green stink bug *Nezara viridula* (Heteroptera, Pentatomidae). Cent. Eur. J. Biol. 3: 189–197.

Polajnar, J., D. Svenšek and A. Čokl. 2012. Resonance in herbaceous plant stems as a factor in vibrational communication of pentatomid bugs (Heteroptera: Pentatomidae). J. R. Soc. Interface 9: 1898–1907.

Polajnar, J., A. Eriksson, A. Lucchi, G. Anfora, M. Virant-Doberletc and V. Mazzoni. 2015. Manipulating behaviour with substrate-borne vibrations–potential for insect pest control. Pest Manag. Sci. 71: 15–23.

Polajnar, J., A. Eriksson, A. Lucchi, M. Virant-Doberlet and V. Mazzoni. 2016. Mating disruption of a grapevine pest using mechanical vibrations: from laboratory to the field. J. Pest Sci. on line DOI 10.1007/s10340-015-0726-3. 13 p.
Potamitis, I., T. Ganchev and D. Kontomidas. 2009. On automatic bioacoustic detection of pests: the cases of *Rhynchophorus ferrugineus* and *Sitophilus oryzae*. J. Econ. Entomol. 102: 1681–1690.
Raman, D.J., R.R. Gerhardt and J.B. Wilkerson. 2007. Detecting insect flight sounds in the field: implications for acoustical counting of mosquitoes. Agric. Bios. Eng. 50: 1481–1485.
Rohde, B., T.M. Paris, E.M. Heatherington, D.G. Hall and R.W. Mankin. 2013. Responses of *Diaphorina citri* (Hemiptera: Psyllidae) to conspecific vibrational signals and synthetic mimics. Ann. Entomol. Soc. Am. 106: 392–399.
Saxena, K.N. and H. Kumar. 1980. Interruption of acoustic communication and mating in a leafhopper and a planthopper by aerial sound vibrations picked up by plants. Experientia 36: 933–6.
Scheffrahn, R.H., W.P. Robbins, P. Busey, N.Y. Su and R.K. Mueller. 1993. Evaluation of novel, hand-held, acoustic emissions detector to monitor termites (Isoptera: Kalotermitidae, Rhinotermitidae) in wood. J. Econ. Entomol. 86: 1720–1729.
Seffrin, R.C.A.S., E.C. Costa and S.T.B. Dequech. 2006. Artropodofauna do solo em sistemas direto e convencional de cultivo de sorgo (*Sorghum bicolor* (L.) Moench) na região de Santa Maria, RS. Ciência e Agrotecnologia 30: 597–602.
Silva, C.C.A., R.A. Laumann, J.B.C. Ferreira, M.C. Blassioli-Moraes, M. Borges and A. Čokl. 2012. Reproductive biology, mating behavior, and vibratory communication of the brown-winged stink bug, *Edessa meditabunda* (Fabr.) (Heteroptera: Pentatomidae). Psyche 2012: 1–9.
Sosa-Gomez, D.R. and J.J. Silva. 2010. Neotropical brown stink bug (*Euschistus heros*) resistance to methamidophos in Paraná, Brazil. Pesq. Aropec. Bras. 45: 767–769.
Stone, C.M., H.C. Tuten and S.L. Dobson. 2013. Determinants of male *Aedes aegypti* and *Aedes polynesiensis* (Diptera: Culicidae) response to sound: efficacy and considerations for use of sound traps in the field. J. Med. Entomol. 50: 723–730.
Todd, J.W. and D.C. Herzog. 1980. Sampling phytophagous Pentatomidae on soybean. pp. 438–478. *In*: Kogan, M. and D.C. Herzog (eds.). Sampling Methods in Soybean Entomology. Springer-Verlag, New York, USA.
Vieira, C.R., M.C. Blassioli-Moraes, M. Borges, E.R. Sujii and R.A. Laumann. 2013. *Cis*-Jasmone indirect action on egg parasitoids (Hymenoptera: Scelionidae) and its application in biological control of soybean stink bugs (Hemiptera: Pentatomidae). Biol. Cont. 64: 75–82.
Vieira, C.R., M.C. Blassioli-Moraes, M. Borges, C.S.S. Soares, E.R. Sujii and R.A. Laumann. 2014. Field evaluation of (*E*)-2-hexenal efficacy for behavioral manipulation of egg parasitoids in soybean. BioCont. 59: 525–537.
Vilhelmsen, L., N. Isidoro, R. Romani, H.H. Basibuyuk and L.J. Quicke. 2001. Host location and oviposition in a basal group of parasitic wasps: the subgenual organ, ovipositor apparatus and associated structures in the Orussidae (Hymenoptera, Insecta). Zoomorph. 121: 63–84.
Virant-Doberlet, M., A. Čokl and M. Zorovic. 2006. Use of substrate vibrations for orientation: from behaviour to physiology. pp. 81–97. *In*: Drosopoulos, S. and M.F. Claridge (eds.). Insect Sounds and Communication: Physiology, Behaviour, Ecology and Evolution. CRC Press, Boca Raton, U.S.A.
Virant-Doberlet, M., R.A. King, J. Polajnar and W.O.C. Symondson. 2011. Molecular diagnostics reveal spiders that exploit prey vibrational signals used in sexual communication. Mol. Ecol. 20: 2204–2216.
Wäckers, F.L., E. Mitter and S. Dorn. 1998. Vibrational sounding by the pupal parasitoid *Pimpla* (*Coccygomimus*) *turionellae*: An additional solution to reliability-detectability problem. Biol. Cont. 11: 141–146.

Walker, T.J. 1996. Acoustic methods of monitoring and manipulating insect pests and their natural enemies. pp. 113–123. *In*: Rosen, D., F.D. Bennett and J.L. Campinera (eds.). Pest Management in the Subtropics. Integrated Pest Management: a Florida Perspective. Intercept, Andover, U.K.

Welter, S.C., C. Pickel, J. Millar, F. Cave, R.A. Van Steenwyk and J. Dunley. 2005. Pheromone mating disruption offers selective management options for key pests. Cal. Agric. 59: 1–22.

Zgonik, V. and A. Čokl. 2014. The role of signals of different modalities in initiating vibratory communication in *Nezara viridula*. Central Eur. J. of Biol. 9: 200–2011.

Zorovič, M. and A. Čokl. 2015. Laser vibrometry as a diagnostic tool for detecting wood-boring beetle larvae. M. J. Pest Sci. 88: 107–112.

Žunič, A., M. Virant-Doberlet and A. Čokl. 2011. Species recognition during substrate-borne communication in *Nezara viridula* (L.) (Pentatomidae: Heteroptera). J. Insect Behav. 24: 468–487.

CHAPTER 12

Suggestions for Neotropic Stink Bug Pest Status and Control

Miguel Borges,[1,*] *Maria Carolina Blassioli-Moraes,*[1] *Raul Alberto Laumann*[1] and *Andrej Čokl*[2]

Introduction

Since the first identification of stink bug pheromone in the late 1980s, the information about communication in stink bugs has developed significantly. A simple search on the Web of Science (Thompson Reuters ISI) database using keywords like semiochemicals, pheromone and communication combined with stink bugs, demonstrates that we have jumped from five studies during the 60s and 70s to 200 in the last 16 years (Fig. 12.1). From Fig. 12.1 we can have an idea of the main areas that have evolved and the enormous progress achieved in the last few years. We can also notice that there are gaps in our knowledge about these insects in several areas like the biosynthesis of semiochemicals, their olfaction and visual perception and their evolutionary life cycle in natural and agricultural systems. These are critical subjects that have little to no recorded study in the literature, but actually all areas of study concerning these sucking-feeding insects need more efforts to provide knowledge in order to develop new technologies aiming to achieve more sustainable agricultural practices.

[1] Embrapa Recursos Genéticos e Biotecnologia, Parque Estação Biológica - W5 Norte (final), CEP: 70770-017, Brasília, Brazil.
Emails: carolina.blassioli@embrapa.br; raul.laumann@embrapa.br
[2] Department of Organisms and Ecosystems Research, National Institute of Biology, Večna pot 111, SI-1000 Ljubljana, Slovenia.
Email: andrej.cokl@nib.si
* Corresponding author: miguel.borges@embrapa.br

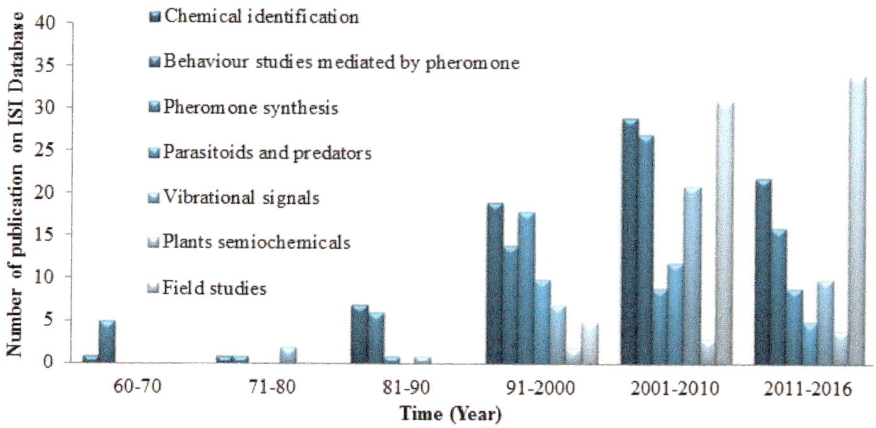

Figure 12.1. Number of studies published on the Thompson Reuters ISI database using the following keywords: semiochemicals, pheromone and communication combined with stink bugs.

Stink bugs are serious pests in different crops in the Neotropical region. *Euschistus heros* (F.) causes serious economic damage to soybean crops (Borges et al. 1998, 2011), and in the U.S., stink bugs pests like *Nezara viridula* (L.) and *Euschistus servus* (Say) have become a severe problem in *Bt* cotton, due to the absence of primary pests, which are the target of *Bt* technology, like *Helicoverpa zea* (Boddie) and *Heliothis virescens* (Fabricius) (Zeilinger et al. 2011). The outbreak of non-target species in *Bt* crops, mainly in cotton and maize, has been reported in different parts of the world where this technology has been applied for more than three years (Zhao et al. 2011, Catarino et al. 2015). The *Bt*-crops were developed to control the main lepidopteran pests in agriculture like *Spodoptera frugiperda* (J.E. Smith), *H. zea* and *H. virescens*, with the purpose of decreasing the insecticide applications and, consequently, the production costs. However, this advent opened a niche for secondary pests which resulted in the decrease of broad spectrum insecticide applications and the absence of competitors due to the *Bt*-toxin, and as a result more insecticides are been applied to control the stink bugs (Zhao et al. 2011, Catarino et al. 2015). In addition, although the studies did not show a direct negative effect of *Bt*-crops on beneficial insects, there is an indirect negative effect, since the *Bt* agricultural system presents simplified food webs (Moraes et al. 2010). Non-tillage agricultural systems also contribute to the population rise in these insects (Chocorosqui and Panizzi 2004, Seffrin et al. 2006) and climate change with global warming allows species migration and the colonization of temperate regions as seen in the cases of some tropical species invading U.S. crops (Panizzi 2015) (see Chapter 11 for details).

The biorational control of stink bugs requires an agricultural system using less chemical insecticides and with a less negative anthropogenic influence on the environment. Plants produce a series of secondary metabolites that give resistance to biotic stress (insects and disease), and several insecticides in use nowadays, have their molecules based on natural products, like spinosad (Birkett and Pickett 2014). Thus, the genes for the biosynthesis of natural insecticides are available in nature, waiting to be explored by genetic engineering. The development of genetically-modified plants with the ability to release pheromones or secondary metabolite compounds can provide a new class of GMOs to manage pests with direct non-toxic effects on beneficial insects (Birkett and Pickett 2014).

Pheromones, plant volatiles and the vibrational signals used by stink bugs in their communication are very feasible components that can generate more biorational methods to control these pests in field conditions.

A search on the ISI database with keywords such as: songs, vibrational communication combined with stink bug shows that since 1970 only 51 studies on stink bug vibrational communication have been published and that these studies covered only 19 species. Recently, Shestakov (2015) described vibrational communication in 16 species (see Chapter 6) (this study is not in the ISI database, 03/06/2016). The vibrational communication of insects is species-specific, presenting very special characteristics of signal modulation in order to avoid noise from different sources, including noise from the conspecifics and heterospecifics emitting vibrational signals (see Chapters 3, 6 and 11). These signals are essential in order for the mates to find each other in the plant environment, and any noise, biotic or abiotic, can interfere in this communication affecting the mating success of stink bugs. Furthermore, it disrupts the third level of communication as parasitoids are able to recognize stink bug vibratory signals produced during mating (Laumann et al. 2007, 2011). Therefore, developing traps that emit signals to attract natural enemies, or even devices which emit signals that can interfere in this communication, can keep the stink bug population below the economic threshold level. Indirect monitoring with devices that allow the recording and processing of the natural signals in the field provides the possibility and opportunity to estimate pest population density in the field in real time through user-friendly monitoring technology.

The worldwide species *N. viridula* was the first stink bug species with an identified sex-pheromone (Brennan et al. 1977, Pavis and Malosse 1986, Aldrich et al. 1987, Baker et al. 1987) and elucidated mating behaviour (Borges et al. 1987). *N. viridula* produce three male-specific compounds α-bisabolene, (4*S*)-*cis*-(*Z*)-bisabolene epoxide and (4*S*)-*trans*-(*Z*)-bisabolene epoxide (Baker et al. 1987, Borges et al. 1987 and Aldrich et al. 1987). Olfactometry bioassays showed that α-bisabolene was not important for female attraction, but that the two bisabolene-epoxides must be present in the blend in a similar ratio as produced by the males to attract females (Borges 1995). The same compounds were later also identified in different Nearctic and Neotropical *Chinavia* sp. species (McBrien et al. 2001, Blassioli-Moraes et al. 2012).

Ever since the first sex pheromone was isolated from *N. viridula* (Brennan et al. 1977, Pavis and Malosse 1986, Aldrich et al. 1987, Baker et al. 1987), semiochemicals have represented an important part of the integrated pest management strategy focused on stink bugs.

From this first description to the present day, forty-five stink bugs species have had their sex or sex-aggregation pheromones described in the literature and from among them only those of *E. heros*, Neartic species of *Euschitus, N. viridula, Halyomoprha halys* (Stal), *Piezodorus hybneri* and (Gmelin) *Plautia stali* (Scott) are commercially available. Most studies involve only chemical identification and laboratory bioassays; most of these 45 species are serious pests of grain and fruit (see Chapter 5).

However, because of different reasons, the practical use of stink bugs' pheromones has not materialized as quickly as expected. In general, the stink bugs' sex pheromones show high volatility and instability, which presents a low risk for the environment and non-target species. On the other hand, this can complicate its deployment. In addition, a series of field bioassays need to be conducted in order to prove sex-pheromone efficiency before it becomes commercially available; such tests are time consuming and can take several years to be completed. The compound (2*R*,6*R*,10*S*)-methyl-2,6,10-trimethyltridecanoate of the economically important Neotropical brown stink bug *E. heros* was the second identified sex-pheromone (Aldrich et al. 1994, Borges and Aldrich 1994). The molecule is relatively stable, has high volatility and is highly attractive to *E. heros* females at very low doses. This sex-pheromone compound has all the traits necessary in order to be used for monitoring and controlling insects in the field. Laboratory bioassays showed that the racemic mixture of methyl-2,6,10-trimethyltridecanoate was as attractive as the isomer produced by the males. The first field experiment was conducted in 1998 (Borges et al. 1998), using a racemic mixture provided by Dr. Kenji Mori. The pheromone traps captured females of *E. heros*, adults of other stink bug species, and the natural enemies of the family Platygastridae (Hymenoptera). In the periods between 2004 and 2007 and 2011/2012 new field experiments were conducted in soybean crops in order to determine parameters like the efficiency of capture, the amount of pheromones needed, the distance and trap-positioning (Borges et al. 2011, Silva et al. 2014). Since these first studies, several other stink bugs pheromones have been identified and evaluated in field conditions (*P. stali, H. halys, Chinavia hilaris*, Say, and *Thyanta perditor*, F.) (see Chapters 5 and 10). These studies showed that these stink bugs' sex pheromones have the potential to be used for monitoring in field conditions and also as an alternative tool of integrated pest management measures.

In order to enhance the number of commercially available pheromone lures and to improve their quality, it is necessary to conduct specific studies of stink bug mating behaviour, field population distribution and migrations. These studies can provide relevant information about when and how to apply the lures in the field (see Chapter 10). The amount of pheromone released from the lures represents a key point. Few studies on stink bugs' identified sex and sex-aggregation pheromones provide data on pheromone dose-response. Consequently, researchers just estimate the pheromone release rate from the lure in order to keep it at least for one month in the field. Different works have shown the importance of the amount of pheromone released from the lure for an effective capture of insects, whereby the wrong amount of lure can result in low or no capture at all (see Chapters 5 and 10). Another important point, for pheromone field application is the exact timing of placing the pheromone traps in the field. If the insects mate outside the crop area, and stay in it only to feed and lay eggs, then there is a chance that the pheromone traps will not catch adults. Since, in the first studies the authors suggested that the sex and sex-aggregation pheromones attract insects to meet their partners on the same plants where males and females afterwards use an extensive repertoire of substrate-borne vibrational signals to locate and recognize the mate (Borges et al. 1987, Čokl and Millar 2009). In addition, to these signals, stink bugs also use other cues to locate the plant host.

The ability of a phytophagous insect to locate a host plant depends upon a series of issues; such as the chemical and physical features of the plant (Chew and Renwich 1995) which may act as an attractant or a repellent (see Chapters 2 and 8). Therefore, the landscape influences of the foraging behaviour of stink bugs, and can also be manipulated in order to be used to control stink bugs in agriculture systems (see Chapters 2 and 8; Ehler 2000). By understanding the function of the secondary metabolites in the agriculture system, we can provide information to develop GMO plants with higher resistance to herbivores and with the ability to attract beneficial insects, as explained above.

It is well-demonstrated that stink bug communication during mating is conducted via signals of different modalities used in a sequence: first plant volatiles and visual signals are used to locate the host plants, afterwards chemical (pheromone) and vibrational signals are used by mates to reach each other and finally short-range chemicals, like cuticular hydrocarbons, visual and contact mechanical signals are used for final recognition and stimulation for copulation. Stink bugs can restart the sequence at any moment if necessary (Borges et al. 1987). The interplay among the different modalities of communication used by stink bugs has not been sufficiently demonstrated and further studies are needed to show that vibrational communication combined with trap pheromones has the potential to monitor or control insects in the field.

Conclusion

In the near future two important problems need to be solved in order to efficiently use the basic knowledge on stink bug communication for biorational control in the field: new formulations for the slow release of pheromone molecules are necessary and have to be settled together with the technology to playback vibratory signals in the traps.

Like Lepidoptera species, stink bugs have pheromone blends with two or more components, but the molecules in the stink bugs' blends are characterized by higher structural diversity, containing molecules with very different physical–chemical properties (Moraes et al. 2008). The latter can complicate their slow release from lures such as the rubber septum in the right ratio of the components at the required concentration. The advance of nanotechnology allows nanoencapsulation and nanoemulsion of volatile molecules but new research on volatile molecules is needed in order to evaluate the possibility of their slow release.

Advanced technology needs to be developed in order to record and reproduce vibrational signals in large areas, for example, like soybean or cotton fields. For effective monitoring in the field, technological development needs to consider devices which can record the vibratory signals in crops, discriminate them from the environmental noise and recognize their species and sex origin. Playback of synthesized signals has to consider the mechanical properties of the host plants that change their temporal, amplitude and frequency characteristics which determine the communication signals' species-specificity. The development of new material with characteristics similar to those of host plants is needed in order to use it for long distance vibration transmission in the field.

Further new advances in the use of the basic knowledge on communication will definitely provide new tools so that their economically relevant use in the biorational control of pest species can be achieved sooner. It is important to note that natural enemies also explore the stink bugs' communication system, including the parasitoid attraction to plants concurrently stressed by water reduction and phytophagous infestations (see Chapter 8). All these aspects should be carried out simultaneously and with fine synchronicity in order to develop effective target-oriented technology adapted for robust use in field conditions.

Acknowledgments

We thank the National Council for Scientific and Technological Development (CNPq), the Brazilian Corporation of Agricultural Research (EMBRAPA), the Research Support Foundation of the Federal District (FAP-DF) and the Slovene Research Agency.

References

Aldrich, J.R., J.E. Oliver, W.R. Lusby, J.P. Kochansky and J.A. Lockwood. 1987. Pheromone strains of the cosmopolitan pest *Nezara viridula* (Heteroptera: Pentatomidae). J. Exp. Zool. 244: 171–175.

Aldrich, J.R., J.E. Oliver, W.R. Lusby, J.P. Kochansky and M. Borges. 1994. Identification of male-specific volatiles from Neartic and Neotropical stink bugs (Heteroptera: Pentatomidae). J. Chem. Ecol. 20: 1103–1111.

Baker, R., M. Borges, N.G. Cooke and R.H. Herbert. 1987. Identification and synthesis of (Z)-(1′S,3′R,4′S)-(–)-2-(3′,4′-epoxy-4′-methylcyclohexyl)-6-methylhepta-2,5-diene, the sex pheromone of the southern green stink bug, *Nezara viridula* (L.). J. Chem. Soc. D 6: 414–416.

Blassioli-Moraes, M.C., R.A. Laumann, M.W.M. Oliveira, C.M. Woodcock, P. Mayon, A. Hooper et al. 2012. Sex pheromone communication in two sympatric Neotropical stink bug species *Chinavia ubica* and *Chinavia impicticornis*. J. Chem. Ecol. 38: 836–845.

Birkett, A.M. and J.A. Pickett. 2014. Prospects of genetic engineering for robust insect resistance. Curr. Opin. Plant Biol. 19: 59–67.

Borges, M., P.C. Jepson and P.E. Howse. 1987. Long-range mate location and close range courtship behavior of the green stink bug, *Nezara viridula* and its mediation by sex pheromones. Entomol. Exp. Appl. 44: 205–212.

Borges, M. and J.R. Aldrich. 1994. Attractant pheromone for Nearctic stink bug, *Euschistus obscurus* (Heteroptera: Pentatomidae): insight into a Neotropical relative. J. Chem. Ecol. 20: 1095–1102.

Borges, M. 1995. Attractant pheromone of the southern green stink bug, *Nezara viridula* (L.) (Heteroptera: Pentatomidae). An. Soc. Entomol. Brasil 24: 215–225.

Borges, M., F.G.V. Schmidt, E.R. Sujii, M.A. Medeiros, K. Mori, P.H.G. Zarbin et al. 1998. Field responses of stink bugs to the natural and synthetic pheromone of the neotropical brown stink bug, *Euschistus heros* (Heteroptera: Pentatomidae). Physiol. Entomol. 23: 202–207.

Borges, M., M.C.B. Moraes, M.F. Peixoto, C.S.S. Pires, E.R. Sujii and R.A. Laumann. 2011. Monitoring the neotropical brown stink bug *Euschistus heros* (F.) (Hemiptera: Pentatomidae) with pheromone-baited traps in soybean fields. J. Appl. Entomol. 135: 68–80.

Brennan, B.M., F. Chang and W.C. Mitchell. 1977. Physiological-effects on sex-pheromone communication in southern green stink bug, *Nezara-viridula* (Hemiptera-Pentatomidae). Environ. Entomol. 6: 169–173.

Catarino, R., G. Ceddia, F.J. Arak and J. Park. 2015. The impact of secondary pests on *Bacillus thurigiensis* (Bt) crops. Plant Biotech. J. 1–12.

Chew, F.S. and J.A.A. Renwick. 1995. Host plant choice in *Pieris* butterflies. pp. 214–238. *In*: Carde, R.T. and W.J. Bell (eds.). Chemical Ecology of Insects. Chapman & Hall, New York.

Čokl, A. and J.G. Millar. 2009. Manipulation of insect signaling for monitoring and control of pest insects. pp. 279–316. *In*: Ishaaya, I. and A.R. Horowitz (eds.). Biorational Control of Arthropod Pests: Application and Resistance Management. Springer, New York, NY.

Ehler, L.E. 2000. Farmscape Ecology of Stink Bugs in Northern California. Entomological Society of Amarica, Lanhan, Maryland, USA.

Gogala, M., A. Čokl, K. Drašlar and A. Blaževič. 1974. Substrate-borne sound communication in Cydnidae (Heteroptera). J. Comp. Physiol. 94: 25–31.

Laumann, R.L., M.C. Blassioli-Moraes, A. Čokl and M. Borges. 2007. Eavesdropping on sexual vibratory signals of stink bugs (Hemiptera: Pentatomidae) by the egg parasitoid *Telenomus podisi*. Anim. Behav. 73: 637–649.

Laumann, R.A., A. Čokl, A.P.S. Lopes, J.B.C. Fereira, M.C. Blassioli-Moraes and M. Borges. 2011. Silent singers are not safe: selective response of a parasitoid to substrate-borne vibratory signals of stink bugs. Anim. Behav. 82: 1175–1183.

McBrien, H.L., J.G. Millar, L. Gottlieb, X. Chen and R.E. Rice. 2001. Male-produced sex attractant pheromone of the green stink bug, *Acrosternum hilare* (Say). J. Chem. Ecol. 27: 1821–1839.

Moraes, M.C.B., M. Pareja, R.A. Laumann and M. Borges. 2008. The chemical volatiles (Semiochemicals) produced by neotropical stink bugs (Hemiptera: Pentatomidae). Neotropical Entomology 37: 489–505.

Moraes, M.C.B., R.A. Laumann, M.F.S. Aquino, D.P. Paula and M. Borges. 2010. Effect of *Bt* genetic engineering on indirect defense in cotton via a tritrophic interaction. Transgenic Res. 20: 99–107.

Pavis, C. and P.H. Malosse. 1986. Mise en evidence d'un attractif sexuel produit par les males de *Nezara viridula* (L.) (Heteroptera: Pentatomidae). C.R. Acad. Sci. Series III 7: 272–276.

Shestakov, L.S. 2015. A comparative analysis of vibrational signals in 16 sympatric species (Pentatomidae, Heteroptera). Entomol. Rev. 95(3): 310–325.

Silva, W.P., M.J.B. Pereira, L.M. Vivan, M.C.B. Moraes, R.A. Laumann and M. Borges. 2014. Monitoramento do percevejo marrom *Euschistus heros* (Hemiptera: Pentatomidae) por feromônio sexual em lavoura de soja. Pesqui. Agropecu. Bras. 49: 844–852.

Zeilinger, A.R., D.M. Olson and D.A. Andow. 2011. Competition between stink bug and heliothine caterpillar pests on cotton at within-plant spatial scales. Entomol. Exp. Appl. 141: 59–70.

Zhao, J.H., P. Ho and H. Azadi. 2011. Benefits of Bt cotton counterbalanced by secondary pests? Perceptions of ecological change in China. Environ. Monit. Asses. 173: 985–994.

Index

9781498732802
#0001 - 240317 - C20 - 229/152/15 [17] - CB - 9781498732802